Mechanisms of Secondary
Brain Damage

NATO ASI Series

Advanced Science Institutes Series

A series presenting the results of activities sponsored by the NATO Science Committee, which aims at the dissemination of advanced scientific and technological knowledge, with a view to strengthening links between scientific communities.

The series is published by an international board of publishers in conjunction with the NATO Scientific Affairs Division

A	**Life Sciences**	Plenum Publishing Corporation
B	**Physics**	New York and London
C	**Mathematical and Physical Sciences**	D. Reidel Publishing Company
		Dordrecht, Boston, and Lancaster
D	**Behavioral and Social Sciences**	Martinus Nijhoff Publishers
E	**Engineering and Materials Sciences**	The Hague, Boston, and Lancaster
F	**Computer and Systems Sciences**	Springer-Verlag
G	**Ecological Sciences**	Berlin, Heidelberg, New York, and Tokyo

Recent Volumes in this Series

Volume 109—Central and Peripheral Mechanisms of Cardiovascular Regulation
edited by A. Magro, W. Osswald, D. Reiss, and P. Vanhoutte

Volume 110—Structure and Dynamics of RNA
edited by P. H. van Knippenberg and C. W. Hilbers

Volume 111—Basic and Applied Aspects of Noise-Induced Hearing Loss
edited by Richard J. Salvi, D. Henderson, R. P. Hamernik,
and V. Colletti

Volume 112—Human Apolipoprotein Mutants: Impact on Atherosclerosis
and Longevity
edited by C. R. Sirtori, A. V. Nichols, and G. Franceschini

Volume 113—Targeting of Drugs with Synthetic Systems
edited by Gregory Gregoriadis, Judith Senior, and George Poste

Volume 114—Cardiorespiratory and Cardiosomatic Psychophysiology
edited by P. Grossman, K. H. Janssen, and D. Vaitl

Volume 115—Mechanisms of Secondary Brain Damage
edited by A. Baethmann, K. G. Go, and A. Unterberg

Volume 116—Enzymes of Lipid Metabolism II
edited by Louis Freysz, Henri Dreyfus, Raphaël Massarelli,
and Shimon Gatt

Volume 117—Iron, Siderophores, and Plant Diseases
edited by T. R. Swinburne

Series A: Life Sciences

Mechanisms of Secondary Brain Damage

Edited by
A. Baethmann
University of Munich
Munich, West Germany

K. G. Go
University of Groningen
Groningen, Holland

and
A. Unterberg
University of Munich
Munich, West Germany

Plenum Press
New York and London
Published in cooperation with NATO Scientific Affairs Division

Proceedings of a NATO Advanced Research Workshop on
Mechanisms of Secondary Brain Damage,
held February 19–23, 1984,
in Mauls/Sterzing, Italy

Library of Congress Cataloging in Publication Data

NATO Advanced Research Workshop on Mechanisms of Secondary Brain
Damage (1984: Mules, Italy)
    Mechanisms of secondary brain damage.

    (NATO ASI series. Series A, Life sciences; v. 115)
    "Proceedings of a NATO Advanced Research Workshop on Mechanisms of
Secondary Brain Damage, held February 19–23, 1984, in Mauls/Sterzing, Italy"—
T.p. verso.
    Includes bibliographies and index.
    1. Brain damage—Congresses. 2. Brain—Wounds and injuries—Complica-
tions and sequelae—Congresses. 3. Cerebrovascular disease—Complications
and sequelae—Congresses. 4. Cerebral ischemia—Congresses. 5. Cerebral
edema—Congresses. I. Baethmann, A. II. Go, K. G. III. Unterberg, A. IV. Title. V.
Series. [DNLM: 1. Brain Damage, Chronic—congresses. 2. Brain Edema—com-
plications—congresses. 3. Cerebral Ischemia—complications—congresses. 4.
Head Injuries—complications—congresses. WL 354 N279 1984]
RC387.5.N38   1984   616.8   86-22671
ISBN-13: 978-1-4684-5205-1   e-ISBN-13: 978-1-4684-5203-7
DOI: 10.1007/978-1-4684-5203-7

© 1986 Plenum Press, New York
Softcover reprint of the hardcover 1st edition   1986
A Division of Plenum Publishing Corporation
233 Spring Street, New York, N.Y. 10013

PREFACE

A distinction between primary and secondary brain damage of various origin, particularly in acute lesions, such as head injury and ischemia is not entirely new. The concept is of practical significance, because it is the foremost intention of all clinical efforts to prevent, or at least attenuate the development of secondary sequelae. Primary damage to nervous elements usually cannot be influenced by treatment. Its prevention is the objective of prophylactic measures. The current volume gathered prominent scientists and clinicians from various fields to provide a competent introduction and survey of the various aspects involved in secondary brain damage. It was attempted to provide criteria for the distinction between the primary and secondary phenomena on a morphological and functional level, on the basis of the kinetics involved and, most importantly, regarding the different specific manifestations, such as disturbances of microcirculation, aspects of the blood-brain barrier, and of cellular structure and function at a molecular level. Although it was not expected that a grand unifying hypothesis will be reached reconcilable with the many, occasionally opposing views on such a complex subject, nevertheless, the present volume attains an appropriate result. It can best be described as a mosaic of many different pieces which only as an ensemble reflect the current state of the art. Head injury and cerebral ischemia were selected as the underlying clinical disorders, not only because of their high-ranking significance from a socio-economical view, but also because the ensuing secondary processes often have a greater impact on outcome than the primary lesion.

An important purpose is to arouse awareness of the problem. Progress in prevention of secondary brain damage can now already be accomplished by more consequent implementation of the currently available resources and skills in management, such as concerning organization, handling of the flow of information, methods of transportation and, last but not least, emergency and intensive care protocols. However, clinical and experimental research must progress to advance and refine our knowledge up to a molecular understanding of the mechanisms of secondary brain damage. Yet, the careful reader of the contributions may arrive at the conclusion that the experimental models employed and the clinical examples studied are still too complex to provide a basis for the development of specific and, hence, more effective forms of treatment. But this is where we currently are.

The present collection of articles covers a wide spectrum, from
the basic sciences investigating molecular mechanisms of cell swelling
under the various pathological conditions to the current practice and the
experiences with complex protocols of treatment. Certainly, the clear and
lucid pathomorphological descriptions of what is primary and secondary
in head injury and cerebral ischemia constitute a main theme. Major
elements, such as diffuse axonal injury, the contusion focus, and vascular damage are competently discussed, or the vascular or parenchymatous
changes occurring in ischemia. Since brain edema is a major component
of secondary brain damage both in head injury and ischemia, it is dealt
with on a broad basis ranging from the changes of extracellular homeostasis, the involvement of autotoxic mediator substances which enhance
blood-brain barrier damage or disturbances of the microcirculation and
probably induce cell swelling, to the response of the afflicted brain on
both physiological and biochemical levels. This response is ingreasingly
utilized to refine our diagnostic armament to obtain quantitative information on the actual level of damage which is pertinent for prognosis.

Coverage of several aspects pertaining to cerebral ischemia has
been exhaustive, in particular those dealing with the important concept
of flow thresholds associated with distinct functional and structural de-
ficiencies of the brain, the postischemic recirculation, the phenomenon
of selective vulnerability, or the role of receptors on the cerebrovascular endothelium. On the other hand, equal emphasis is given to the more
mundane clinical issues, such as current mortality and morbidity from
secondary brain damage in head injury together with evaluation of the
potential for improvement. Although, in cerebral ischemia a dividing line
between primary and secondary phenomena is more difficult to draw, useful suggestions have been made for the prevention or mitigation of the
evolving secondary mechanisms. Certainly, this is a constantly moving
field where definite answers have to be given yet. Altogether, the current volume attempts to provide a competent survey and status report
of a subject not only enticing enormous scientific excitement but also
offering pertinent perspectives for the clinical routine. Its purpose
would be fulfilled, if it is used as a basis and initiation for further
experimental and clinical efforts to improve our understanding and
methods to the ultimate benefit of the afflicted patient.

We would like to take the opportunity to express our gratitude to
NATO-Scientific Affairs Division in Brussels, its representative Dr. M.
di Lullo, and to Dr. W. Thiede and Peter Hansen of UCB-Chemie, Kerpen,
FRG. Their generous support made possible this publication. Without the
meticulous editorial and secretarial assistance provided by Ruth Demmer,
Ulrike Goerke, Isolde Juna, and Sylvia Schneider in addition to their
daily routine in the Institute for Surgical Research at the University of
Munich, this work had not arrived at a happy conclusion.

A. Baethmann
K.G. Go
A. Unterberg

CONTENTS

II. Pathophysiological and Biochemical Mechanisms

a) Cerebral Edema

CONTENTS

III. Clinical Aspects and Treatment

a) Head Injury

b) Cerebral Ischemia

PRIMARY BRAIN DAMAGE IN NON-MISSILE HEAD INJURY

J. Hume Adams, D.I. Graham and T.A. Gennarelli*

University Department of Neuropathology, Institute of
Neurological Sciences, Southern General Hospital, Glasgow
G51 4TF, Scotland, U.K., and Division of Neurosurgery*
University of Pennsylvania, Philadelphia, USA

There have been various approaches to the classification of brain
damage resulting from a non-missile head injury. One is that being used
in this Symposium, namely primary and secondary . Primary damage occurs
at the moment of injury and consists mainly of contusions and diffuse
axonal injury, while secondary damage can be considered as a complication
of the original injury and includes intracranial haematoma, brain damage
secondary to raised intracranial pressure, shift and herniation of the
brain, brain swelling, and hypoxic brain damage (Adams, Gennarelli, and
Graham, 1982). Since it is now known as a result of early CT-scans and
experimental studies (see below) that some types of so-called secondary
brain damage, such as haematoma and swelling, may be present very soon
after the injury, the distinction between primary and secondary damage is
becoming somewhat blurred. There is therefore much to commend classi-
fying brain damage resulting from a head injury as being focal or
diffuse (Adams and Graham, 1984).

There are three principal types of primary brain damage in non-
missile head injury. Two of these - cerebral contusions and diffuse axonal injury - are encountered frequently in patients who have survived
their injury long enough to be admitted to hospital. The third is most
frequently seen in patients who die very soon after their injury and
consists of multiple small haemorrhages throughout the brain. Other less
common types of brain damage that occur as primary events are ponto-
medullary rents and some types of intracranial haematoma.

Further light has been shed on the pathogenesis of all types of
brain damage in head injury from the experimental studies on subhuman
primates undertaken by Gennarelli and his group in Philadelphia (Gennarelli, 1983; Adams, Graham and Gennarelli, 1983) using devices referred
to as Penn 1 and Penn 2. The Penn 1 device was based on inertial, i.e.

1

non-impact, controlled angular acceleration of the head through 60° in
the sagittal plane: with the Penn 2 device the acceleration rate was
slower and the head could be moved in oblique and coronal (lateral)
planes as well as sagittal.

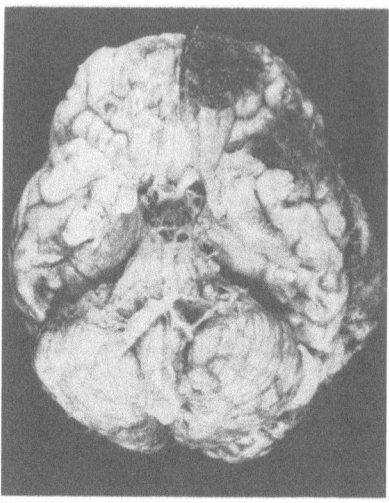

Fig.1: Recent contusions in the frontal and temporal lobes. (Reproduced
with permission from Adams, Graham and Gennarelli, 1983).

CEREBRAL CONTUSIONS

Cerebral contusions have long been considered the classical features of a non-missile head injury, but fatal brain damage may occur in
an individual who sustains a head injury or in a subhuman primate subjected to angular acceleration of the head without there being a trace of a
conventional contusion (Adams, Gennarelli and Graham, 1982). Contusions
are a type of focal brain damage brought about mainly by contact phenomena when the surface of the brain impacts on bony protuberances in the
base of the skull. They have a highly characteristic appearance and
distribution (Adams et al., 1980a). They occur principally at the crests
of gyri and in the acute stages are haemorrhagic (Fig. 1). With the passage of time they come to be represented by shrunken brown areas on the
surface of the brain (Fig 2). Recent quantitative studies using a contusion index (Fig. 3) which is based on the extent and depth of contusions
in any region of the brain (Adams et al., 1980b; Adams, Gennarelli and
Graham, 1982) have confirmed that in man and in subhuman primates
subjected to angular acceleration of the head using the Penn 1 device
contusions are most severe in the frontal and in the temporal lobes.
Contusions not obvious on external examination of the brain are frequently found on sectioning in the cortex above and below the Sylvian
fissures (Fig. 4).

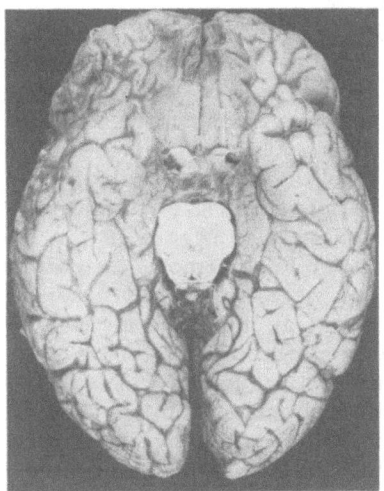

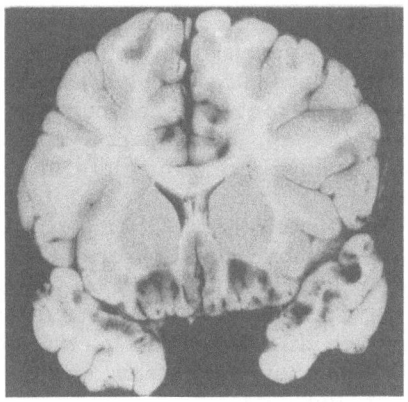

Fig.2: Old contusions in the frontal and temporal lobes: patient made a complete recovery from a head injury
sustained some 40 years
prior to death. (Reproduced
with permission from Adams,
Gennarelli and Graham, 1982).

Fig.4: Recent contusions above and
below the Sylvian fissures
(Reproduced with permission
from Adams, 1975)

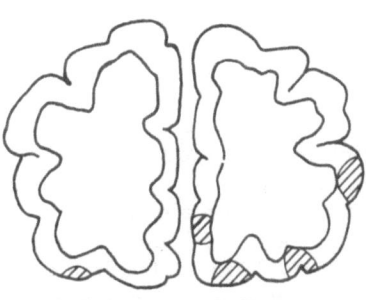

L. Frontal   R. Frontal
C I = 1 x 1 = 1   C I = 2 x 3 = 6

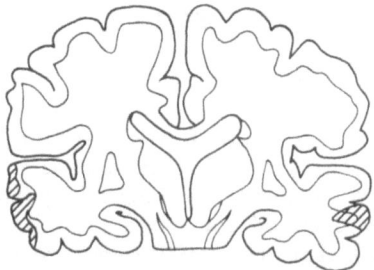

L. Temporal   R. Temporal
C I = 2 x 2 = 4   C I = 3 x 1 = 3

Fig.3: Two representative line diagrams to show how the contusion index
(C.I.) is derived for different regions in the brain. It is obtained by multiplying the depth of a contusion (1 = superficial cortex;
2 = full thickness of cortex; 3 = cortex plus white matter) by the
extent (1 = localised; 2 = moderately extensive; 3 = extensive).
(Reproduced with permission from Adams et al. 1980b).

Contusions have been intensively studied by many pathologists, particularly by Lindenberg and Freytag (1957, 1960), who suggested that contusions should be classified as coup (contusions occurring at the point of impact), contrecoup (contusions contralateral to the site of impact) and intermediate coup (contusions in the deeper structures of the brain). An analysis of 30 patients with localising fractures of the skull, however, has shed considerable doubt on the significance of contrecoup contusions as being of help in defining the site of injury (Adams et al., 1980b) since contusions were not more severe in the cerebral hemisphere contralateral to the fracture. Furthermore, in patients with frontal or occipital fractures, contusions were always more severe in the frontal lobes. Gliding contusions occur along the superior border of the cerebral hemispheres (Fig. 11a) and are often restricted to the deeper layers of the cortex and the adjacent white matter. They tend to be particularly prominent in patients and in subhuman primates that have sustained diffuse axonal injury, and have been attributed by Dawson et al (1980) to shear and strain stresses where the brain is relatively adherent to the dura.

Since contusions are a focal type of brain damage, it is not surprising that patients with quite severe cerebral contusions may make a very satisfactory recovery from their injury. They are significantly more severe in patients with a fracture of the skull than in those who do not have a fracture; there is no difference in the severity of contusions between patients who do and do not experience a lucid interval after head injury; and they are less severe in patients with diffuse axonal injury than in patients who do not have this type of brain damage (Adams et al., 1982).

DIFFUSE AXONAL INJURY

It is becoming increasingly apparent that this is the most important type of brain damage that may be sustained as a result of a head injury in that even if there are no complications, the outcome in patients with severe diffuse axonal injury is poor. It is the commonest structural abnormality found in patients who have remained vegetative or severely disabled for more than four weeks after their injury (Graham et al. 1983). This type of damage was first clearly defined by Strich (1956) and since that time has been referred to by a variety of names (see Adams and Graham, 1984). From the outset Strich (1956) attributed the damage to axons to shearing injury affecting nerve fibres at the moment of the original injury. Many have endorsed this view but others have suggested that the damage to nerve fibres is more often secondary to hypoxic brain damage or to damage to the brain stem resulting from an intracranial expanding lesion (Jellinger, 1977; Peters and Rothemund, 1977). The situation has recently been clarified and Strich's contention vindicated in that structural abnormalities identical to those seen in man have been produced in subhuman primates subjected to non-impact controlled angular acceleration of the head in circumstances in which the

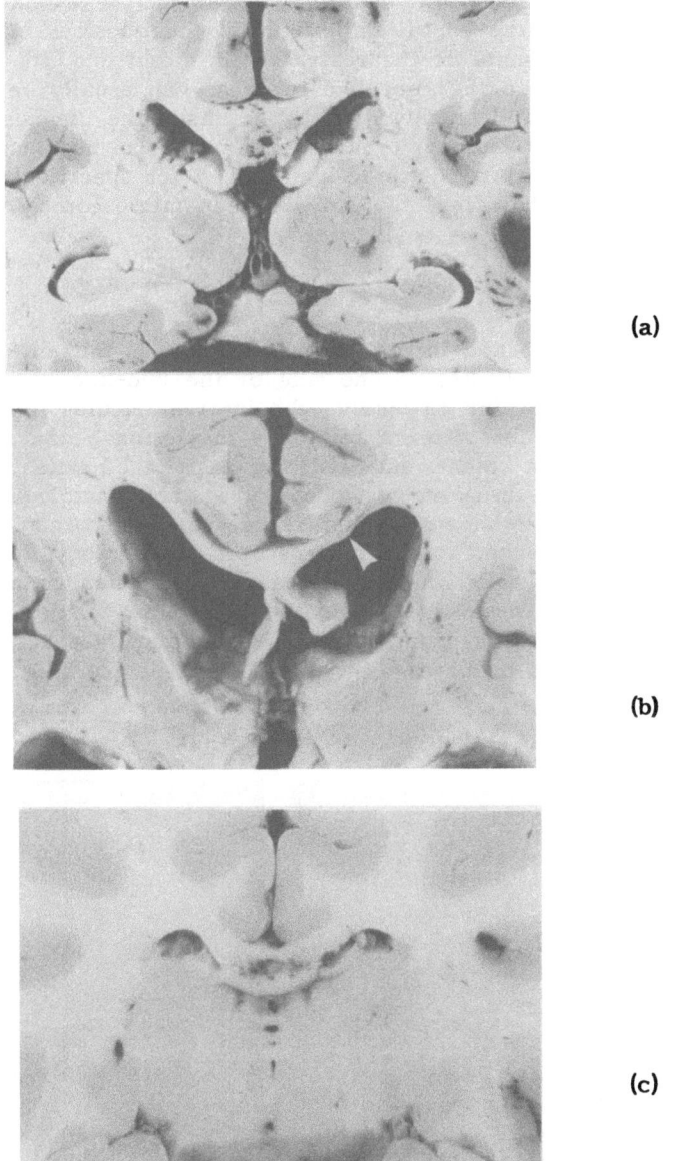

Fig.5: Diffuse axonal injury: lesion in corpus callosum. (a) recent
haemorrhagic lesion in man; (b) old lesion (arrow) in man;
(c) recent haemorrhagic lesion in subhuman primate. (Reproduced
with permission from Adams, Graham and Gennarelli, 1983).

axonal damage could not be attributed to an increase in intracranial pressure or an episode of hypoxia (Gennarelli et al., 1982). There are three distinctive structural abnormalities in diffuse axonal injury; (i) a focal lesion in the corpus callosum; (ii) focal lesions in one or both dorsolateral quadrants of the rostral brain stem; and (iii) diffuse damage to axons. Since the first two of these can usually be identified macroscopically, it is rarely difficult to make the diagnosis of diffuse axonal injury post mortem provided the brain has been properly fixed prior to dissection and these lesions looked for specifically. Diffuse injury to axons, however, can only be seen microscopically.

Corpus Callosum

The focal lesion in the corpus callosum is initially haemorrhagic (Fig. 5a) and usually lies to one side of the mid-line. It may, however, extend to the mid-line to involve the interventricular septum and the pillars of the fornix. After a few days the lesion becomes rather granular and brown in colour, and with the passage of time comes to be represented by a shrunken cystic scar (Fig. 5b). Histological examination in the early stages shows varying degrees of haemorrhage and rarefaction necrosis, and adjacent to the macroscopic abnormality there are numerous axonal retraction balls. In the late stages the lesion comes to be represented by a shrunken fenestrated glial scar. Similar abnormalities have been produced by controlled non-impact angular acceleration of the head in subhuman primates using the Penn 2 device (Fig. 5c).

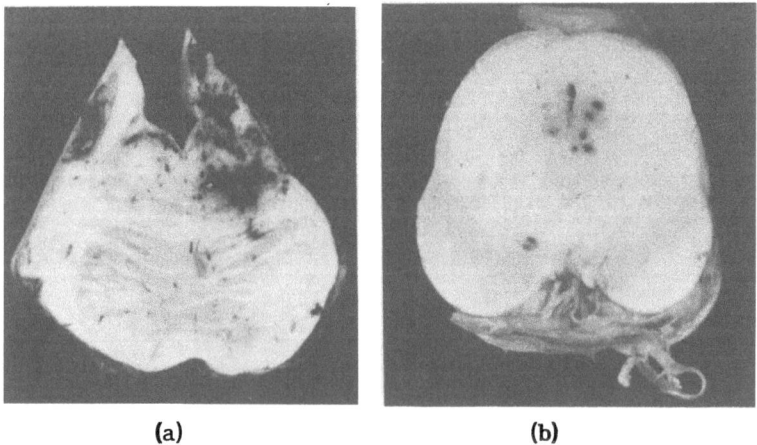

(a)   (b)

Fig.6: Diffuse axonal injury - recent haemorrhagic lesions in the rostral brain stem in (a) man and (b) subhuman primates. (Reproduced with permission from Adams et al. 1982).

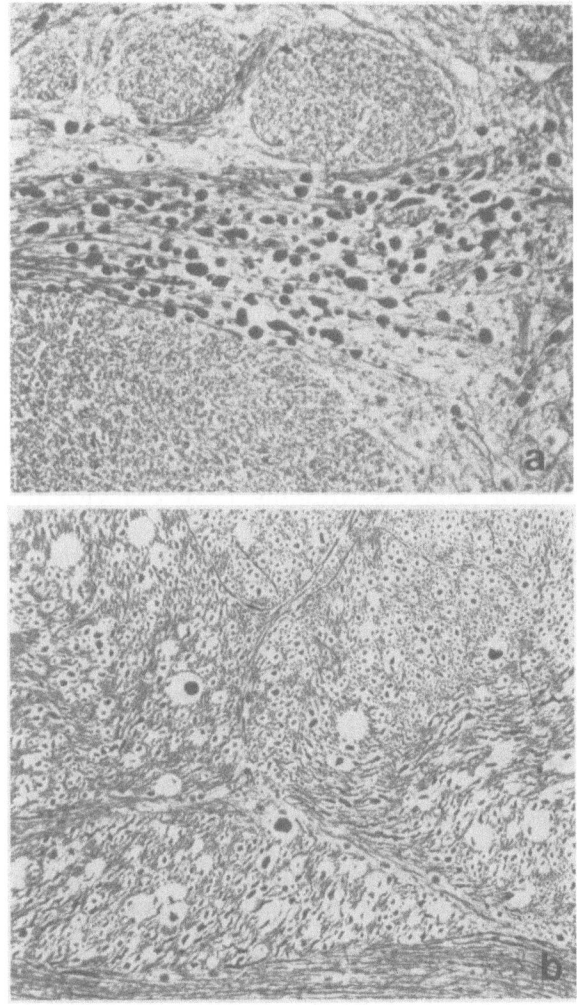

Fig.7: Diffuse axonal injury. Numerous retraction balls are present on
axons; (a) man; (b) subhuman primate. Palmgren: both x 100. (Reproduced with permission from Adams,Gennarelli and Graham,1982)

Brain Stem

The lesions in the dorsolateral quadrants of the rostral brain stem (Fig. 6a) follow a similar time course to those in the corpus callosum. They vary greatly in size and after a few days become brown and granular: with the passage of time they come to be represented by shrunken rarefied scars. Similar lesions have been produced in subhuman primates with the Penn 2 device (Fig. 6b).

Diffuse Damage to Axons

This takes forms that are related to the duration of the survival of the patient, and the structural abnormalities may be difficult to recognise unless appropriate stains are used and appropriate regions examined. In patients of short survival (days) there are large numbers of axonal retraction balls in the white matter (Fig. 7a). Their distribution is not uniform but they occur particularly in the subcortical parasagittal white matter, in the corpus callosum remote from the focal lesions referred to above, in various tracts in the brain stem, and in the cerebellum. Axonal retraction balls with a similar distribution to that seen in cases of diffuse axonal injury in man have been produced in the brains of subhuman primates with the Penn 2 device (Fig. 7b).

After a few weeks the most striking feature is the presence of large numbers of small clusters of microglia throughout the white matter of the cerebral hemispheres, the cerebellum and the brain stem (Fig. 8). These have also been produced in subhuman primates with the Penn 2 device. After some months, long tract degeneration stainable by the MAR-CHI method (Strich, 1968) becomes apparent. This is often particularly conspicuous in the medial lemnisci, in the central tegmental tracts, and in the pyramidal tracts (Fig. 9) throughout the brain stem and the spinal cord but it can also be identified in the white matter of the cerebral hemispheres including the internal capsules (Adams, 1975). By this time there is usually some reduced bulk and increased consistency of the white matter of the cerebral hemispheres, thinning of the corpus callosum, and compensatory enlargement of the ventricular system.

Pathogenesis

As indicated above, this has always been a controversial issue but the evidence from clinico-pathological studies in man and experiments on subhuman primates has firmly established that this type of brain damage occurs at the moment of injury. Thus in man in a series of 45 patients with diffuse axonal injury (Adams et al., 1982) not one of the patients experienced a lucid interval, while in the subhuman primates with diffuse axonal injury there was nothing to suggest that this was a secondary type of brain damage - it certainly could not be attributed to raised intracranial pressure or hypoxia. Furthermore, the subhuman primates that

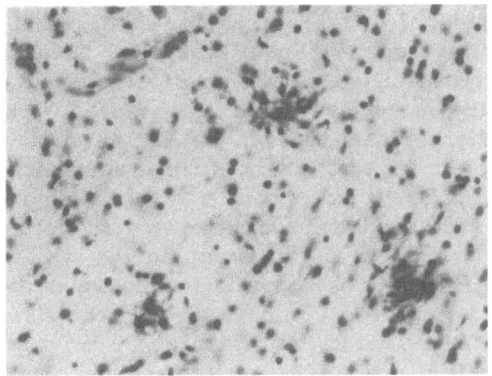

Fig.8: Diffuse axonal injury. Multiple clusters of microglia in the cerebral white matter. Cresyl violet x 250. (Reproduced with permission from Adams, Gennarelli and Graham, 1982).

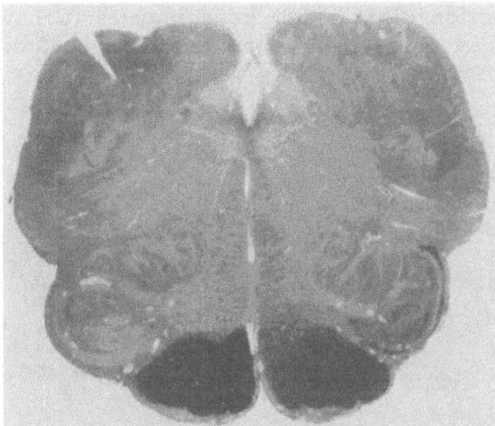

Fig.9: Diffuse axonal injury. There is established degeneration in the pyramidal tracts. Marchi preparation. (Reproduced with permission from Adams and Graham, 1984).

showed the classical features of diffuse axonal injury in man remained in
persistent coma until the time of sacrifice (Gennarelli et al., 1982),
while those with less severe brain damage showed varying degrees of
neurological dysfunction from the moment of acceleration and this
persisted until the time of sacrifice.

The production of diffuse axonal injury in subhuman primates by
non-impact controlled acceleration of the head (Gennarelli et al., 1982)
has allowed of a quantitative definition of various grades of diffuse
axonal injury. In the most severe grade - grade 3 - there are focal
abnormalities in the corpus callosum and in the dorso-lateral quadrant of
the rostral brain stem in addition to diffuse damage to axons. In grade 2
there is diffuse damage to axons and a focal lesion in the corpus callosum. In grade 1 structural abnormalities are restricted to axonal
retraction balls and clusters of microglia in the white matter. In these
experiments, there was a close correlation between the severity of
diffuse axonal injury and the clinical state of the subhuman primates
(Gennarelli et al., 1982). Animals with grade 1 or grade 2 diffuse axonal
injury were moderately disabled after their injury, while those with
grade 3 remained severely disabled or in persistent coma until death. A
so far unpublished analysis of our experience with fatal head injuries in
man has shown that minor degrees of diffuse axonal injury are not uncommon in any type of fatal head injury in man. Oppenheimer (1968) has also
commented on the occurrence of clusters of microglia after an apparently
mild head injury, and it may well be that some of the failure of recovery
of intellectual function after minor head injuries (Gronwall and
Wrightson, 1974) may be related to minor degrees of diffuse axonal
injury.

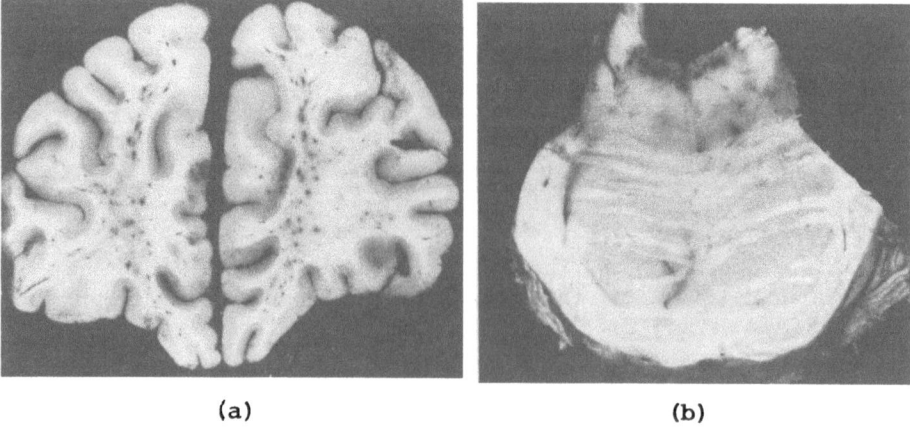

(a)   (b)

Fig.10: Multiple petechial haemorrhages in the brain in patients who died
some 30 minutes after a head injury. (Reproduced with permission
from Adams and Graham, 1984).

MULTIPLE PETECHIAL HAEMORRHAGES IN THE BRAIN

The importance of this type of brain damage in head injury has been reviewed and emphasised by Tomlinson (1970). In his experience the damage is seen in patients who die within a few hours of injury and the haemorrhages are particularly prominent in the rostral brain stem, particularly in the floor of the aqueduct and the rostral part of the fourth ventricle. Our experience in this Institute is similar in that one of the commonest findings in patients who die very shortly after a non-missile head injury is the presence of multiple petechial haemorrhages, not only in the brain stem (Fig. 10b) but also in the white matter of the frontal (Fig. 10a) and temporal lobes. These lesions can be remarkably striking in patients who are reported to have died instantaneously. Histological examination shows that the haemorrhages are small collections of blood in perivascular spaces. The pathogenesis of this type of brain damage has not been established and it can only be concluded that certain acceleration/deceleration impulses selectively damage arterioles, capillaries and venules in the brain.

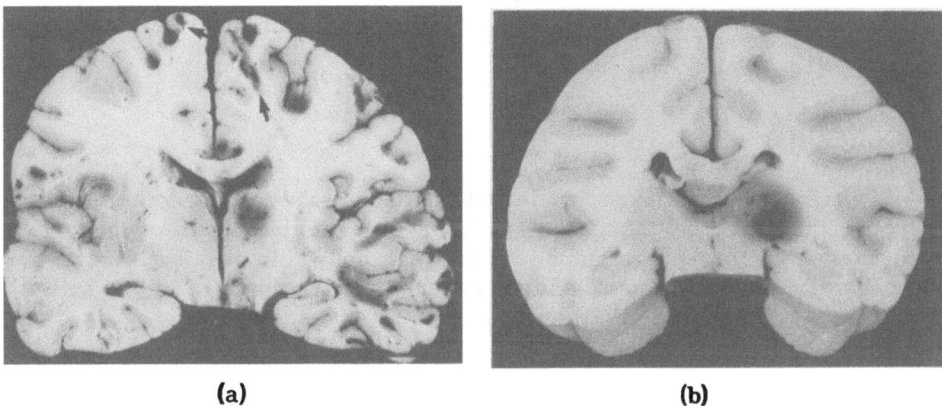

(a)   (b)

Fig.11: Haematomas in basal ganglia. (a) a patient with diffuse axonal injury: there are also bilateral parasagittal gliding contusions (arrows); (b) a subhuman primate with diffuse axonal injury.

OTHER TYPES

Traumatic intracranial haematoma is usually thought of as a secondary event and therefore a complication of the original injury, but there seems little doubt that haemorrhage from damaged blood vessels commences at the time of the injury. Certainly in subhuman primates subjected to angular acceleration of the head, fatal acute subdural haematoma can occur within minutes of the injury (Adams, Graham and Gennarelli, 1982). Furthermore, we are becoming increasingly aware of deeply placed haematomas in and adjacent to the basal ganglia in fatal non-missile head injuries (Fig. 11a), and haematomas have been observed in subhuman primates with diffuse axonal injury (Fig. 11b). In an as yet unpublished analysis of eight patients (age range 4 to 25 years) there is no doubt that these haematomas in the basal ganglia were the result, and not the cause, of the injury. Not one of the eight cases experienced a lucid interval, in four there were also the structural abnormalities of diffuse axonal injury, and in six the injury was the result of a road traffic accident.

The so-called ponto-medullary rent (Hardman, 1979) is widely recognised, and is usually associated with a ring fracture in the base of the skull. This type of damage has also been produced in subhuman primates subjected to angular acceleration at very high acceleration levels (Adams, Graham and Gennarelli, 1982). Patients with such rents are usually dead on admission to hospital (Lindenberg and Freytag, 1970) but Pilz, Strohecker and Grobovschek (1982) have recently described patients with subtotal ponto-medullary rents who survived for periods ranging from 8 to 26 days after their injury. This type of brain damage appears to be the only type of primary damage to the brain stem that can occur in isolation. Other intracranial structures that may be damaged at the time of injury include cranial nerves, the hypothalamus and the pituitary stalk.

CONCLUSIONS

Primary brain damage by definition occurs at the moment of injury and could therefore be thought of as being irreversible. It can only be hoped that a better understanding of the pathophysiological basis of this type of brain damage in man and in experimental animals coupled with early CT-scans might however help in patient management.

REFERENCES

Adams JH, The neuropathology of head injuries, in: "Handbook of clinical neurology", Vol. 23, Injuries of the brain and skull, Part 1, Vinken PJ, Bruyn GW, eds., American Elsevier, New York (1975).

Adams JH, Gennarelli TA, Graham DI, Brain damage in non-missile head injury: Observations in man and subhuman primates, in: "Recent advances in neuropathology", 2, Smith WT, Cavanagh JB, eds., Churchill Livingstone, Edinburgh (1982).

Adams JH, Graham DI, Diffuse brain damage in non-missile head injury, in: "Recent advances in histopathology", 12, Anthony PP, Mac Sween RNM, eds., Churchill Livingstone, Edinburgh (1984).

Adams JH, Graham DI, Gennarelli TA, Neuropathology of acceleration-induced head injury in the subhuman primate, in: "Head injury: Basic and Clinical Aspects", Grossmann RG, Gildenberg PL, eds., Raven Press, New York (1982).

Adams JH, Graham DI, Gennarelli TA, Head injury in man and experimental animals - neuropathology, Acta Neurochirg Suppl. 32:15 (1983).

Adams JH, Graham DI Murray LS, Scott G, Diffuse axonal injury due to non-missile head injury in humans, Ann Neurol 12:557 (1982).

Adams JH, Graham DI, Scott G, Parker LS, Doyle D, Brain damage in non-missile head injury, J Clin Pathol 33:1132 (1980a).

Adams JH, Scott G, Parker LS, Graham DI, Doyle D, The contusion index: a quantitative approach to cerebral contusions in head injury, Neuropathol Appl Neurobiol 6:319 (1980b).

Dawson SL, Hirsch CS, Lucas FV, Sebek BA, The contrecoup phenomenon: reappraisal of a classic problem, Human Pathol 11:155 (1980).

Gennarelli TA, Head injury in man and experimental animals - clinical aspects, Acta Neurochirg Suppl. 32:1 (1983).

Gennarelli TA, Thibault LE, Adams JH, Graham DI, Thompson CJ, Marcincin RP, Diffuse axonal injury and traumatic coma in the primate, Ann Neurol 12:564 (1982).

Graham DI, McLellan D, Adams JH, Doyle D, Kerr A, Murray LS, The neuropathology of severe disability after head injury, Acta Neurochirg Suppl. 32:65 (1983).

Gronwall D, Wrightson P, Delayed recovery of intellectual function after minor head injuries, Lancet 2:605 (1974).

Hardman JM, The pathology of traumatic brain injuries, in: "Advances in neurology", Vol. 22, Thompson RA, Green JR, eds., Raven Press, New York (1979).

Jellinger K, Pathology and pathogenesis of apallic syndromes following closed head injuries, in: "The apallic syndrome", Ore GD, Gerstenbrand F, Lücking CH, Peters G, Peters UH, eds., Springer Verlag, Berlin (1977).

Lindenberg R, Freytag E, Morphology of cerebral contusions, Arch Pathol 63:23 (1957).

Lindenberg R, Freytag E, A mechanism of cerebral contusions: a patho-logic-anatomic study, Arch Pathol 69:440 (1960).

Lindenberg R, Freytag E, Brain stem lesions of traumatic hyperextension of the head, Arch Pathol 90:504 (1970).

Oppenheimer DR, Microscopic lesions in the brain following head injury, J Neurol Neurosurg Psychiat 31:229 (1968).

Peters G, Rothemund E, Neuropathology of the traumatic apallic syndrome, in: "The apallic syndrome", Ore GD, Gerstenbrand F, Lücking CH, Peters G, Peters UH, eds., Springer Verlag, Berlin (1977).

Pilz P, Strohecker J, Grobovschek M, Survival after ponto-medullary tear, J Neurol Neurosurg Psychiat 45:422 (1982).

Strich SJ, Diffuse degeneration of the cerebral white matter in severe dementia following head injury, J Neurol Neurosurg Psychiat 19:163 (1956).

Strich SJ, Notes on the Marchi method for staining degenerating myelin in the peripheral and central nervous system, J Neurol Neurosurg Psychiat 31:110 (1968).

Tomlinson BE, Brain-stem lesions after severe head injury, in: "The pathology of trauma", Sevitt S, Stoner HB, eds., J Clin Pathol 23, Suppl.4:154 (1970).

DIFFUSE AXONAL INJURY - A NEW CONCEPTUAL APPROACH

TO AN OLD PROBLEM

Thomas A. Gennarelli,[*] J.H. Adams,[**] D.I. Graham[**]

Department of Neurosurgery, University of Pennsylvania
Philadelphia, Pennsylvania, USA [*] ; Department of Neuropathology, University of Glasgow, Scotland [**]

Neurosurgeons commonly teach their trainees and explain to families
that patients with prolonged traumatic coma unaccompanied by mass lesions
have suffered "brain damage" as a result of their trauma. Despite this
frequently held belief, there is little to document the nature of the
"brain damage" that results from the immediate effects of trauma. Consequently, brain swelling, edema, ischemia or other secondary mechanisms
are offered in explanation for the clinical symptoms and the outcome from
injury. Evidence from our own investigations coupled with that of others
supports the concept that primary injury to axons is responsible for
many, if not most, cases of traumatic coma that is not due to mass lesions. We have named the condition **Diffuse Axonal Injury (DAI).** The term
DAI, therefore, is descriptive of both the clinical complex of prolonged
traumatic coma and the pathologic substrate that underlies that complex.
This paper reviews the genesis of the concept of DAI and discusses the
implications of DAI as an approach to prolonged traumatic coma.

THE CLINICAL SPECTRUM OF DIFFUSE AXONAL INJURY

Diffuse brain injuries form a distinct subset of head injuries.
These are characterized by a general disturbance of neurological function
that begins exactly at the moment of injury, the lack of a mass lesion,
and the absence of visible morphological change except for the most
severe variety. These injuries comprise a continuum from mild to severe,
beginning with concussion and extending through the various forms of DAI
(Fig. 1). Hence, there is no sharp division between various gradations of
severity. The fact which has caused nomenclature conflicts in the literature is, in our opinion, best resolved by definition. It is well recog-

15

TABLE 1

DIFFUSE AXONAL INJURY:

DAI

Prolonged traumatic coma of immediate onset
and unaccompanied by mass lesions

Brain Stem Contusion

Central Cerebral Trauma

Cerebral Concussion (Severe, Prolonged, Grade V)

Closed Head Injury

Contusion Cerebri

Diffuse Brain Damage of Immediate Impact Type

Diffuse Degeneration of White Matter

Diffuse Injury

Diffuse Neuronal Injury

Diffuse White Matter Shearing Injury

Edema

Inner Cerebral Trauma

Shearing Injury

Strich Injury

Unspecified Head Injury

nized that six hours of coma are associated with a significantly less favorable outcome than shorter periods. Therefore, six hours of coma is a convenient divisor between concussion and DAI. Traumatic coma of less than six hours can be defined as classical cerebral concussion, while coma greater than six hours as DAI. Within the category of DAI or prolonged traumatic coma, gradations in clinical status and outcome are recognized (1,2). Three degrees of DAI are distinguished as follows:
- **mild DAI:** coma of 6 to 24 hours duration,
- **moderate DAI:** coma longer than 24 hours without prominent brain stem motor signs,
- **severe DAI:** coma longer than 24 hours regularly accompanied by brain stem motor signs (decerebration, decortication).

Other distinctions of DAI may be considered, but we have obtained useful clinical experience by employment of these criteria. E.g., DAI of mild, moderate and severe type as defined above have mortality rates of 15, 24 and 51% respectively (1,2).

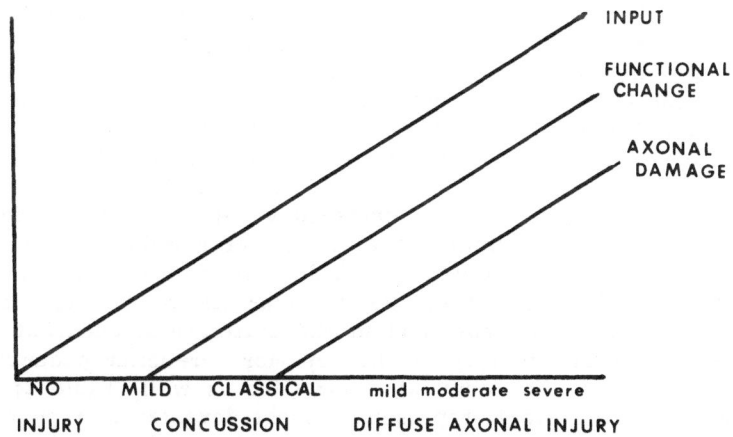

Fig.1: The clinical spectrum of the diffuse brain injuries is presented diagramatically. As the severity of injury increases from no injury to severe DAI, the amount of axonal damage grows. The acceleration input to the head, the cause of the diffuse brain injuries also increases. The relationships are depicted as linear for convenience since quantitative correlations are not yet available.

The nomenclature of the severe form of DAI has been problematic despite a clear recognition of this frequently occurring syndrome. Table 1 gives many of the synonyms used for severe DAI. This syndrome is characterized by immediate deep traumatic coma that lasts days to months and is accompanied by initial or delayed decerebration, brain stem reflex abnormalities, and autonomic dysfunction (hypertension, hyperhidrosis, hyperpyrexia). Death is frequent and is not due to increased intracranial pressure, but rather to complications of prolonged coma. Survivors become vegetative, die then or pass through the vegetative state before recovering further and, almost always, incompletely. Ample evidence that the axon is the primary locus of injury justifies the name severe DAI, in contradistinction to the numerous terms shown in Table 1 that reflect the pathophysiology less precisely.

THE PATHOLOGIC SPECTRUM OF DIFFUSE AXONAL INJURY

The continuum of clinical injury is shown in Figure 1 as a spectrum of increasingly severe pathological change. Documentation of the pathologic continuum has awaited the development of appropriate animal models especially of the spectrum of DAI. Human pathological material of the less frequently fatal types of injuries is unavailable. The longstanding debate of whether cerebral concussion is associated with pathological abnormality has been subjected to renewed studies by experimentalists using recently developed techniques. These studies have demonstrated axonal damage that is sparse but qualitatively similar to abnormalities seen in DAI (3,4,5,6,7). Our own work in primates where DAI is produced by angular acceleration demonstrates that as DAI worsens, the amount and distribution of axonal damage increases (8). Recently, this clinical-pathological spectrum has been verified in man as well (9). It now appears quite clear that axonal damage is strongly related to diffuse brain injury from concussion through the severe forms of prolonged coma. The severe form of DAI deserves a special discussion, since this entity has received considerable attention in the literature (cf. 8,10,11). In addition to the axonal damage that is scattered widely throughout the brain and only detectable by the microscope, small yet histologically visible hemorrhages are frequent. Biomechanical data suggest that these are tissue tears due to high levels of tensile or shear strain. Although usually occurring in the corpus callosum and in the dorsolateral quadrant of the rostral brainstem in the vicinity of the superior cerebellar peduncle (12-17) they are frequently found at other sites as well (Table 2; 18). These tissue tears, therefore, represent a most devastating type of damage where axons and vascular structures are torn by the force of the injury.

As the severity of injury increases, the sequence of axonal damage can best be summarized as follows. At first, only few axons are damaged requiring special neuroanatomic techniques for detection. However, axonal injury may be inferred from the presence of central chromatolysis as a reflection of the cell body response to axonal injury (19). As severity

TABLE 2

DIFFUSE AXONAL INJURY

TISSUE TEARS

Axonal and Vascular Disruption

95% Midbrain-Pons Tegmentum

92% Corpus Callosum

90% Third Ventricle Choroid Plexus

88% Parasagittal (Gliding Contusions)

88% Hippocampus

83% Periventricular

80% Septum Pellucidum

77% Lateral Ventricle, Choroid Plexus

62% Fornix

61% Cingulum

56% Thalamus

38% Midbrain-Pons Base

17% Basal Ganglia

from Grcevic (18)

increases, larger numbers of axons are damaged, yet injury is not macroscopically apparent. However, at the appropriate time injured axons can be detected by reduced silver impregnation stains. The most severe variety of DAI is characterized by larger numbers and a wider distribution of injured axons throughout the brain. In addition, tears in brain tissue cause multifocal areas of combined axonal and vascular damage that can be macroscopically recognized as small hemorrhagic lesions in characteristic locations.

AXONAL CHANGES IN DIFFUSE AXONAL INJURY

Morphology of Axonal Damage

The morphological changes associated with DAI appear to be qualitatively similar in mild or severe cases, while severity is determined by the number of injured axons. With silver impregnation two types - disruptive and non-disruptive axonal injury (Fig. 2) - are seen in humans with severe DAI and animals with all grades of DAI. Disruptive damage implies structural discontinuity of the axon (axotomy) that occurs at the time of injury (Fig. 2F and G). Club-, bulb- and ball-shaped dilatations occur without apparent axonal continuity beyond them. The changes are similar to those described by Cajal in experimental transections (20). These dilatations are classically named retraction balls in their most developed form. Retraction balls appear several days after injury, during which time they progressively enlarge. We have detected club-shaped axonal enlargements as early as six hours after injury but not sooner. In non-disruptive axonal damage axotomy cannot be verified, rather, injury-in-continuity or internal axonal damage is present (Fig. 2A and B). Axons have single or multiple bulbous dilatations with obvious continuity on both sides of the abnormalities. Unlike those disruptively damaged, these axons can be followed long distances in a histological section without finding terminal accumulations.

Axonal Morpho-Physiology in Diffuse Axonal Injury

Movement of intracellular material from the cell body down the axon and from the nerve terminals to the cell body is a general property of neurons called axoplasmic transport. Due to its complexities, axoplasmic transport has only recently allowed an understanding, e.g. of the mechanisms, the type of material involved, and the manner in which transport occurs (cf. 21-24). In general, materials are transported rapidly (100-400 mm/day), or slowly (1 mm/day). Rapidly transported materials are usually membrane-bound and include various organelles and vesicles. Horseradish peroxidase (HRP) is rapidly transported and thus been useful as a neuroanatomical marker (cf. 25). Cytoskeletal proteins are transported in the slow phase.

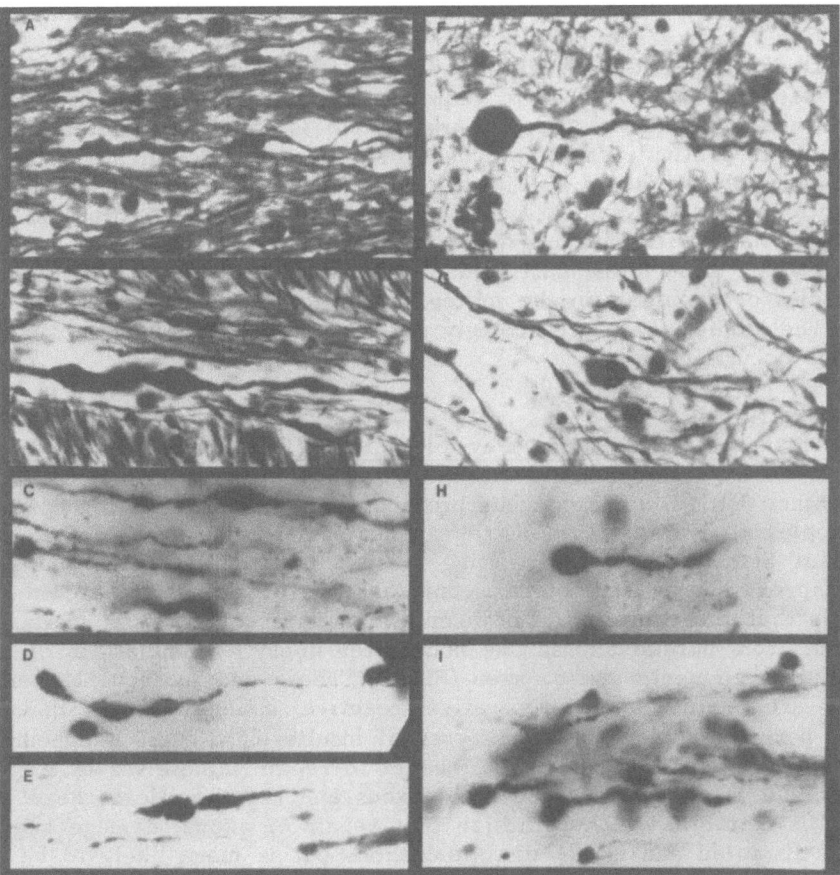

Fig.2: Photomicrographs A, B, F and G demonstrate axonal abnormalities
in reduced silver stains. A and F (Bodian's method) are from a
patient who died of severe DAI after head trauma, while B and G
(Palmgren's method) are from a monkey with severe DAI. In C, D,
E, H and I abnormal axonal profiles from monkeys with DAI are
shown by the HRP method. Two types of axonal abnormalities are
recognized which are identical in appearance in man and monkey
in silver and HRP-preparations. Fusiform swelling occurs in axons
with continuity maintained (non-disruptive damage, A-C), while terminal accumulations at sites beyond where axonal continuity can
be demonstrated (disruptive damage, D-I). All microphotographs are
similarly enlarged (x200).

Rapid axonal transport of HRP appears to be hardly influenced by the deep coma of severe DAI except at sites of axonal damage where HRP accumulates. The accumulation of axoplasm containing HRP results in dilatations that are identical in appearance to the terminal (retraction-) balls seen in silver stains (Fig. 2 H). Both disruptive (Fig. 2D, E, H, I) and non-disruptive (Fig. 2 C) axonal damage is present in HRP studies. From the silver impregnation studies referred to above, it can be argued that axotomy is present but is too far from the site of dilatations-in-continuity that it cannot be followed histologically. This argument is not valid for HRP transport studies, since HRP injected into the motor cortex is found in non-disruptive damage in the corpus callosum, ventrolateral thalamus and corticospinal tract and as distal as mid-pons. All these sites must remain in structural continuity with the cell bodies or terminals in the motor cortex for HRP to have reached them. The studies therefore strongly suggest that non-disruptive axonal injury is occurring indeed.

Ultrastructural Axonal Morpho-Physiology in Diffuse Axonal Injury

Since HRP is an exogenous protein, it cannot be used to determine the axoplasmic transport of endogenous materials in DAI, other than to document that transport is occurring to sites of axonal damage. Our recent, preliminary electronmicroscopic analysis of axonal damage in DAI suggests that the composition of axoplasm at sites of injury is markedly different from normal axons. One hour after injury accumulations of membranous elements can be seen (Fig. 3). These and subsequent changes (Fig. 4) are reminescent of reparative (reactive) changes present in central or peripheral axons in other types of insults (26). Thus, it appears that central axons damaged in DAI begin to repair themselves by a process that is a general property of axons and not specific to head injury. It is attractive to hypothesize that repair of axonal damage begins in both disrupted and in non-disrupted axons but is more likely to be successful in non-disruptive injury. The severity of the clinical syndrome and the outcome from DAI depends not only on the total number of damaged axons and their location but also on the proportion of disrupted to non-disrupted axons. However, it is well known that secondary axonal damage can occur in non-disrupted axons (27-29). Axoplasm can continue to be transported to a site of internal axonal injury where the transport mechanism is impaired. There, the resulting accumulation of transported material progressively enlarges, thins and finally ruptures the axon. Local chemical influences may similarly result in the conversion of a structurally continuous but internally damaged axon to a disrupted axon, thereby lowering the chances of successful functional repair.

A brief review of the microarchitecture of the axon and the dynamic aspects of axoplasmic transport provides insight into the internal axonal damage caused in DAI and the difficulty of successful axonal repair. Current concepts are indicative of an extremely complex axonal structure. An endomembranous system - the axoplasmic reticulum - similar to and pro-

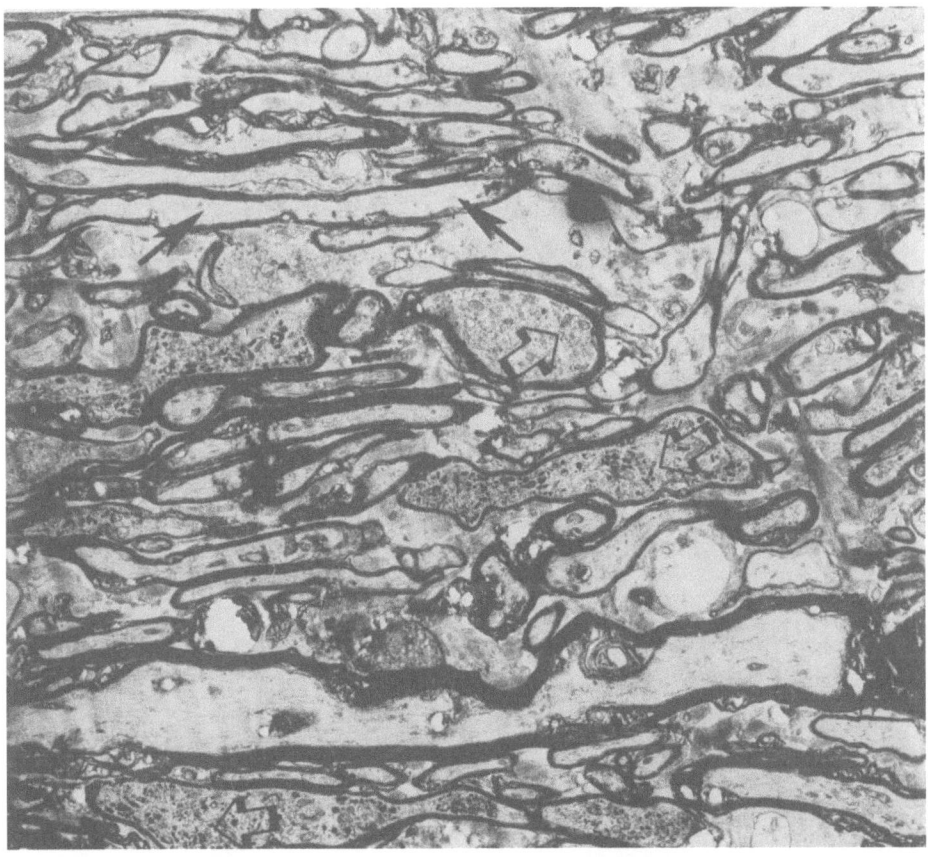

Fig.3: Low power (x5000) electronmicrograph of the corpus callosum from an animal with severe DAI one hour after injury. Several damaged axons are seen with accumulations of membranous material (open arrows). A normal axon is indicated by the solid arrow.

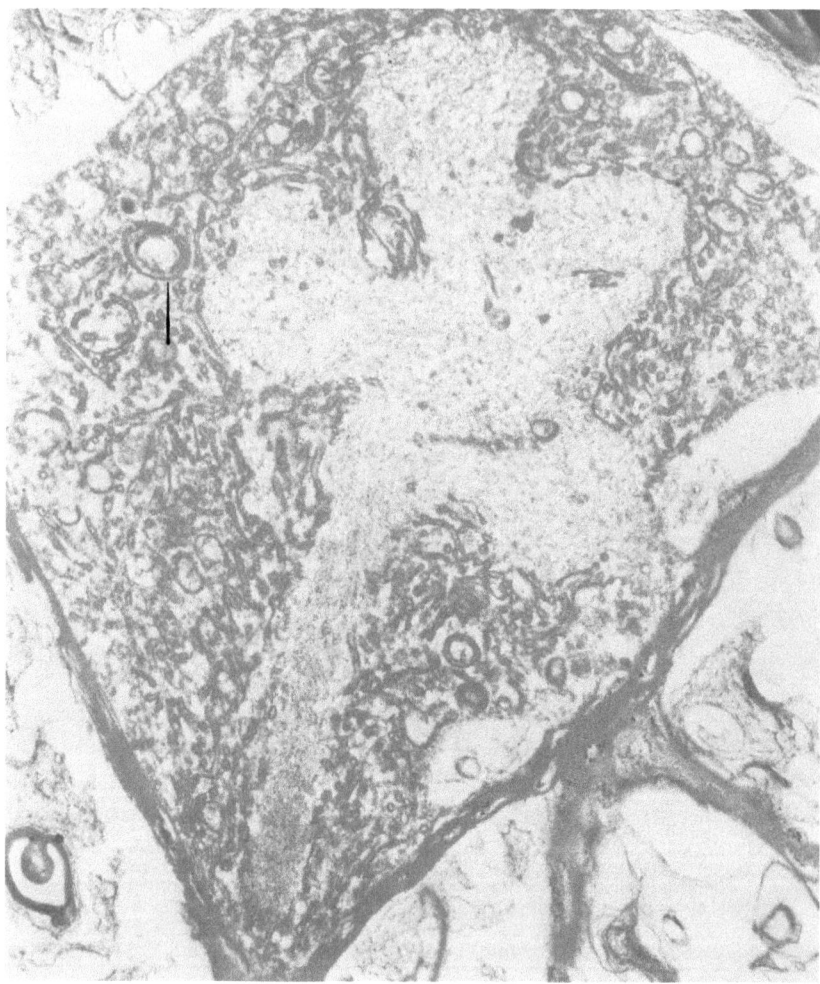

Fig.4: This markedly swollen axon ball was found in the corpus callosum 24 hours after severe DAI in a monkey. A small portion of the proximal appears at the bottom of the ball demonstrating the enormous degree of swelling. The ball is packed with membraneous organelles subjected to rapid axoplasmic transport. The organelles have been displaced peripherally by filamentous material which presumably has arrived later by slower transport (magnification x11,000).

bably derived from the smooth endoplasmic reticulum (SER), transcends the axon from the region of the hillock to the vicinity of the synapse. Interposed between the axoplasmic reticulum and the axolemma is a microtrabecular array of cross-linked neurofilaments and neurotubules that comprise the cytoskeleton of the axon. The materials involved in axoplasmic transport pass through this dense matrix. Proteins and other compounds synthesized in the cell body are subjected to anterograde transport from the cell body and are destined for use elsewhere in the axon. Proteins, i.e. enzymes or neural transmitters which are synthesized in the rough endoplasmic reticulum of the soma pass through the SER and Golgi apparatus. They are encapsulated there in a membrane to form vesicles. The vesicles are then transported in association with the axoplasmic reticulum by a mechanism that progressively dissociates and reforms the cross-linkages of the microtrabecular matrix. Other vesicular elements, such as mitochondria, lysosomes, peroxisomes are also transported by this mechanism. Non-vesicular transport of proteins, usually of cytoskeletal components such as tubulin and neurofilament proteins synthesized in free ribosomes in the soma, is not associated with the axoplasmic reticulum. It is subjected to the slow axoplasmic transport. Retrograde transport of materials toward the cell body is principally vesicular and includes phagosomes, multivesicular bodies and vesicles containing products of pinocytosis, endocytosis, or phagocytosis.

Obviously, the axon has an extremely complex microstructure through which membranous and non-membranous elements are transported in retrograde and anterograde directions. As a result there is considerable transaxonal traffic. It is therefore understandable that internal axonal injury upsets this delicately balanced finely tuned system. It is tempting to postulate that mechanical trauma, especially the tensile and shear strains causing DAI, produces distortion of and non-disruptive damage to the axoplasmic-reticulum-cytoskeletal complex at its weakest point. This is the node of Ranvier. Here, impairment of axoplasmic transport due to the localized injury results in accumulation of transported axoplasm proximal and distal to the damaged area. This influences profoundly the composition of the materials transported to the cell body and signals the presence of injury to the soma which then initiates synthesis of proteins needed for repair (19). Electronmicroscopy suggests that, first, membrane proteins and, later, neurofilamentous and microtubular proteins arrive at the site of axonal injury (Fig. 3,4). Further studies are clearly necessary to better understand the specific events that follow internal axonal damage and to identify the microarchitectural elements within the axon that are the locus of primary injury in DAI. Then, strategies may become available that promote internal axonal repair in non-disruptive axonal damage and prevent secondary axotomy.

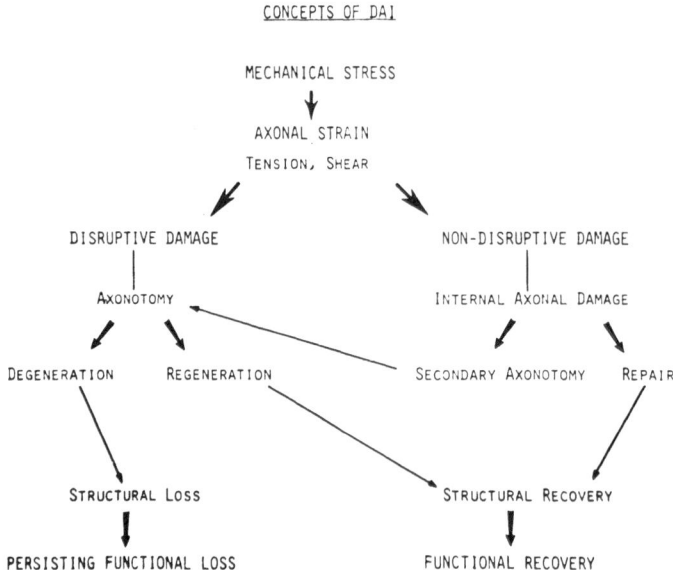

Fig.5.: Schematic concept of DAI of our laboratory. Further research is necessary to determine the exact proportion of disruptive to non-disruptive damage and the success of repair and regeneration.

SUMMARY

A comprehensive concept is shown in Figure 5. Diffuse axonal injury appears to be fundamentally different from brain damage which occurs in traumatic mass lesions. It is now clear that in traumatic coma there is a spectrum from mild to severe injury which correlates with the number of damaged axons. Two types of lesions occur in DAI, disruptive (axotomy) and non-disruptive damage (internal axonal damage). Repair of axonal damage is attempted in both types of axonal injury but is more likely to be successful in non-disruptive injury. Thus the clinical severity and outcome depend not only on the total number of damaged axons, but also on the proportion of disrupted to non-disrupted axons. However, this proportion can change, if secondary axonal damage occurs in non-disrupted axons causing them to become disrupted. The nature of non-disruptive axonal damage remains illusive but further investigations on the microarchitecture and the dynamic aspects of axonal transport may provide a better understanding. An improved understanding of the basic mechanisms of DAI is required for the development of specific therapeutic interventions aimed at promoting axonal repair and eliminating secondary axotomy to improve the outcome of DAI.

REFERENCES

1. Gennarelli TA, Head injury in man and experimental animals: clinical aspects, Acta Neurochirurg Suppl 32:1 (1983).
2. Gennarelli TA, Spielman GM, Langfitt TW, Gildenberg PL, Harrington T, Jane JA, Marshall LF, Miller JD, Pitts LH, Influence of the type of intracranial lesion on outcome from severe head injury: A multicenter study using a new classification system, J Neurosurg 56:26 (1982).
3. Gennarelli TA, Jane J, Thibault LE, Steward O, Axonal damage in mild head injury demonstrated by the Nauta method, in: "Advances in neurotraumatology," Villani R, et al, eds., Excerpta Medica, Amsterdam (1983).
4. Jane J, Timel RW, Pobereskyn LH, Tyson GW, Steward O, Gennarelli TA, Outcome and pathology of minor head injury, in: "Seminars in neurological surgery," Grossman RG, ed., Raven Press, New York (1982).
5. Jane J, Steward O, Gennarelli TA, Rimel R, Pathology of minor head injury, Proc Am Assoc Neurol Surg 127: (1981).
6. Povlishock JT, Becker DP, Cheng CLY, Vaughan GW, Axonal change in minor head injury, J Neuropath and Exp Neurol 42:225 (1983).
7. Barron KD, Axon reaction and its relevance to CNS trauma, in: "Head injury: Basic and clinical aspects," Grossman RG, Gildenberg PL, eds., Raven Press, New York (1982).
8. Gennarelli TA, Thibault LE, Adams JH, Graham DI, Thompson CJ, Marcincin RP, Diffuse axonal injury and traumatic coma in the primate, Ann Neurol 12:564 (1982).
9. Pilz P, Axonal injury in head injury, Acta Neurochirg, Suppl 32:119 (1983).
10. Adams JH, Graham, DI, Murray S, Scott G, Diffuse axonal injury due to nonmissile head injury in humans: An analysis of 45 cases, Ann Neurol 12:557 (1982).
11. Adams JH, Gennarelli TA, Graham DI, Brain damage in nonmissile head injury: Observations in man and in subhuman primates, in: "Recent advances in neuropathology," Smith R, Cavanagh J, eds., Edinburgh, Churchill, (1982).
12. Strich SJ, Diffuse degeneration of the cerebral white matter in severe dementia following head injury, J Neurol Neurosurg Psychiatry 19:163 (1956).
13. Strich SJ, Shearing of nerve fibers as a cause of brain damage due to head injury, Lancet 2:443 (1961).
14. Strich SJ, The pathology of brain damage due to blunt head injuries, in: "The late effects of head injury," Walker AE, Caveness WF, Critchley M, eds., Springfield, Il, Thomas, (1969).
15. Strich SJ, Lesions in the cerebral hemispheres after blunt head injury, J Clin Pathol 23 (Suppl 4):166 (1970).
16. Adams JH, The neuropathology of head injuries, in: Handbook of clinical neurology. Vol 23, "Injuries of the brain and skull", Part I. Vinken PJ, Bruyn GW, eds., American Elsevier, New York (1975).

17. Adams JH, Mitchell DE, Graham DI, Doyle D, Diffuse brain damage of immediate impact type, Brain 100:489 (1977).
18. Grcevic N, Topography and pathogenic mechanisms of lesions in "inner cerebral trauma," Rad Jazu (Med) 4:265 (1982).
19. Grafstein B, The nerve cell body responses to axonotomy, Exp Neurol 48:32 (1975).
20. Cajal S, Degeneration and regeneration of the nervous system, Oxford University Press, London, (1928).
21. Weiss DG, "Axoplasmic transport," Springer-Verlag, Berlin (1982).
22. Weiss DG, Gorio A, "Axoplasmic transport in physiology and pathology," Springer-Verlag, Berlin (1982).
23. Weiss DG, General properties of axoplasmic transport, in: "Axoplasmic transport in physiology and pathology," Weiss DG, Gorio A, eds., Springer-Verlag, Berlin (1982).
24. Grafstein B, Forman DS, Intracellular transport in neurons, Physiol Rev 60:1167 (1980).
25. Mesulam MM, "Tracing neural connections with horseradish peroxidase", John Wiley, Chichester (1982).
26. Lampert PW, A comparative electronmicroscopic study of reactive degenerating, regenerating, and dystrophic axons, J Neuropathol and Exper Neurol 26:345 (1967).
27. Kapeller K, Mayor D, An electron microscopic study of the early changes proximal to a constriction in sympathetic nerves, Proc Roy Soc B 172:39 (1969).
28. Kapeller K, Mayor D, An electron microscopic study of the early changes distal to a constriction in sympathetic nerves, Proc Roy Soc B 172:39 (1969).
29. Kao CC, Spinal cord cavitation after injury, in: "The spinal cord and its reaction to traumatic injury," Windle WF, ed., Marcel Dekker, Inc New York, Chapter 16 (1980).

HISTOPATHOLOGY AND COMPUTERIZED TOMOGRAPHY OF HUMAN TRAUMATIC CEREBRAL SWELLING

Raymond A. Clasen, P. Guariglia, R.J.
Stein, S. Pandolfi and J.J. Lobick

Departments of Pathology and Diagnostic Radiology
Rush-Presbyterian-St. Luke's Medical Center, and
the Office of the Medical Examiner
County of Cook, Chicago, Il. 60612

The Relationship between Cerebral Swelling and Edema

Any discussion of cerebral edema must begin with a definition of terms. This relates to the question of whether a fundamental distinction can be drawn between cerebral swelling and cerebral edema, or whether these terms simply describe the gross and microscopic picture of what is essentially the same process (1-3). It has been proposed that a gross distinction can be made between swelling and edema. In the former the cut surface of the brain is dry and in the latter wet. Those holding the opposing view argue that this difference is quantitative, i.e. that the wet brain is more edematous than the dry brain. The resolution of this controversy depends upon the definition of the histopathologic basis of cerebral swelling. This has been a contribution of electron microscopy.

In our experience, the swelling associated with cerebral infarction is dry. Electron microscopic studies have shown this to be predominantly an intracellular process (4,5). The swelling associated with the cryogenic cerebral lesion in the rhesus monkey is generally, but not always, wet. This is predominantly an extracellular process (6). It has been proposed that in cerebral swelling the process is intracellular and in cerebral edema extracellular (1). In our experience, the swelling associated with brain tumors is generally dry but may in a significant number of instances be wet. Ultrastructural studies have shown that this process is predominantly extracellular (7,8). This is, we believe, conclusive evidence that the dry and wet cut surface of the brain does not define an intra- and extracellular process, respectively.

If there is no fundamental difference between cerebral swelling and edema, then the relationship between them becomes a matter of definition. Reichardt in 1905 defined cerebral swelling as an increase in brain volume, not caused by hyperemia, excess of fluid (excluding edema), or a space occupying process (9). We prefer not to exclude the latter since the gross signs by which we recognize an increase in brain volume may be caused by a hemorrhage or tumor in the absence of edema, and we would like to use the term to describe these gross signs. Cerebral swelling is defined in this paper as an abnormal increase in the volume of cerebral tissue. Cerebral edema is defined as a form of cerebral swelling in which there is an increase in tissue fluid within brain substance. These definitions reflect current usage in the United States (10).

Traumatic Cerebral Swelling and Edema

Few would argue with the dictum that cerebral swelling is important in the pathology of cerebral trauma but its histopathologic substrate has remained elusive. In an early description of traumatic cerebral edema, Courville described two gross forms, a generalized type and a focal form occurring around contusions (11). His subsequent descriptions of the histologic basis of this edema in terms of perivascular and pericellular space enlargement and swelling of oligodendroglia, are by the standards of the post-electron-microscopic era artifacts of tissue preparation (12). The description of three stages of cerebral swelling as tumefaction, edema, and liquefaction (13,14) has not been confirmed by others and finds no support in electron microscopic studies (15). Feigin (16) has described staining of the white matter around contusions in a jaundiced patient with acute cerebral trauma, but he gives no histologic description of the changes in this brain and the staining could be confined to necrotic tissue. His interpretation of the extensive atrophy of white matter in old traumatized brains as being secondary to edema, rather than primary traumatic necrosis, is open to question. Hardman (17) described focal traumatic edema as being vasogenic but also did not provide a histologic description nor is one given in his references. In Lindenberg's review (18), there is no histologic description of either focal or generalized traumatic edema. He describes only gross swelling of necrotic tissue. In the review of Adams (19), less than one-half page is devoted to post-traumatic cerebral edema and there is only one gross illustration. He speaks of diffuse swelling of both hemispheres as having a predilection for children and of showing pallor of the white matter and early reactive gliosis. He further states that edema of the vasogenic type is frequent in the white matter adjacent to contusions. In ordinary histologic sections, vasogenic edema is recognized by the staining of the proteins in the edema fluid either by the periodic acid-Schiff (PAS) method or by the collagen dye in a trichrome stain (1). Neither technique was used in these studies. In another review, in which both staining techniques were used, vasogenic edema is not even mentioned (20). The authors report three patients who died as the result of herniation from generalized cerebral edema, the histologic basis of which was not defined. They described 22 patients with diffuse homogenizing necrosis of neurons who showed severe swelling implying that post-traumatic swelling may be simply ischemic swelling.

Computerized Tomography of Traumatic Cerebral Edema

The advent of computerized tomography (CT) has provided the means for both observing and measuring cerebral edema in the living patient and invites a reassessment of cerebral edema in head trauma based on correlative CT and histologic studies.

Cerebral edema in the form of hypodensity of white matter with mass effect occurs in about 5% of patients with traumatic epidural hemorrhages and implies an unfavorable prognosis (21). In acute and subacute subdural hematomas, edema occurs in about 15% of patients and is described as being hypodense and homogeneous. The histopathology in relation to CT of neither form has been described. Cerebral edema is not seen in chronic subdural hematomas.

The pathology of subdural hematoma is illustrated in the following case report: This 88 year old female was admitted to an outside hospital in a confused state following a fall in her home. A CT scan revealed an acute right subdural hematoma with compression of the underlying

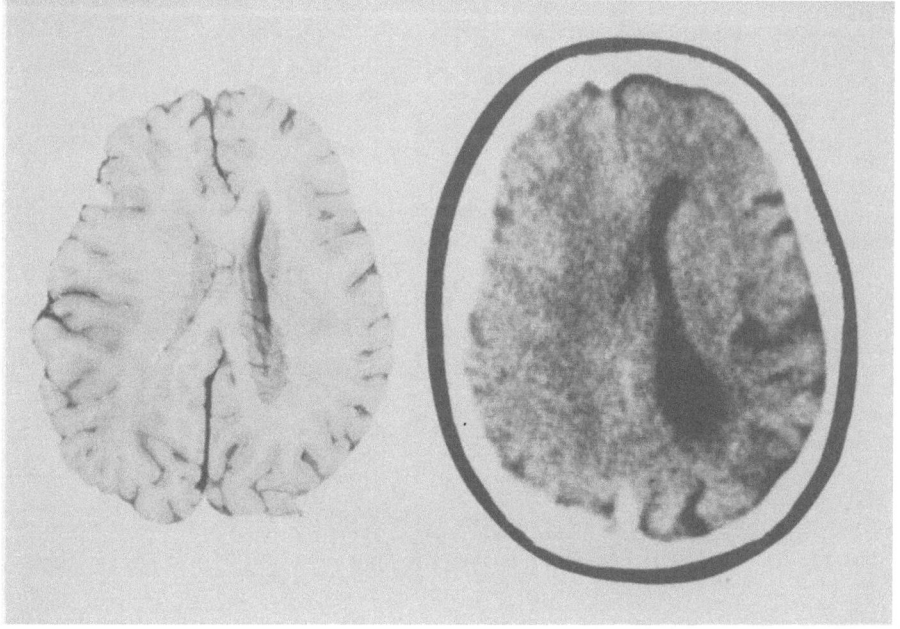

Fig.1: Cross section of brain and CT image obtained 7 days earlier

hemisphere but no areas of hypodensity (Figure 1). A technetium scan showed flattening of activity over the right hemisphere but the blood flow to the hemispheres was symmetrical. The patient was given only medical management. She became semicomatose and expired on the seventh hospital day. Examination of the fixed brain revealed a small old contusion of the right temporal lobe. The right hemisphere showed marked distortion (Figure 2) with ventricular collapse and signs of herniation but in spite of this the fixed right hemisphere weighed 462.0 gms as compared to 485.5 for the left. This 4.8% decrease in weight is significant (22). There was no histologic evidence of significant edema in either hemisphere. We interpret the weight difference as a reflection of atrophy related to the old contusion and as further evidence that the right hemisphere was not swollen. We suggest that in the absence of areas of hypodensity in the CT scan, patients with extra-axial hemorrhages do not have swelling of the underlying hemisphere. Unfortunately, such swelling may develop when the hematoma is surgically evacuated and blood flow restored to the hemisphere (21). The basis of this swelling is probably cerebral ischemia. The problem is discussed in a previous publication (26). This postsurgical swelling would probably not have occurred in our patient since the blood flow was not diminished in the underlying hemisphere.

Focal traumatic cerebral edema is seen in association with contusions. Three types of cerebral contusions have been described on the basis of CT scans (21). Type I contusions consist of focal areas of hypodensity and comprize 12% of contusions. They occur most often in white matter and are reversible in 2-3 weeks. The histopathology of this

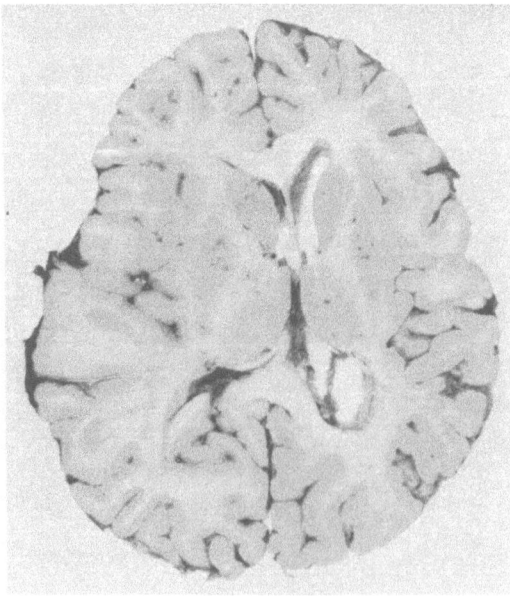

Fig.2: Cross section of above brain at a lower level

lesion has been described in one patient (23). Grossly there was no evidence of contusion but microscopically the grey matter showed rare petechial hemorrhages and mild acute inflammation. The neuropil was not swollen and the histologic basis for the CT hypodensity was separation of the nerve fibers in the underlying white matter. Nerve fiber abnormalities were not seen and there was no evidence of stainable edema fluid between the fibers.

Type II contusions contain areas of hemorrhage in the hypodense foci producing a mottled or salt and pepper appearance. Contre-coup contusions are classified as Type III. Hemorrhages occurring in the focal hypodense areas which are of sufficient size to be visualized on CT are designated as contusional hemorrhages (21).

In the acute phase, Type II and III contusions show swelling of the neuropil with interspersed dense areas (23,24). The neurons generally have pycnotic nuclei and cytoplasm which stains red with eosin. These are changes which are characteristics of ischemia. The histologic changes in contused white matter are more variable ranging from pallor of nerve fibers with the luxol fast blue stain to areas of necrosis associated with pools of PAS-positive edema fluid (23). By the definition given above this is edema but one could quarrel with its designation as vasogenic edema since the latter was originally described as occurring in viable tissue (25). Vasogenic edema is also seen in the white matter adjacent to the contusion, at least when significant hemorrhage is present (23). Cerebral hemorrhages of non-traumatic etiology are also associated with vasogenic edema (26).

Generalized traumatic cerebral edema presents on CT as a diffuse area of hypodensity which may involve all or part of a hemisphere and may be unilateral or bilateral (27,28). The CT numbers (CT#) in these hemispheres are low (29). This lesion is relatively rare occurring in 6-7% of abnormal scans in patients with severe head trauma. When this form of edema is associated with small parenchymal hemorrhages in the region of the corpus callosum, and subarachnoid hemorrhage, the condition has been designated as shearing injury of the white matter (30). These patients present in deep coma and generally show decerebrate posturing. The prognoses for recovery is poor but some do survive.

The following case report illustrates diffuse traumatic cerebral edema: The patient was an 18 month old male black infant. According to the mother, the patient fell from his crib on the evening of 3/23/82. He was alert but had a bruise on the forehead. He was placed back in the crib and subsequently went to sleep. On the following morning he was lethargic and febrile. The patient was admitted to an outside hospital with a diagnosis of fever and dehydration. He was transferred to our institution on 3/25 in a comatose state with reactive pinpoint pupils. The eyes deviated to the right and he had nystagmus with the fast component going to the right. He was flaccid with bilaterally positive Babinski signs. An area of ecchymosis was noted on the right side of the

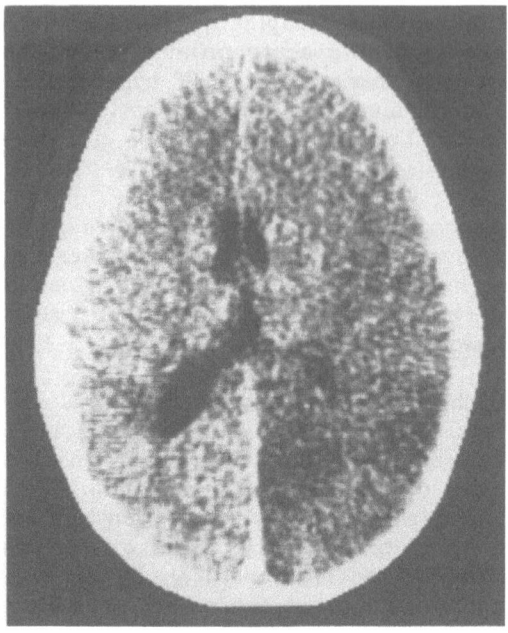

Fig.3: CT image of 18 month old infant two days after head injury

face. The rectal temperature was 102° . A CT scan, obtained on the
Siemens Somatom 2, showed diffuse edema of the right hemisphere
(Figure 3). The mean CT# of the right hemisphere was 27.8 Hounsfield
units (HU) with a standard deviation (SD) of 3.5 and a standard error
($S\bar{x}$) of 0.1 (n= 3,564). The lower portion of the left hemisphere measured 30.9 HU (SD 3.8, $S\bar{x}$ 0.1, n= 1,925). This difference is statistically significant at the 1% level of probability.

A pressure monitor was placed and a pentobarbital coma induced.
The intracranial pressure measured 20 cm of water (15.4 mmHg). That
evening the right pupil became larger than the left and was fixed. The
doll's eyes reflex was negative. The patient was given steroids, mannitol, and furosemide. He was also placed on a hypothermic blanket. The
right pupil decreased in size and became reactive. On 3/26, bilateral
papilledema was noted and the right pupil again became fixed and dilated.
On 3/27 both pupils were fixed and dilated and there was no response to
painful stimuli. On 3/29, there was no response to cold caloric stimulation. A nuclide angiogram showed perfusion of the brain. A CT scan
showed an increase in the edema of the right hemisphere with involvement
of the left frontal lobe (Figure 4). The edema in the right hemisphere
did not involve the basal ganglia or the white matter. The mean CT# of
different parts of the brain are given in Table I for white matter, the
anterior value was obtained from the frontal lobe and the posterior value
from the parietal lobe adjacent to the ventricle. The values for cerebral

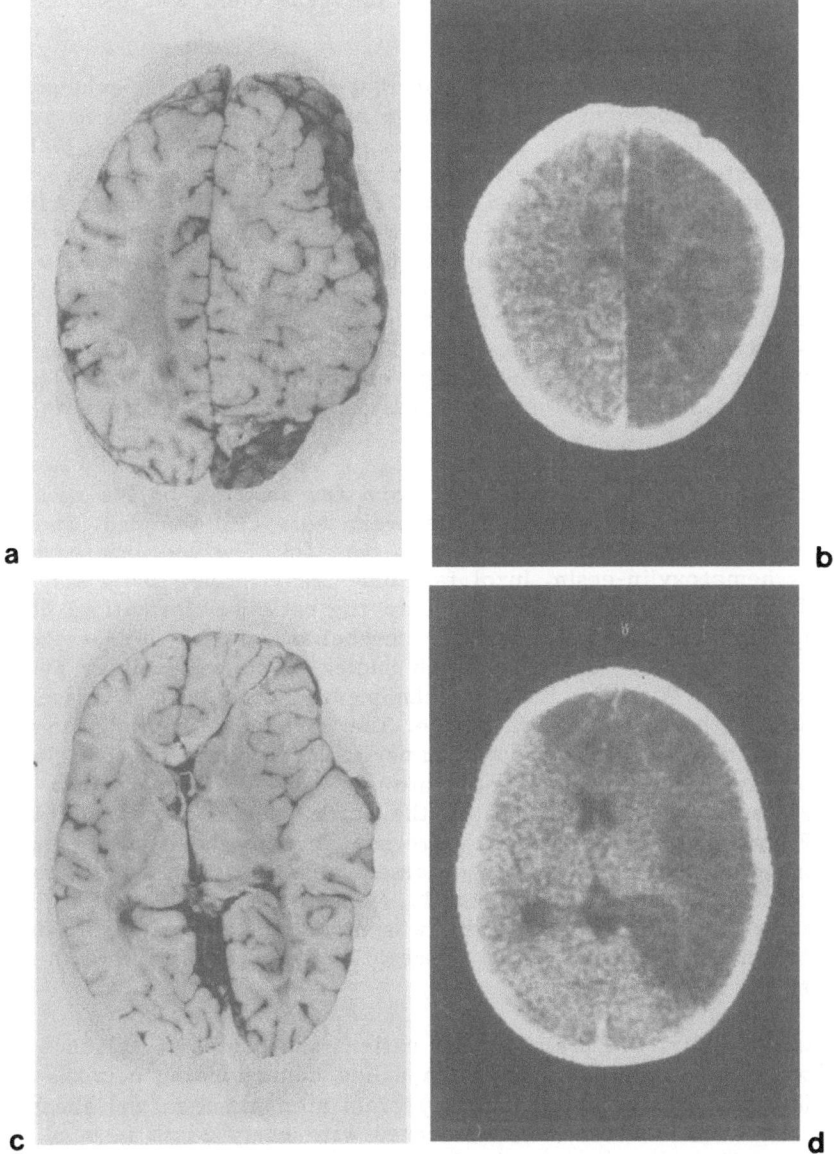

Fig.4: CT images of infant (cf. Fig. 3) obtained 6 days after insult
     compared with gross specimens of the brain

cortex were obtained from the insular region. Values for the left hemisphere were taken from comparable areas. No enhancing lesions were
seen.

On 3/30, the pentobarbital drip was discontinued. On 4/1, the
nuclide angiogram showed no blood flow and the electroencephalogram was

flat. There were no spontaneous respirations. The child was pronounced dead and the respirator disconnected.

Autopsy revealed subgaleal hemorrhage. There was no evidence of extracerebral or subarachnoid hemorrhage and no contusions were seen. The external surface of both hemispheres was flattened and the pial vessels congested, more on the right than the left. The right hippocampal gyrus was markedly herniated as were the cerebellar tonsils. After fixation in 20% formalin buffered with 1% acetic acid, the left hemisphere weighed 430.5 and the right 464.0 g. The right hemisphere weighed 7.8% more than the left but this is not an accurate assessment of the mass increment since the left hemisphere was also swollen on CT. The brain was sectioned in the plane of the CT scan (Figure 4). The right hemisphere showed subfalcine herniation and ventricular collapse. The left ventricle was dilated. The tissue was friable but there were no focal lesions.

Labeled sections were prepared from the surfaces of the specimens shown in Figure 4. Other areas of the brain were also sampled. These were slowly dehydrated and embedded in paraffin. The sections were stained by hematoxylin-eosin, luxol fast blue-cresyl violet and PAS: Scattered ischemic neurons were found in the reticular formation, both hippocampi, the basal ganglia, and the cerebellar cortex. These cells had homogenous eosinophilic cytoplasm with nuclei which were either pycnotic or stained poorly. With the Kluver technique the cytoplasm of these neurons stained blue rather than purple. The cortex of the right hemisphere and left frontal lobe showed numerous necrobiotic neurons, loss of neurons, and marked capillary proliferation (Figure 5B). Spongioform change was seen in many sections at the border between grey and white matter (Figure 5A). These changes were seen in both hemispheres but were much more pronounced in sections obtained from the right hemisphere and left frontal lobe. The slides were easily separated by blind sorting. The white matter in the right hemisphere was more pale than the left and contained numerous myelin basket figures. These were quite rare in sections from the left hemisphere.

Histologically, the brain of this patient showed clear evidence of diffuse ischemic change. This has been called homogenizing necrosis of neurons (20) and is, along with primary brain stem damage and shearing injury of the white matter (30), associated with early death in head injury patients. The changes seen in the edematous cortex of our patient are those of ischemia, but greatly accentuated. In a similar reported case widespread ischemic necrosis was also seen (31). Without the CT scan, the myelin pallor and numerous basket figures seen in our patient would be interpreted as ischemic swelling in the underlying white matter.

In anoxic-ischemic encephalopathy white matter is hypodense on CT both in the adult and the infant and the basal ganglia are involved (32). In infarction due to vascular occlusion there is hypodensity of both white and grey matter after the second day in both children (33)

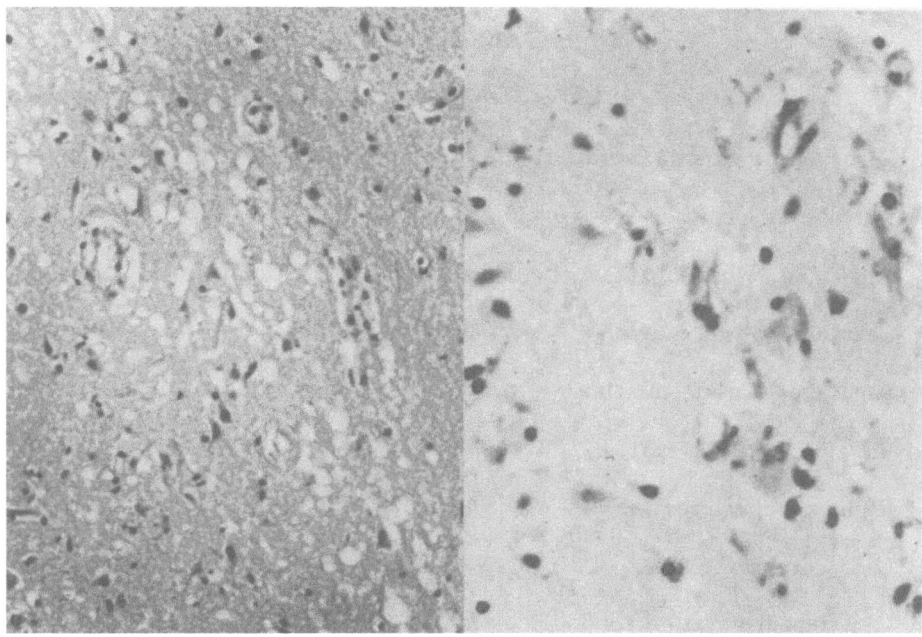

Fig.5: Left section of cerebral cortex stained by hematoxylin-eosin.
Right section of cerebral cortex stained by the Kluver technique.

and adults (34). The initial CT scan in our case (Figure 3) showed
diffuse edema of the right hemisphere without demarcation between white
and grey matter. The second scan (Figure 4) shows this demarcation but
with reversal of the normal white-grey matter density relationship. In
the four day time period between the scans, the CT# of the grey matter
decreased by 3.3 HU while the CT# of the white matter increased by 2.5
HU. In the second scan the CT# in the right posterior white matter is
significantly higher than the CT# of the left posterior white matter
(Table 1). We interpret this sequence as indicating worsening of edema in
the grey matter with compression of the white matter and obliteration of
the extracellular space.

While it is possible that the histologic changes which we observed
in the white matter of the right hemisphere occurred in the two day
interval between the time the scan was obtained and the autopsy was
performed, we doubt that this happened. Myelin pallor with histologic
stains is evidence of edema (or atrophy) only if it is due to separation
of the fibers. We suggest that the white matter was compressed at the
time of autopsy, that the pallor reflects an intrinsic abnormality of the
nerve fiber probably related to anoxia, and that the myelin basket figures reflect a change which is insufficient to cause gross swelling.
These views are supported by the fact that while the brain showed anat-

TABLE 1

	$\bar{x}$	SD	$S\bar{x}$	n
Right basal ganglia	35.3	3.6	0.2	489
Left basal ganglia	35.3	3.8	0.2	489
Right ant. white matter	30.5	3.4	0.3	332
Left ant. white matter	31.4	3.6	0.3	184
Right post. white matter	30.3	3.9	0.3	172
Left post. white matter	28.0	9.3	0.7	197
Right cerebral cortex	24.5	3.4	0.2	228
Left cerebral cortex	35.4	4.1	0.3	285

Attenuation numbers in Housfield units from the lower image shown in Figure 4.
$\bar{x}$ = mean, SD = standard deviation, $S\bar{x}$ = standard error, n = number of observations.

omic evidence of swelling, there is no evidence of the cut sections (Figure 4) that the white matter is swollen.

Diffuse traumatic cerebral edema is not simply an ischemic lesion. It does not occur in the distribution of a single vessel and the basal ganglia are not involved. It is an ischemic lesion with a unique distribution determined by pathogenetic factors associated with physical trauma probably related to changes in local blood flow.

Computerized Tomography of Traumatic Cerebral Swelling

There is also a traumatic space occupying lesion in which the brain is not hypodense. This has been designated as traumatic cerebral swelling (TCS) rather than edema. It usually involves both hemispheres. It has been reported to occur in 13% of abnormal scans in trauma patients generally (21) and in 28% in the pediatric age group (35,36). In adults the lesion is isodense (21). In children there is a measurable increase in CT# of white matter (35) and a subsequent decrease to normal levels as the condition resolves over a one week period (36). This form of cerebral swelling may be confined to one hemisphere where it has been described as occurring within 20-30 min after trauma (37,38).

TCS has been attributed to vascular congestion (35,39,40). In the living young adult, the mean ventricular volume of the brain is 30.9 ml (41) and the total spinal fluid volume, 76 ml (42). Although volumetric measurements have not, to our knowledge, been published, there is no indication from linear CT measurements that the spinal fluid volume in children is appreciably different (43-45). The blood volume of the brain as a whole, as measured by emission CT of technetium labeled red blood cells, is 4.34 ml/100 g (46). The blood volume of human white matter, as measured by x-ray absorption CT following intravenous iodinated contrast material, is 2.35 ml/100 g and the ratio of the blood volume in white to grey matter is 0.44 (47). Absorption rather than emission values were used to compute these values because the spatial resolution of the former is superior. The ratio of the volume of white to grey matter in the human cerebral hemisphere at age 20 is 0.77 and the mean hemisphere weight, 526 g (48). Based on the above data, the supratentorial blood volume as a whole is 45.7 ml and for white matter 10.8 ml. If congestion resulted in a doubling of cerebral blood volume (CBV), ventricular collapse could result only if the remaining cerebrospinal fluid was not displaced but TCS also includes obliteration of the cisterns. If the congestion (swelling) is confined to white matter, the volume increment could clearly not account for ventricular collapse. If congestion resulted in tripling of CBV, an increment of 91.4 ml would result. This could theoretically account for ventricular collapse.

In the rhesus monkey there is a linear relationship between $PaCO_2$ and CBV (49). In that study the maximum reported CBV increase with elevation of $PaCO_2$ was 51.4% of normal. In a series of experiments in cats the maximum increase in CBV was produced by a combination of nembutal anesthesia, anoxia, and hypercapnia (50). This was a 100% increase. In two patients with pseudotumor cerebri increases in CBV of 62 and 110% have been reported (51). It is interesting to note that the authors did not attempt to explain the accompanying cerebral swelling on the basis of vascular congestion. In one study of experimental cerebral trauma in cats a 289% increase in CBV was reported (52). The experimental method used sometimes produces intracerebral hemorrhage (53) and this was not ruled out in that study. We conclude that the maximal allowable increase in cerebral blood volume, due to congestion, under pathologic conditions, is about 100% and therefore, vascular congestion cannot account for the ventricular collapse seen in TCS.

Initial studies of TCS reported an increase in blood flow (54). Four children with this condition have had CBV measured by emission CT (46). All patients showed CT signs of cerebral swelling at the time that the measurements were made. Three patients showed a decrease, not an increase, in CBV. The mean was 12% and the decrease was greater in grey than white matter. One patient showed an increase in CBV of 21%.

TCS is associated with an increase in the CT# of white matter of about 3 HU (35). There is a linear relationship between the hematocrit and CT# of whole blood; a 50% hematocrit corresponds to 60 HU (55).

Assuming that the CT# of whole brain is the sum of the vascular and extravascular components, doubling the blood volume of white matter would increase its CT# from 30.0 to 30.7 HU. Clearly, congestion can not explain the increase in CT# of white matter in TCS.

In the original description of diffuse traumatic edema (27), it was suggested that if this lesion were bilateral, it could not be recognized visual examination of the CT scan but would be manifest only by ventricular collapse and obliteration of the cisterns which is the CT picture of TCS. We propose this as the best explanation for TCS which can be offered at the present time. This also offers an explanation for the white matter hyperdensity observed in children. This could represent compression of white matter as is demonstrated in our second case report. The extracellular space is larger in the immature brain (56) and such compression would be more likely to occur in children. In a reported study of CT and histopathology, two patients with bilateral isodense swelling without coexisting contusions were described (57). Both showed severe cortical ischemic damage.

The mean weight of supratentorial grey matter is 595 g (48) of which 298 g are basal ganglia (58). An intracranial volume increment of 80 gms could theoretically collapse the ventricles (41,42). The addition of 80 g of Ringer's solution to 297 g of cerebral cortex with a water content of 85.3% (59) would increase the water content to 88.1%. For plasma the increase would be 86.6%. This would result in a decrease in CT# of 6.3 HU for Ringers and 2.9 HU for plasma (26). The actual composition of the fluid which accumulates in ischemic encephalopathy is intermediate in composition but is probably closer to Ringer's solution. Five HU would be a reasonable estimate. A decrease of this magnitude if spread over the entire cerebral cortex might not be visible in relation to the basal ganglia but should be measurable if normals can be established.

The occurrence of TCS within minutes after trauma is not consistent with an ischemic lesion (37,38). We propose that both of these patients are examples of acute extra-axial traumatic hemorrhages. This is fairly apparent on one of the scans (38) and at least suggestive on the other (37).It is also consistent with both clinical courses. This is also a second possible explanation for TCS. Undetected hemorrhage occurring at the base or apex of the skull, with brain compression could produce the CT signs of bilateral TCS. It may also be noted in our first case report that what appears on CT to be a relatively small amount of blood (Figure 1) may produce severe distortion of the brain (Figure 2). Surgical intervention was withheld in this patient because it was thought that there was not a significant volume of subdural blood and the distortion was due to isodense cerebral swelling.

Classification of Traumatic Cerebral Edema

Traumatic cerebral edema is not a single entity. It includes both cytotoxic and vasogenic components (25) as well as unclassifiable elements. It occurs within cerebral contusions. In the cerebral cortex this takes the form of an intracellular edema which has the histology of an ischemic lesion. Anoxic edema is generally classified as cytotoxic edema but this is a poor term for lack of oxygen since it implies that the actual swelling is due to something else like lactic acid. It might be better to separate anoxic edema from cytotoxic edema and make it a separate entity (10). Diffuse traumatic edema is not confined to an area of contusion but belongs to the same general category.

Traumatic cerebral edema in the contused white matter takes the form of collections of protein rich edema fluid in areas of tissue necrosis. This may be reasonably classified as vasogenic edema. The edema which develops outside of the area of contusion is clearly of the vasogenic type. This is the reversible component of cerebral contusions demonstrated on serial CT scans in head injury patients (60).

The histologic changes seen in association with the CT lesion designated as a Type I contusion are difficult to classify. The process is confined to white matter, like vasogenic edema, but unlike that entity, there is no evidence that the edema fluid contains plasma proteins. It is clearly not cytotoxic, anoxic, hydrocephalic, osmotic, or hydrostatic, but simply traumatic. Must we create yet another category or is it time for the paradigm to change?

ACKNOWLEDGEMENTS

This work was supported by a grant from the National Institute of Neurological and Communicative Disorder and Stroke, NS-03677.

REFERENCES

1. Zülch KJ, Neuropathological aspects and histologic criteria of brain edema and brain swelling, in: "Brain edema," Klatzo I and Seitelberger F, eds., Springer-Verlag, New York (1967).
2. Greenfield JG, The histology of cerebral edema associated with intracranial tumors (with special reference to changes in the nerve fibers of the centrum ovale), Brain 62:129 (1939).
3. Small RG, and Krehl WA, Cerebral swelling and cerebral edema, J Neuropath Exp Neurol 11:192 (1952).
4. Hartmann JF, Becker RA, and Cohen MM, Cerebral ultrastructure in experimental hypoxia and ischemia, in: "Biochemistry, ultrastructure and physiology of cerebral anoxia, hypoxia, and ischemia," Cohen MM, ed., S. Karger, New York (1973).
5. Garcia JH, Lossinsky AS, Kauffman FC, Conger KA, and Mena H,

Fine structure and biochemistry of brain edema in regional
cerebral ischemia, in: "Cerebrovascular diseases," Price TR and
Nelson E, eds., Raven Press, New York (1979).

6. Bakay L, and Lee JC, "Cerebral edema," Charles C. Thomas,
   Springfield (1965).

7. Long DM, Hartmann JF, and French LA, The ultrastructure of human
   cerebral edema, J Neuropath Exp Neurol 15:373 (1966).

8. Aleu F, Samuels S, and Ransohoff J, The pathology of cerebral edema
   associated with gliomas in man, report based on ten biopsies,
   Amer J Path 48:1043 (1966).

9. Perret GE, and Kernohan JW, Histopathologic changes of the brain
   caused by intracranial tumors (so-called edema or swelling of
   the brain), J Neuropath Exp Neurol 2:341 (1943).

10. Katzman R, Clasen RA, Klatzo I, Meyer JS, Pappius HM, and Waltz
    AG, Brain edema in stroke, Stroke 8:512 (1977).

11. Courville CB, "Pathology of the central nervous system", Pacific
    Press Publishing Association, Mountain View, California (1937).

12. Courville CB, Structural changes in the brain consequent to traumatic
    disturbances of intracranial fluid balance, Bull Los Angeles
    Neurol Soc 7:55 (1942).

13. Scheinker IM, Cerebral swelling, histopathology, classification, and
    clinical significance of brain edema, J Neurosurg 4:255 (1947).

14. Evans JP, "Acute head injury," 2nd edition, Charles C. Thomas,
    Springfield, Illinois (1963).

15. Hirano A, The fine structure of brain in edema, in: "The structure
    and function of nervous tissue", Bourne GH, ed., Academic
    Press, New York (1969).

16. Feigin I, Sequence of pathological changes in brain edema, in: "Brain
    edema", Klatzo I, Seitelberger F, eds., Springer-Verlag, New
    York (1967).

17. Hardman JM, The pathology of traumatic brain injuries, in: "Complica-
    tions of nervous system trauma", Thompson RA and Green JR,
    eds., Raven Press, New York, Advances in Neurology 22:15
    (1979).

18. Lindenberg R, Trauma of the meninges and brain, in: "Pathology of
    the nervous system", 2, J. Minckler, ed., McGraw-Hill, New
    York (1971).

19. Adams JH, The neuropathology of head injury, in: "Handbook of
    clinical neurology", Vinken PJ and Bruyn GW, eds., Elsevier
    Publishing Company, New York (1975).

20. Clifton GL, McCormick WF, and Grossman RG, Neuropathology of
    early and late deaths after head injury, Neurosurg 8:309 (1981).

21. Lanksch W, Grumme T, and Kazner E, "Computed tomography in
    head injuries", Springer-Verlag,, New York (1979).

22. Clasen RA, Huckman MS, VonRoenn KA, Pandolfi S, Laing I, and
    Clasen JR, Time course of cerebral swelling in stroke: a
    correlative autopsy and CT study, in: "Brain edema", Cervos-
    Navarro J and Ferszt R, eds., Raven Press, New York (1980).

23. Penn RD and Clasen RA, Traumatic brain swelling and edema, in:
    "Head injury", Cooper PR, ed., Williams & Wilkins, Baltimore

(1982).

24. Rand CW, Courville CB, Histologic studies of the brain in cases of fatal injury to the head. VI Cyto-architectonic alterations, Arch Neurol Psychiat 36:1277 (1936).

25. Go KG, The classification of brain edema, in: "Brain edema", de Vlieger M, deLange SA, and Beks JWF, eds., John Wiley & Sons, New York (1981).

26. Clasen RA, Huckman MS, VonRoenn KA, Pandolfi S, Laing I, and Lobick JJ, A correlative study of computed tomography and histology in human and experimental vasogenic cerebral edema, J Comput Assist Tomogr 5:313 (1981).

27. DeVillasanta JM and Taveras JM, Computerized tomography (CT) in acute head trauma, Amer J Roent 126:765 (1976).

28. Koo AH and LaRogue RL, Evaluation of head trauma by computed tomography, Radiol 123:345 (1977).

29. Cooper PR and Moody S, Neurodiagnostic studies and the management of head injury, Comput Tomogr 2:197 (1978).

30. Zimmerman RA, Bilaniuk LT, and Gennarelli T, Computed tomography of shearing injuries of the cerebral white matter, Radiol 127:393 (1978).

31. Miller JD, Gudeman SK, Kishore PS, and Becker DP, Computed tomography in brain edema due to trauma, in: "Brain Edema", Cervos-Navarro J and Ferszt R, eds., Raven Press, New York (1980).

32. Lee SH and Rao KCVG, "Cranial computed tomography", McGraw-Hill, New York (1983).

33. Hammock MK and Milhorat TH, "Cranial computed tomography in infancy and childhood", Williams & Wilkins, Baltimore (1981).

34. Valk J, "Computed tomography and cerebral infarctions", Raven Press, New York (1980).

35. Bruce DA, Alavi A, Bilaniuk LT, Dolinskas C, Obrist W, and Kuhl D, Diffuse cerebral swelling following head injuries in children: The syndrome of malignant brain edema, J Neurosurg 54:170 (1981).

36. Zimmerman RA, Bilaniuk LT, Bruce DA, Dolinskas C, Obrist W, and Kuhl D, Computed tomography of pediatric head trauma: Acute general cerebral swelling, Radiol 126:403 (1978).

37. Kobrine AI, Timmins E, Rajjoub RK, Rizzoli HV, and Davis DO, Demonstration of massive traumatic brain swelling within 20 minutes after injury, J Neurosurg 46:256 (1977).

38. Waga S, Tochio H, and Sakakura M, Traumatic cerebral swelling developing within 30 minutes of after injury, Surg Neurol 3:191 (1979).

39. Jennett B, Clinical brain swelling: Edema or engorgement, in: "Brain edema", de Vlieger M, de Lange SA, and Beks JWF, eds., John Wiley & Sons, New York (1981).

40. Miller JD and Corales RL, Brain edema as a result of head injury: Fact or fancy? in: "Brain edema", de Vlieger M, de Lange SA, and Beks JWF, eds., John Wiley & Sons, New York (1981).

41. Brassow F and Baumann K, Volume of brain ventricles in man

determined by computer tomography, <u>Neuroradiol</u> 16:187 (1978).

42. Sager WD, Gell G, Ladurner G, and Ascher PW, Calculation of cerebral tissue and cerebrospinal fluid spaces from computer tomograms, <u>Neuroradiol</u> 16:176 (1978).

43. Haug G, Age and sex dependence of the size of normal ventricles on computed tomography, <u>Neuroradiol</u> 14:201 (1977).

44. Pederson H, Gyldensted M, and Gyldensted C, Measurement of the normal ventricular system and subarachnoid space in children with computed <u>Neuroradiol</u> 17:231 (1979).

45. Pelicci LJ, Bedrick AD, Cruse RP, and Vannucci RC, Frontal ventricular dimensions of the brain in infants and children, <u>Arch Neurol</u> 36:852 (1979)

46. Kuhl DE, Alavi A, Hoffman EJ, Phelps ME, Zimmerman RA, Obrist WD, Bruce DA, Greenberg JH, and Uzzell B, Local cerebral blood volume in head-injured patients, determination by emission computed tomography of 99mTc-labeled red cells, <u>J Neurosurg</u> 52: 309 (1980).

47. Ladurner G, Zilkha E, Sager WD, Iliff LD, Lechner H, and DuBoulay GH, Measurement of regional cerebral blood volume using the EMI 1010 Scanner, <u>Brit J Radiol</u> 52:371 (1979).

48. Miller AKH, Alston RL, and Corsellis JAN, Variation with age in the volumes of grey and white matter in the cerebral hemispheres of man: measurements with an image analyser, <u>Neuropath Appl Neurobiol</u> 6:119 (1980).

49. Grubb RL, Raichle ME, Eichling JO, and Ter-Pogossian MM, The effects of changes in $PaCO_2$ on cerebral blood volume, blood flow, and mean transit time, <u>Stroke</u> 5:630 (1974).

50. White JC, Verlot M, Silverstone B, and Beecher HK, Changes in brain volume during anesthesia: The effects of anoxia and hypercapnia, <u>Arch Surg</u> 44:1 (1942).

51. Mathew NT, Meyer JS, and Ott EO, Increased cerebral blood volume in benign intracranial hypertension, <u>Neurol</u> 25:646 (1975).

52. Lewis HP, Ramirez R, and McLaurin RL, Intracranial blood volume after head injury, <u>Surg Forum</u> 29:433 (1968).

53. Tornheim PA, Liwnicz BH, Hirsch CS, Brown DL, and McLaurin RL, Acute responses to blunt head trauma, experimental model and gross pathology, <u>J Neurosurg</u> 59:431 (1983).

54. Obrist WD, Dulinskas CA, Gennarelli TA, and Zimmerman RA, Relation of cerebral blood flow to CT scan in acute head injury, <u>in:</u> "Neural Trauma," Popp AJ, Bourke RS, Nelson LR, and Kimelberg HK, eds., Raven Press, New York (1979).

55. New PFJ, and Aronow S, Attenuation measurements of whole blood and blood fractions in computed tomography, <u>Radiol</u> 121:635 (1976).

56. Bondareff W, Myers RE, and Brann AW, Brain extracellular space in monkey fetuses subjected to prolonged partial asphyxia, <u>Exp Neurol</u> 28:167 (1970).

57. Snoek J, Jennett B, Adams JH, Graham DI, and Doyle D, Computerized tomography after recent severe head injury in patients without acute intracranial hematoma, <u>J Neurol Neurosurg</u>

Psychiat 42:215 (1979).
58. Blinkov SM and Glezer II, "The Human Brain in Figures and Tables," Plenum Press, New York (1968).
59. Katzman R and Pappius HM, "Brain Electrolytes and Fluid Metabolism", Williams and Wilkins, Baltimore (1973).
60. Kishore PRS, Radiography evaluation, in: "Head Injury," Cooper PR, ed., Williams and Wilkins, Baltimore (1982).

THE TEMPORAL GENESIS OF PRIMARY AND SECONDARY BRAIN DAMAGE IN EXPERIMENTAL AND CLINICAL HEAD INJURY

Donald P. Becker

Division of Neurological Surgery Medical College of Virginia
Richmond, Virginia, U.S.A.

Head injury in both the clinical and laboratory setting is a complex and multifactorial process which may involve many interactive primary and secondary insults. Not only is direct mechanical injury a factor but ischemia/hypoxia, brain swelling and edema, arterial and intracranial hypertension, subarachnoid hemorrhage, to name but a few, are frequently involved. Studies from our head injury center during the past ten years have attempted to identify and document primary and secondary factors, examine their temporal relationships and study their interaction in the pathophysiology of traumatic brain injury. It will be the purpose of this presentation to summarize the data from our head injury center regarding some of these factors. This will include both clinical and laboratory studies aimed at distinguishing between primary and secondary effects and their temporal sequence. It is our fundamental hypothesis that much of the damage incurred in head trauma is due to secondary effects and that areas of brain tissue which are rendered marginally dysfunctional can recover if the proper milieu is provided. Clinical and laboratory studies will be presented in the order of occurrence following trauma.

THE CNS AND SYSTEMIC RESPONSES AT IMPACT WITH FLUID PERCUSSION INJURY

The systemic response to injury in the fluid percussion model with ventilated animals involves a sympatho-adrenal surge characterized by arterial hypertension, brady- and tachycardia, an increase in serum catecholamines, and an increase in glucose levels. This response is graded and proportional to the severity of injury. These responses are transient in nature with the exception of elevated serum glucose which may persist for hours following injury (1). In addition, post-traumatic arterial hypotension may follow with higher levels of injury which may be mediated by the release of endogenous opiates (2). It has also been demonstrated that alpha blockade may prevent many of the above mentioned post injury cardiopulmonary changes (3). Thus, the sympatho-adrenal response appears to play a major role in the acute systemic reaction to fluid percussion injury. Additionally, other factors such as endogenous opioids may participate.

In addition to the systemic features of the acute impact a number of direct CNS effects have been documented in laboratories of our center. Recent data generated by HAYES and KATAYAMA have advanced our understanding of certain neurophysiological changes associated with experimental injury. Low levels of injury produce an initial, brief generalized areflexia followed by a reversible comatose state associated with postural muscle hypotonia and flaccidity. Additional experiments have demonstrated that during the initial period of generalized areflexia, afferent input transmission was depressed although the excitability of motor neuron pools was increased. In contrast, during periods of flaccidity, spinal cord somatomotor functions were depressed while transmission of afferent inputs was recovering. Behavioral and EEG data indicated that the flaccid comatose states were probably not attributable to depression of the reticular activating system (4). Systemic transection of the brainstem showed that activity within structures lying between collicular and midpontine levels is necessary to produce this latter condition. Cholinergic activation of pontine inhibitory areas within this same region of the rostral pons can produce profound descending inhibitory influences on postural somatomotor function in conjunction with other features of coma including suppression of eye-opening responses. Such effects occur without EEG slow waves. Moreover, other data indicate that local rates of glucose utilization within this pontine inhibitory area increase following concussive head injury (5). Thus, it is possible that a predominance of activity within this pontine inhibitory area could provide at least one neural basis for the reversible comatose state following concussive head injury characterized by close association between flaccidity and other indices of coma. Finally, other data from our laboratories (6) have provided preliminary evidence that flaccid comatose states may be mediated not only by cholinergic systems but also by neuropeptides.

Furthermore, numerous changes in cerebral blood flow resulting in impaired autoregulation, CO_2 reactivity, and response to hypoxia have been documented with a central parietal impact with this model (7,8) Early CBF responses are characterized by hyperemia followed by near normal CBF values after 30 to 60 minutes if cerebral perfusion pressure is maintained. (9)

Numerous acute vascular functional, metabolic, and structural changes have been observed by POVLISHOCK, KONTOS and ELLIS in the cortex and brainstem. In our consideration of the vascular response to head injury we have focussed on blood-brain barrier integrity, intraparenchymal vascular structure and pial microvessels. By using intravascularly administered horseradish peroxidase (HRP) coupled with scanning (SEM)- and transmission (TEM) electron microscopic analyses, we have evaluated traumatically induced intraparenchymal vascular changes.

Additionally, with comparable morphological studies, we have assessed traumatically induced changes in the pial arterioles, which were viewed and measured through a cranial window thereby providing information regarding the pial arteriolar diameter and responsiveness to normal physiological stimuli. Moreover, the cranial window offered the added advantage of providing a portal for infusing agents assumed to mimic, prevent or exacerbate those vascular phenomena seen with injury.

With brain injury uncomplicated by intraparenchymal hemorrhage, foci of increased vascular permeability to HRP were recognized in the brainstem (10). The duration of this altered permeability was brief (approximately 60 minutes) and was not associated with an initial increase in brain water (11). TEM suggested that the altered vascular permeability occurred without cleaving of the interendothelial tight junctions and it appeared that an increased endothelial vesicular activity was spatially and temporally linked to the HRP passage (10). Although we have advocated that increased vesicular activity was the mechanism for the transendothelial passage of the protein, recent studies in our laboratory have suggested that this concept may not be entirely correct (12). In addition to these forward and transient blood-brain barrier alterations, head injury was also seen with SEM and TEM to cause endothelial changes throughout the intraparenchymal vasculature (13). Typically, such widespread endothelial change was reflected by endothelial balloons and craters, which constituted foci of endothelial damage. Comparable endothelial change was also observed in the pial arterioles which was correlated with sustained vasodilation and a loss of responsiveness to hypocapnia (14). When the normally occurring post-traumatic hypertensive episode was prevented, no morphologic or functional abnormalities were recognized in either the pial or the intraparenchymal vasculature (14). However, such manipulation had no effect on the increased permeability to HRP of the brainstem vasculature. This suggested that the increased vascular permeability to HRP was a direct result of the trauma, while the pial and widespread intraparenchymal change was the result of a vasoactive substance triggered by the post-traumatic hypertension. The observation of pial dilation and of formation of endothethial lesions not stimulating platelet aggregation suggested that a vasodilator anti-aggregatory agent was responsible. Since prostaglandin I_2 has such properties, we explored the possibility that brain injury caused increased prostaglandin synthesis. ELLIS measured prostaglandin levels in frozen cerebral cortex of cats following fluid percussion injury. He found that there was a transient increase in PGE_2 at 8 minutes after injury, with PG levels returning to control by one hour after injury. These results tend to support the hypothesis that increased prostaglandin synthesis and generation of free radicals causes pial arteriolar lesions after trauma. Pretreatment of the animals with drugs which inhibited prostaglandin synthesis prevented both the morphological and functional abnormalities previously noted in the pial vasculature (15).

The mechanism by which increased prostaglandin synthesis produced the observed vascular abnormalities was also considered. Since it is known that certain steps in the normal biosynthetic pathway of prostaglandins release free oxygen radicals, we studied the role of oxygen radicals in the observed vascular change. Via the topical application of various radical scavengers under the cranial window, we were able to examine the role of oxygen radicals in the genesis of those abnormalities seen with head injury. Such topical applications blocked or significantly diminished those vascular changes seen with trauma (15) and suggested that oxygen radicals were directly involved. We are presently continuing our investigation of oxygen radicals and their effect on the cerebral vasculature (16). It is hoped that such an approach will provide for an enhanced appreciation of those mechanisms at work in head injury.

The possible contribution of the lipoxygenase enzyme system which also forms free radicals to brain injury is unknown. More generally, the capacity for lipoxygenase metabolism of arachidonic acid in the brain is unknown. ELLIS has recently found that the cerebral lipoxygenase enzyme system is capable of making several hydroxylated derivatives of arachidonic acid. The most abundant product which he has found to date is 12-HETE (12-hydroxy-eicosatetraenoic acid). The formation of 12-HETE by brain tissue from various species and the possible role of 12-HETE in the brain's acute and chronic response to trauma is currently under investigation.

EARLY SECONDARY EFFECTS INVOLVING CEREBRAL ISCHEMIA AFTER MILD AND SEVERE HEAD INJURY

The importance of early secondary ischemia following head injury has been demonstrated both in the clinic and the laboratory at our center. Laboratory studies have demonstrated that even reversible levels of trauma and ischemia when combined result in irreversible injury. Such evidence confirms suspicions that trauma increases the vulnerability of the brain to cerebral ischemia. (See JENKINS et al., this vol.)

GRAHAM and ADAMS have shown that ischemic brain injury is common in patients who died from severe head trauma (17). Signs of ischemia were most often found in the fronto-parietal watershed areas. Cerebral blood flow (CBF) measurements have been performed in head injured patients since 1972. Yet, despite the score of investigations that have been performed and documented, OVERGAARD and coworkers were the only group who reported a significant number of patients with ischemia. It was shown that patients with occurrence of regional flows less than 17 ml/100g min did poorly. However, only regional ischemia was reported in these patients and whole brain CBF appeared to be within a normal physiologic range (18).

The definition of ischemia must include consideration of the needs of the tissue. CBF values alone are not sufficient because the needs of the brain are greatly reduced in coma. In studies of severely head injured patients, we measure the arteriovenous difference in oxygen content $(AVDO_2)$. The normal $AVDO_2$ is in the range of 6 to 7 ml O_2 /100 ml. If $AVDO_2$ is clearly above this value, blood supply to the brain is not sufficient. If $AVDO_2$ is below this value, CBF levels of less than 17 ml/100g min may be adequate.

We have found in severely head injured patients that 6 of the 53 patients studied had CBF levels below the threshold for infarction (i.e. 18 ml/100g min;19). These CBF measurements were obtained within 24 hrs of injury. The timing of CBF measurements is important. In some patients, ischemia could be found at 8 hours post-injury, while CBF values 16 hours post injury were above the ischemic threshold. Our studies of $AVDO_2$ in patients permitted us to document the existence of ischemia 6 hours post injury. We believe it is necessary to obtain $AVDO_2$ as early as possible. In six patients we attempted to reverse ischemia by raising the blood pressure or by hemodilution with mannitol (20). In four patients the reversal of ischemia was reflected in an improvement of the clinical status, multimodality evoked potentials, and/or cerebral O_2 -consumption $(CMRO_2)$. No improvement of CBF was found in 2 patients. In one of these patients, $AVDO_2$ decreased so that $CMRO_2$ remained unchanged.

The implications are that early measurements of both $AVDO_2$ and CBF are necessary to detect cerebral ischemia in patients with severe head injury. It strongly influences the clinical status and outcome and demonstrates that early ischemia may be potentially treatable. The importance of assessing not only blood flow but metabolism is obvious for determining if blood flow is adequate to meet the metabolic needs of the injured brain. We are currently evaluating such relationships in the laboratory model of head injury and have developed a technique for the simultaneous measurement of regional cerebral blood flow and regional glucose utilization using radioactive microspheres and (^{14}C)-2-deoxyglucose (21). Employing this method we have demonstrated a close linear relationship between CBF and metabolism in the normal cat brain. Initial studies show that even after mild trauma the normal coupling between CBF and metabolism is altered in some brain regions.

THE DEVELOPMENT OF BRAIN EDEMA AS A FUNCTION OF INTENSITY AND DIRECTION OF FLUID PERCUSSION IMPACT

Brain swelling, whether due to an increase in brain water and/or an increase in vascular volume, remains one of the most serious sequelae of head injury. The progressive increase of this volume in the closed calvarium ultimately leads to intracranial hypertension and ischemia. The severity of this problem is in part reflected in the statistics. 50 % of patients who succumb to head injury die from high intracranial pressure

(22,23). One objective of our research is to identify the role of brain edema in the swelling process and to determine in a well defined animal model the pathophysiologic sequence of events associated with the edematous process.

In the laboratory, the most frequently used model to study edema of the traumatized brain is local freeze injury to the exposed cortex. The cryogenic model has the advantage of high reproducibility and has been studied extensively. The disadvantage is the etiologic dissimilarity with clinical traumatic brain edema. Experimental models of closed head injury have also been studied extensively but relatively few have obtained direct measures of tissue water content. TORNHEIM et al. measured tissue density following mechanical impact to the closed skull of cats (24).Animals sacrificed 48 hours after trauma demonstrated unilateral fracture of the frontal bone, subarachnoid hemorrhage and the development of significant edema of the contused hemisphere. The lack of morphologic reproducibility due to the thickness and contour of the cranial bones is a confounding factor of this model.

The fluid percussion model obviates the variability introduced by the container properties of bone. In the fluid percussion model the controlled impact is directed to the exposed dura. A severe fluid percussion injury is immediatly followed by a rapid rise in arterial pressure and intracranial pressure. Mean systemic pressure one hour following impact returned to normal level while the intracranial pressure remained slightly elevated one hour after trauma. These events prompted a study of tissue water content by CORALES and MILLER (25) to determine if the rise in intracranial pressure following traumatic injury is due to brain edema. In these studies, adult cats were subjected to a central fluid percussion injury of high intensity (2.6-5.3 atmospheres) and sacrificed at 5 minutes and 30 minutes while ICP was still elevated. They reported no increase in water content.

We recently studied animals subjected to the same level of injury (2.8 atmospheres) with both central and lateral percussion and observed that all animals developed significant brain edema at 24 hours following impact. With central injury, the edema was maximum in the brain stem reaching 80 % water per gm tissue. Water content of the supratentorial compartment was increased form a control level of 68 % to a mean of 75 %. The water content of cortex was essentially normal with a slight increase from control levels of 80.9 % to 83 % in the region directly under the percussion cylinder. The fact that the water content of cortex was normal and the increase confined to the white matter suggests that the edema is of vasogenic origin. With central injury, edema develops symmetrically above the tentorium. With lateral injury, the edema is maximum in the impacted hemisphere. The water content of the brainstem is also elevated but less than that seen with central injury. We are now proceeding with experiments which will quantitatively describe the edema profile representative of both types of impact.

In summary, we feel that the preliminary evidence clearly demonstrates that mechanical fluid percussion injury results in the development of brain edema. Edema develops within 24 hours from time of injury and is restricted to the white matter. These findings and the concomitant reduction in spatial buffering capacity suggests that the development of brain edema must be studied if we are to gain further insight into the pathophysiology of head injury.

THE DEVELOPMENT OF BRAIN TISSUE ACIDOSIS FOLLOWING FLUID PERCUSSION INJURY

During the past several years, we have become increasingly focussed on detrimental effects of tissue metabolism that occur following head injury.This interest is a natural extension of our hypothesis that in brain injury many cells are rendered dysfunctional but are not mechanically disrupted. We contend that the cells may recover if the milieu is favorable. If it is unfavorable, they may die. This leads to the challenge of defining the milieu in which the injured cell finds itself and defining an ideal milieu for recovery. The interest in metabolic disturbances was strengthened by observations that of all our patients who died after injury, 30 % had evoked potentials early after injury, which were indicative of a good cerebral recovery. Some of the patients died with normal ICP and adequate CBF and no identifiable systemic complication. What caused the slow progressive brain dysfunction in the absence of elevated ICP and seemingly adequate cerebral blood flow?

It seemed reasonable to suspect that brain tissue acidosis following injury could be a major cause. The evidence is strong that acidosis occurs in brain tissue following injury, and that acidosis is not an ideal milieu for tissue healing. Experiments were performed in our animal intensive care unit to test the hypothesis that tromethamine, an agent that causes hypocapnic alkalosis and which may have intracellular buffering effects, reduces mortality in brain injured cats. The study provided evidence supporting this hypothesis. Mortality was significantly reduced. Thus, our laboratory experience and assessment of the literature has strengthened our postulate that cerebral acidosis is a major culprit in the progressive brain dysfunctional state. These are not new ideas. Metabolic acidosis of the CNS following traumatic injury has been reported by numerous investigators over the last 15 - 20 years and attributed to increased levels of lactic acid. CSF lactic acidosis has also been observed following cerebral hypoxia-ischemia (26,27). In head injured patients ENEVOLDSON et al. (28) reported that lactate levels correlated with outcome. However, while CSF lactic acidosis has been associated with injury, cause and effect remains to be determined. Obviously, the possibility exists that incrased CSF- or tissue-lactate levels may only reflect tissue damage already incurred and may not be harmful in and of itself. This issue has been addressed most thoroughly in the cerebral ischemia literature.

Fig. 1a: Light microphotograph of a plastic thick section of a segment of the red nucleus 7 days after a 2.2 atm insult. The blocked out areas demonstrate reactive axonal swellings in proximity to both normal and altered neuronal somata. These areas are enlarged below (1b, c) for better visualization.

Fig. 1b: Two reactive swellings (RS) approximate a neuronal soma (arrow) and its dendrites (D). Note that the neuronal soma is enlarged and manifests change consistent with chromatolysis.

Fig. 1c: Here a reactive swelling (RS) approximates a neuronal soma (arrow) and one of its proximal dendrites (D). Note that both the soma and dendrite appear unaltered, with the nucleus found in a central position and Nissl substance intact.

Fig. 1d: Light microphotograph of a plastic thick section of the cerebral peduncle 7 days after a 2.5 atm injury showing a reactive axonal swelling (arrow). Note that the swelling is continuous with the proximal axonal segment, while separated from its distal segment. Note that despite the presence of this reactive change, adjacent axons (A) are inconspicuous.

Fig. 1e: Light microphotograph which demonstrates the axonal segment distal to that shown above in Fig. 1d. Note the collapsed and fragmented myelin (arrows) as well as the normal axonal profiles (A).

Fig. 1f: Peroxidase-laden reactive swelling (block arrow) seen 48h after injury. Note the regenerative sprouts (arrow heads) which arise from the apex of the reactive swelling.

Fig. 1g: With light microscopy, a large peroxidase-laden reactive swelling (block arrow) with a regenerative sprout (arrow head) is seen 72h after injury.

Fig. 1h: By the end of the first post-traumatic week regenerative sprouts (arrow heads) can be seen moving in a directed fashion through the substance of the cerebral peduncle.

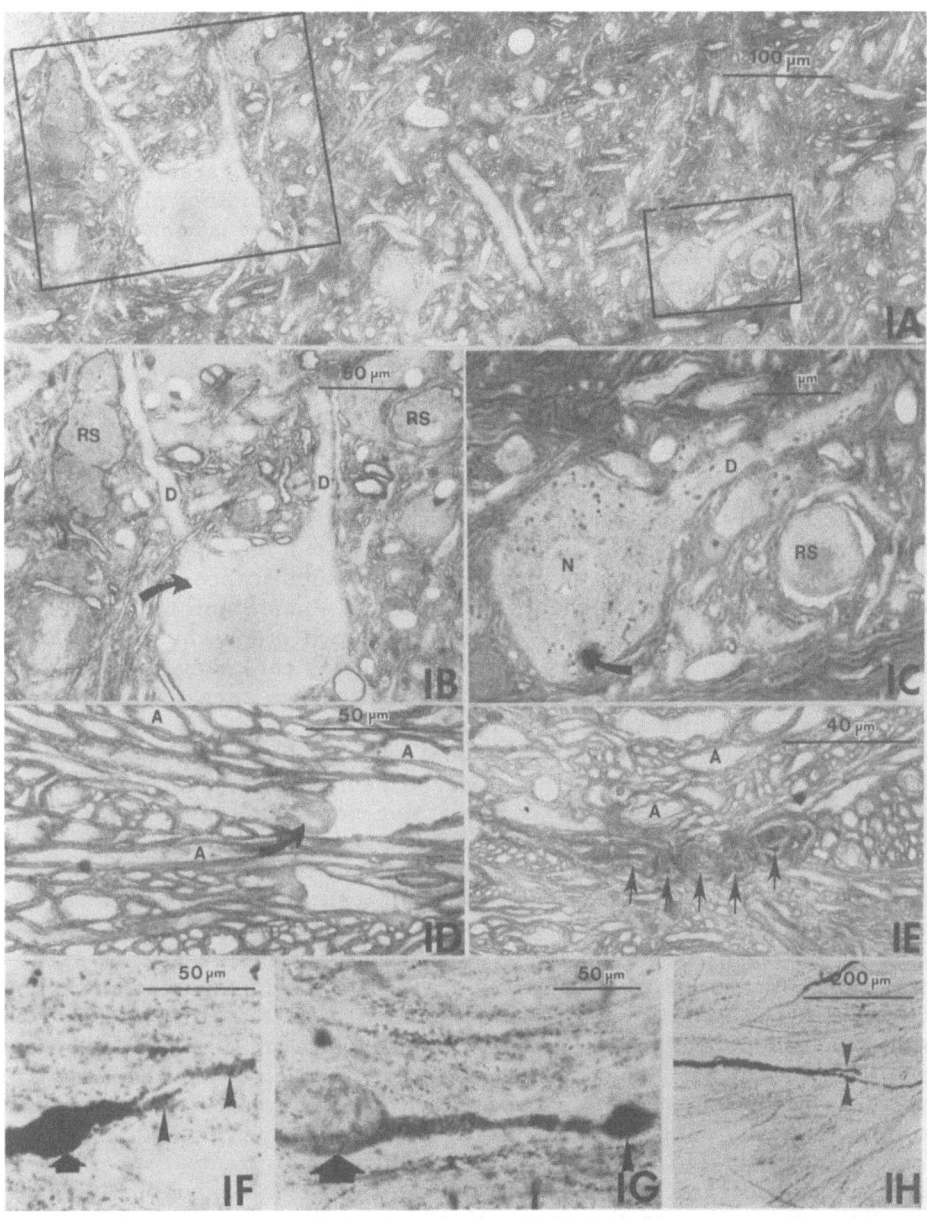

Our laboratories are now engaged in the study of brain tissue acidosis associated with fluid percussion injury. These studies by MARMAROU and JENKINS have shown that significant brain tissue acidosis develops within 8 hours of impact. Brain tissue pH was measured in the impacted and contralateral hemisphere and cerebrospinal fluid of adult cats using pH depth electrodes. Animals were subjected to a lateral injury of 2.8-3.0 atmospheres. The brain tissue pH of the impacted hemisphere fell from a control level of 7.29 (+ 0.183 SD) to 6.65 (+ 0.280 SD) 8 hours following injury. Similar reductions were observed in the contralateral hemisphere. After a 1 to 2 hour delay, the pH of the cerebrospinal fluid followed the reduction of pH observed in the tissue. Serum pH was unaffected. Studies are now in progress to confirm the reduction of pH seen by the electrodes. For that purpose tissue metabolites are employed to calculate intracellular pH using the creatine kinase and bicarbonate carbonic acid methods. Additional CBF studies are under way to determine if the acidosis is flow related.

THE CHRONIC COURSE OF SELECTIVE AXONAL AND SOMATIC CHANGE FOLLOWING MILD HEAD INJURY

In the preceeding passages emphasis has been placed on secondary insults such as ischemia, brain edema, acidosis and the role they play in the pathobiology of traumatic injury.While the significance of these secondary insults is not to be underestimated, it is important to remember that the secondary insults exert their effect on a brain which has sustained varying degrees of primary morphopathological change. In this context, it is relevant that recently much attention has been focussed on the occurrence of primary axonal damage in both minor (29,30) and severe (31,32,33) head injuries. The amount and localization of such axonal alteration has been linked to the observed neurological dysfunction (34) through the identification of axon retraction balls, microglial proliferation, and demyelinization in head injured animals and man. Despite the significance attached to reactive axonal change, little is known regarding the initial genesis or long term fate of such injured axons. Moreover, their overall significance in the pathobiology of head injury remains to be clarified. For example, with regard to the genesis of axon retraction balls, it has been assumed that shear- and tensile strains of the traumatic episode immediately disrupt axons so that they retract and extrude a ball of axoplasm. Furthermore, although great significance has been attached to any axonal abnormalities, it is unclear whether reactive axonal change occurs either in concert with or in isolation from focal neuronal somatic, dendritic, or synaptic alteration. Such lack of clarification regarding the pathobiological significance of traumatically induced axonal change has led to the inference that the axons are preferentially vulnerable to the traumatically generated forces. It is thereby implied that widespread axonal disruption rather than diffuse neuronal somatic and dendritic damage is the most relevant insult sustained by the injured brain.

Recently, we have critically evaluated the genesis of traumatically induced axonal change in minor head injury and have rejected previously held beliefs regarding retraction ball formation (35). Through the use of anterograde axonal transport of horseradish peroxidase studied at the light microscopic- and transmission electron microscopic level we have noted that the traumatic insult does not immediately disrupt axons to cause axoplasmic extrusion. Rather, a subtle focal impairment of axoplasmic transport is elicited, which over 12 hours causes progressive axonal swelling and disruption of axonal continuity resulting in formation of the "retraction ball". Furthermore, the development of such axonal abnormalities were also recognized with injuries of increased severity (unpublished findings) known to elicit both neuronal somatic alteration (34) and activation (5) coupled with blood-brain barrier- (10,35) as well as other forms of vascular perturbation (14,12). Therefore, it would seem that reactive axonal change is but one component of an avalanche of cellular abnormalities resulting from the traumatic insult. We have initiated the following studies to further investigate the overall significance of traumatically induced axonal change, and to correlate its relationship or lack thereof to concomitant neuronal alteration.

Cats were subjected to a fluid percusssion injury ranging in severity from 1.3 to 2.5 atmospheres. The animals were allowed to survive for either 15 minutes, 1h, 6h, 24h, 48h, 72h, 1 week or 2 week periods when they were transcardially perfused with aldehydes. Two days prior to sacrifice the animals received cerebral cortical and cerebellar injections of horseradish peroxidase (HRP) to study anterograde axonal transport in the major cerebral cortical and cerebellar efferents. At the above designated post-traumatic survival times, the brains were removed and processed for light microscopic (LM)- and transmission electron microscopic (TEM) visualization of the HRP reaction product.

Consistent with our previous findings (1), there was no support of the contention that trauma caused axonal shearing and retraction with axoplasmic extrusion. Rather, irrespective of its intensity, the injury caused an initial intra-axonal HRP pooling which was associated with an accumulation of organelles. Such pooling was recognized in cerebral and cerebellar efferents coursing the cerebral peduncles, red and vestibular nuclei, reticular formation and the superior cerebellar peduncles. Over a 24h course HRP pooling and organelle accumulation increased in size. There lobulation of the axonal profile occurred with formation of a ball-like swelling (Figs: 1a, b, c, d, 2a) which ultimately became detached from the axon segment distal to its origin. By 24h the reactive axonal swellings were maximal, showing no increase in either size or organelle content. By 48h through the second post-traumatic week those axonal segments distal to the reactive swellings demonstrated myelin collapse indicative of WALLERIAN degeneration (Fig. 1e). The reactive swellings and proximal axonal segments manifested various forms of reactive change. Occasional reactive swellings showed complex lobulation (Fig. 2b,c) with neurofilamentous hyperplasia and/or increased electron density. Such changes were consistent with a degenerative response.

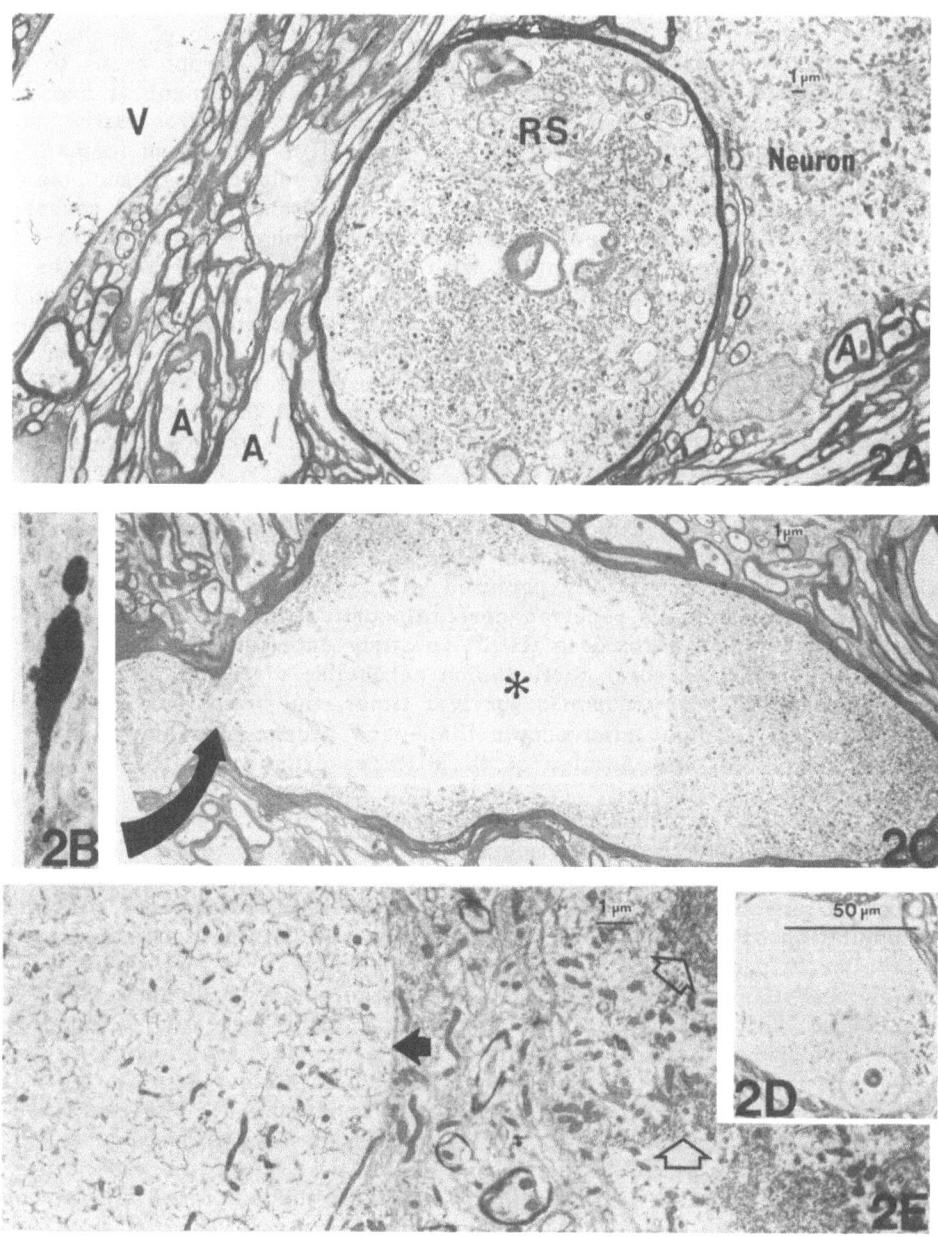

Fig. 2a: Electron micrograph of a large reactive swelling (RS) which approximates a neuron, axons (A) and a venule (V) all of which appear unaltered.

Fig. 2b: Peroxidase-laden axon visualized by light microscopy which demonstrates lobulation suggestive of degeneration.

Fig. 2c: With electronmicroscopy, lobulation is also recognized. Note that the reactive swelling which contains a neurofilamentous core (asterisk) demonstrates segmentation (arrow) in its profile.

Fig. 2d: Ligth microphotograph of a neuron undergoing chromatolysis 1 week after a 2.2 atm fluid percussion injury.

Fig. 2e: Electron micrograph demonstrating a portion (solid arrow) of the neuron shown in Figure 2d. Note the clearly atypical neuronal soma which manifests changes consistent with chromatolysis. An adjacent neuron demonstrates no alteration. Its profiles of the rough endoplasmic reticulum (open arrows) appear unaltered.

Macrophages could be identified within the investing myelin sheath directly attached to the swollen axolemma. Some reactive axonal swellings manifested no change over the 2 week period post-trauma, whereas others demonstrated regenerative sprouts which coursed in a directed fashion through the brain parenchyma (Figs. 1f, g, h). These regenerative sprouts had the ultrastructural characteristics of growth cones (36).

When the brain parenchyma associated with the above described reactive axonal swellings was assessed for any index of neuronal somatic or dendritic change, it seemed remarkable that in all injury ranges reactive axons could be found in direct proximity to both neuronal somata and dendritic profiles which had no abnormality (Figs. 1a, c, 2a). Central chromatolytic changes were occasionally found within neuronal somata focally related to altered axonal profiles at 48 h through the second post-traumatic week (Figs. 1a, b, 2d, e). These neurons were swollen with excentric nuclei, dispersion of clustered ribosomes, and occasional neurofilamentous hyperplasia (Fig. 2d, e).

The results of this investigation are of interest from several perspectives. The observation of reactive change in injuries of varying severity reinforces previous concepts regarding axonal vulnerability in head injury. Moreover, the observation that the reactive axonal swellings are not an immediate consequence of trauma, but rather the result of an initial disruption of axoplasmic transport which with time becomes progressively severe supports our previous findings. It remains unclear, however, why over a two week post-traumatic course some reactive axonal swellings are unchanged, whereas others degenerate or give rise to regenerative sprouts. Perhaps, the proximity of the axonal damage to the neuronal cell body and/or the functional status of that cell body may determine the axon's ultimate response. The regenerative sprouting seen in relation to the reactive swellings is a previously unreported phenomenon. The directed growth of regenerative sprouts through the brain parenchyma is intriguing. In conditions where trauma causes diffuse axonal injury with neither overt disruption of the brain parenchyma or any marked gliosis, it would appear that the brain's microenvironment may be relatively unaltered, thereby providing a conduit for directed growth (37). Our observation that reactive axonal change can be found both in isolation from or in concert with focal axonal alteration poses some unique conceptual problems. Although our studies demonstrate that reactive axonal change can occur in the same anatomical loci in which neurons remain unaltered, the findings of altered axons adjacent to chromatolytic neurons would support that both elements are simultaneously damaged by the trauma. Obviously, much remains to be clarified regarding the significance of these altered neuronal somata. Yet, their presence would suggest that they constitute important components of the brain's response to injury.

SUMMARY

Our common thesis is that treatment of brain injury can improve outcome. Such treatments will be based upon an understanding of both the reversible and irreversible processes induced by injury. We have described in brief terms the temporal genesis of primary and secondary brain damage in experimental and clinical head trauma. The various projects attack these problems with the tools of different disciplines from different viewpoints and with different insights. All have the same goal - to understand the fundamental basis of neuronal loss after injury and, hopefully, to reverse this process.

ACKNOWLEDGEMENTS

This report was prepared by incorporating the contributions of the Neuroscience Research Staff of the Medical College of Virginia, Richmond. I wish to acknowledge the scientific contributions of Douglas S. DeWitt, Ph.D., Earl F. Ellis, Ph.D., Ronald L. Hayes, Ph.D., Larry W. Jenkins, Ph.D., Hermes S. Kontos, M.D., Ph.D., Harry Lutz, Ph.D., Anthony Marmarou, Ph.D., J. Paul Muizelaar, M.D., Polly Newlon, MSc., Ph.D., John T. Povlishock, Ph.D., John D. Ward, M.D. and Mildred Yang, Ph.D. I also wish to thank Ms. Linda Brooks for her assistance in preparation of the manuscript. This research was supported in part by NIH grants NS12587, NS20193, as well as Biomedical Research Support Grants 2SO7RR-05724-05430 and -05697 and the Division of Neurosurgery, Medical College of Virginia.

REFERENCES

1. Rosner MJ, Newsome HH, and Becker DP, Mechanical brain injury: The sympathoadrenal response, J Neurol Surg (in press).
2. Hayes RL, Kulkarni P, Galinat BJ, and Becker DP, Effects of naloxone on systemic and cerebral responses to experimental concussive injury in the cat, J Neurosurg 58:720 (1983).
3. Millen JE, Glauser FL, and Zimmerman M, Physiological effects of controlled concussive brain trauma, J Applied Physiol 49:856 (1980).
4. Katayama Y, Glisson JD, Becker DP, and Hayes RL, Concussive head injury producing suppression of sensory transmission within the lumbar spinal cord in cats, J Neurosurg 63:97 (1985).
5. Hayes RL, Pechura CM, Katayama Y, Povlishock JT, Giebel ML, and Becker DP, Activation of pontine cholinergic sites implicated in unconsciousness following cerebral concussion in the cat, Science 223:301 (1984).

6. Hayes RL, Katayama Y, Hellgeth MC, and Becker DP, Behavioral suppression produced by carbachol microinjections into the rostral pons of the cat: Antagonism by naloxone and thyrotropin-releasing hormone (in preparation).

7. Lewelt W, Jenkins LW, and Miller JD, Autoregulation of cerebral blood flow after experimental fluid-percussion injury of the brain, J Neurosurg 53:500 (1980).

8. Lewelt W, Jenkins LW, and Miller JD, Effects of experimental fluid percussion injury of the brain on cerebrovascular reactivity to hypoxia and to hypercapnia, J Neurosurg 56:332 (1982).

9. DeWitt DS, Jenkins LW, Lutz H, Wei EP, Kontos HA, Miller JD, and Becker DP, Regional cerebral blood flow following fluid percussion injury, J Cereb Blood Flow Metabol 1, Suppl. 1: 579 (1981).

10. Povlishock JT, Becker DP, Sullivan HG, and Miller JD, Vasculature permeability alterations to horseredish peroxidase in experimental brain injury, Brain Res 152:223 (1978).

11. Miller JD, and Corales RL, Brain edema as a result of head injury: Fact or Fallacy? in: "Brain Edema," de Vlieger M, deLange S, and Beks JWF, eds., John Wiley and Sons, New York (1981).

12. Povlishock JT, and Kontos HA, The pathophysiology of pial and intraparenchymal vascular dysfunction, in: "Head injury: Basic and clinical aspects", Grossmann RS, and Gildenberg PL, eds., Raven Press, New York (1982).

13. Povlishock JT, Kontos HA, Wei EP, Rosenblum WI, and Becker DP, Changes in the cerebral vasculature after hypertension and trauma: A combined scanning and transmission electron microscopic analysis, in: "The cerebral microvasculature", Eisenberg HM, and Suddith RL, eds., Plenum, New York (1980).

14. Wei EP, Dietrich WD, Povlishock JT, Navari RM, and Patterson JL, Functional, morphological and metabolic abnormalities of the cerebral microcirculation after concussive brain injury in cats, Circ Res 46:37 (1980).

15. Wei EP, Kontos HA, Dietrich WD, Povlishock JT, and Ellis EF, Inhibition by free radical scavengers and by cyclooxygenase inhibitors of pial arteriolar abnormalities from concussive brain injury in cats, Circ Res 48:95 (1981).

16. Wei EP, Christman CW, Kontos HA, and Povlishock JT, Effects of oxygen radicals on cerebral arterioles, Am J Physiol (In press).

17. Graham DI, and Adams JH, Ischemic brain damage in fatal head injuries, Lancet 1:265 (1971).

18. Overgaard J, Mosdal C, and Tweed WA, Cerebral circulation after head injury. Part 3: Does reduced regional cerebral blood flow determine recovery of brain function after blunt head injury? J Neurosurg 55:63 (1981).

19. Jones TH, Morawetz RB, Crowell RM, Marcoux FW, FitzGibbon SJ, DeGirolami U, and Ojemann RG, Thresholds of focal cerebral ischemia in awake monkeys, J Neurosurg 54:77 (1981).

20. Muizelaar JP, Wei EP, Kontos HA, and Becker DP, Mannitol causes compensatory cerebral vasoconstriction and vasodilation in response to viscosity changes, J Neurosurg 59:822 (1983).

21. DeWitt DS, Becker DP, and Hayes RL, A technique for the simultaneous, quantitative measurement of blood flow and glucose utilization within tissue samples from the feline brain, (in preparation).
22. Becker DP, Miller JD, Ward JD, Greenberg RP, Young HF, and Sakalas R, The outcome from severe head injury with early diagnosis and intensive management, J Neurosurg 47:491 (1977).
23. Marshall LF, Smith RW, Shapiro HM, The outcome with aggresive treatment in severe head injuries. I. The significance of intracranial pressure monitoring, J Neurosurg 50:20 (1979).
24. Tornheim PA, McLaurin RL, and Thorpe JF, The edema of cerebral contusion, Surg Neurol 5:171 (1976).
25. Corales RL, Miller JD, and Becker DP, Intracranial pressure and brain water content in acute graded experimental cerebral trauma, in: "Intracranial pressure, IV", Shulman K, Marmarou A, Miller JD, Hochwald G, Becker DP, and Brock M, eds., Springer Verlag, Berlin (1980).
26. Siesjö BK, Cell damage in the brain: A speculative synthesis, J Cereb Blood Flow Metabol 1:155 (1981).
27. Rehncrona S, Rosen I, Siesjö BK, Brain lactic acidosis and ischemic cell damage. 1. Biochemistry and neurophysiology, J Cereb Blood Flow Metabol 1:297 (1981).
28. Enevoldsen EM, and Jensen FT, Cerebrospinal fluid lactate and pH in patients with acute severe head injury, Clin Neurol Neurosurg 80:213 (1977).
29. Povlishock JT, Becker DP, Cheng CLY, and Vaughan GW, Axonal change in minor head injury, J Neuropath Exp Neurol 42:255 (1983).
30. Gennarelli TA, Jane J, Thibault LE, and Steward O, Axonal damage in mild head injury demonstrated by the Nauta method, in: "Advances in neurotraumatology", Villani R, Papo I, Giovanelli M, Gaini SM, and Tomei G, eds., Excerpta Medica, Amsterdam (1982).
31. Adams JH, Mitchell DE, Graham DI, and Doyle D, Diffuse brain damage of an immediate impact type, Brain 200:489 (1977).
32. Strich SJ, Shearing of nerve fibers as a cause of brain damage due to head injury, Lancet 2:443 (1961).
33. Gennarelli TA, Thibault LE, Adams JH, Graham DI, Thompson CJ, and Marcincin RP, Diffuse axonal injury and traumatic coma in the primate, Ann Neurol 12:564 (1982).
34. Povlishock JT, Becker DP, Miller JD, Jenkins LW, and Dietrich WD, The morphopathologic substrates of concussion, Acta Neuropath 47:1 (1979).
35. Povlishock JT, Becker DP, Kontos HA, and Jenkins LW, Neural and vascular alterations in brain injury, in: "Neural trauma - seminars in neurological surgery", Popp AJ, Bourke RS, Nelson LR, and Kimelberg HK, eds., Raven Press, New York (1979).
36. Povlishock JT, The fine structure of axons and growth cones of the human fetal cerebral cortex, Brain Res 114:379 (1976).

37. Guth L, Barrett CP, Dontai EJ, Deshipande SS, and Albuquerque EX, Histopathological reactions and axonal regeneration in the transsected spinal cord of hibernating squirrels, J Comp Neurol 203:297 (1981).
38. Barrow KD, Dentinger MP, Nelson LR, and Miney JE, Ultrastructure of axonal reaction in red nucleus of cats, J Neuropath Exp Neurol 34:222 (1975).

EVOLUTION OF NEURONAL ISCHEMIC INJURY

Julio H. Garcia and K.A. Conger

Departments of Pathology and Neurology, University of
Alabama, Birmingham, Alabama, USA

INTRODUCTION

A decrease of blood flow to the mammalian brain sufficiently severe to impair neuronal function constitutes the condition of ischemia.
In several situations in human and animals it has been determined that
neuronal activity, i.e. action potentials disappears, if cerebral blood
flow (CBF) falls to a level below 12-15 ml/100g min (Sundt et al, 1981;
Garcia, 1984). However, it has also been observed that this is reversible, as long as blood flow is reestablished within a short period of
time. Different ischemic flow thresholds have been experimentally described for neurons. The flow threshold for release of K^+ is clearly lower than the one resulting in complete electrical failure (Astrup, 1981).

It is well accepted that there is a period of grace during which
ischemic injury is reversible (Heiss, 1983). However, the length of the
reversibility period cannot be established in absolute figures for a variety of reasons. The mechanisms of cerebral ischemia are extremely
heterogeneous. At least five different mechanisms are recognized (Garcia,
1983). The conditions resulting from each of these are not uniform or
predictable. Induction of cerebral ischemia by occlusion of a single
artery decreases CBF to unpredictable levels. The resulting CBF is not
only topographically heterogeneous within the respective arterial territory, but local CBF changes after arterial occlusion are probably varying
with time due to internal adaptations such as changes in the vascular ca-
liber. The heterogeneity in blood flow alterations commonly observed in
animals of the same species is generally attributed to the variable
pattern of collateral connections characteristic of the intracranial
vasculature in most mammals. Thus, the circulatory conditions induced by
single artery occlusion are within the corresponding arterial territory a
veritable patchwork of ischemia, hyperemia, and normal CBF (Yama-

guchi et al, 1971; Garcia, et al, 1983). More predictable and more uniformly distributed areas of injury can be expected, if brain ischemia is
induced in a systemic manner. These are, e.g. lowering of the systemic
blood pressure below the level of autoregulation, or impeding the cranial
circulation by a blood pressure cuff placed around the neck (Rossen et
al, 1943). However, heterogeneity and multifocal injury are also the rule
in those instances (Nemoto et al, 1977). The multifocal nature of injury
in global ischemia is generally explained to result (1) from topographic
(circulatory) factors, and (2) from an intrinsic or selective susceptibility of certain neuronal groups.

In the face of systemic injury, such as hypotension topographic
(circulatory) factors would be responsible for the preferential damage
of brain areas located most distant from the arterial supply. Those
"watershed" territories are commonly referred to as terminal field,
arterial border, or boundary zones (Brierley, 1979). An additional factor
has been invoked to explain why in a given segment of ischemic cerebral
cortex a layer of neurons or even an isolated group of cells may survive
unscathed in the midst of completely necrotic neurons. The property allowing some cells to respond differently to a given level of ischemia has
been called selective vulnerability. Apart from vaguely defined intrinsic
neuronal factors, the mechanisms of selective vulnerability are unknown.
Naturally, understanding the conditions which protect the surviving neurons could be important in developing adequate forms of treatment of
ischemic injury. In addition to the presumed topographical circumstances
of injury and assumed idiosyncrasy of some neurons, numerous other factors which modify outcome of ischemia have been identified. These are:

(1) Age. Young brains withstand ischemia better than old brains,
presumably because the former retain plasticity.

(2) Barbiturates. Partly because of the lowered metabolic activity,
these drugs "protect" the brain from focal ischemic injury (Astrup,
1980). Animals given high doses of barbiturates before occluding a large
cerebral artery had less severe brain injury than paired controls receiving a placebo (Hoff et al, 1975).

(3) Hypothermia (18° C) lowers neuronal metabolic activity, which
has been shown to provide protection in instances where the entire brain
circulation is interrupted. An example is the surgical repair of congeni-
tal heart defects (Treasure et al, 1983).

(4) Lowered serum glucose levels induced by starvation before the
insult result in a considerable decrease of ischemic brain injury when
compared with animals with normal or increased serum glucose levels
during ischemia (Myers, Yamaguchi, 1977). As to serum glucose it has
been suggested that increased levels of lactic acid resulting from anaerobic glycolysis may have deleterious effects on neuronal survival. Some authors have suggested that there is a threshold for lactacidosis of
approximately 20-25 mmol/kg. Accordingly, tissue levels exceeding this

threshold are incompatible with neuronal survival (Rehncrona et al,
1980). Table 1 summarizes some on the known factors capable of influencing outcome of ischemic brain injury and emphasizes the inherent difficulty of comparative studies of ischemia. In most experiments reported to
date, control of the severity of ischemia, i.e. maintainance of a steady
level of perfusion has not been accomplished.

TABLE 1

Factors Influencing Effects of Decreased CBF

1. Duration and Severity
2. Age
3. Collateral Arterial Connections
4. Selective Neuronal Vulnerability
5. Body Temperature
6. Serum Glucose
7. Barbiturates

In the average experiment the investigator may attempt to control duration of ischemia, for example by releasing an arterial ligature.
However, the investigator would have no way of knowing the quality of
the collateral arterial connections which determine local CBF. After
single artery occlusion, multifocal and heterogeneous injury develops
where local CBF can be expected to continuously change as a result of
adjustments of collateral flow and vascular tone.

MECHANISMS OF CEREBRAL ISCHEMIA IN HUMAN

The local circulatory conditions are extremely variable in an individual experiment, and the mechanisms of ischemic injury may considerably differ. Five major types of ischemic injury with slightly varying
structural features have been identified (Garcia, 1983).

(1) Absolute brain ischemia prevails, if an animal is decapitated,
the cerebral blood is replaced by a plasma substitute, or the intracranial pressure is raised above the level of the systolic blood pressure. This type of brain ischemia is complete. In order to differentiate
this from autolytic destruction of cells which occurs only after several
hours, we have proposed that it be designated as postmortem ischemia.
As suggested from the survival of transplanted cadaver organs, e.g.
hearts, or others that were ischemic for up to 120 min (McGiffin, 1983),
this form of ischemia does not immediately induce irreversible tissue damage.

(2) A second variety of ischemic injury develops, if severe hypo-

tension, or even cardiac arrest is superseded by <u>reperfusion of the brain</u> after restoration of normal blood pressure. Similar conditions have been studied in human volunteers (Rossen, 1943), or animals (Nemoto et al, 1977) by inflating a blood pressure cuff placed around the neck to a level above the systolic pressure. As might be anticipated in <u>transient global ischemia,</u> the nature and topographic distribution of brain lesions are entirely different. These depend on the duration of the ischemic episode, the success in re-establishing a normal CBF upon reperfusion, the age of the patient, the anatomic features of the collateral circulation, and the body temperature and serum glucose levels at the time of ischemia. The mean serum glucose levels in patients submitted to the hospital in cardiac arrest who did not regain consciousness were significantly higher, i.e. 341 mg/100 ml than of patients who regained consciousness after circulatory arrest, i.e. 262 mg/100 ml (Longstreth and Inui, 1984). In hypothermia (18° C), barbiturate anesthesia, and hypoglycemia, complete global cerebral ischemia of up to 30 min can be tolerated by gerbils (Meriones unguiculatus) with minimal neuronal necrosis not causing a detectable neurologic deficit (Treasure et al, 1983).

(3) <u>Regional and incomplete brain ischemia</u> is induced by occluding a single cerebral artery distal to the arterial circle at the base of the brain (Hudgins, Garcia, 1970). In the territory of the occluded artery, circulatory and metabolic conditions of marked heterogeneity develop during the first 2-3 hrs. The <u>clearly demarcated classical infarct</u> becomes apparent only after a relatively prolonged period of 2-3 hrs occlusion (Garcia, Kamijyo, 1974). The structural features of an arterial infarction can be modified by reperfusion of a cerebral territory that remained ischemic for up to 6 hrs. This induces grossly visible hemorrhage. Reperfusion of the ischemic territory may also result in selective necrosis of subcortical white matter (Kamijyo et al, 1977).

(4) <u>Occlusion of venous channels</u> can induce brain lesions which are traditionally designated as <u>infarctions.</u> However, as compared to an infarction of arterial origin several differences can be recognized. In addition to the different topographic distribution, <u>venous infarctions</u> are associated with large hemorrhages involving grey and white matter. Moreover, because arterial inflow is initially preserved edema and the leukocytic inflammatory response are more pronounced in venous as compared to arterial infarctions.

(5) <u>Lacunar infarctions.</u> A fifth and different pattern of brain injury results from lesions of intraparenchymal arteries or arterioles that supply the basal ganglia, the thalamic nuclei, and the pontine base. These lesions have been traditionally called lacunes. Two features are typical. They have a well defined volume of 0.2 to 15 mm³ and involve preferentially the basal ganglia and thalamic nuclei (Mohr, 1982).

Examples of all of the above mentioned manifestations of ischemia (Garcia, 1983) are found in human specimens, e.g. illustrated in the following case.

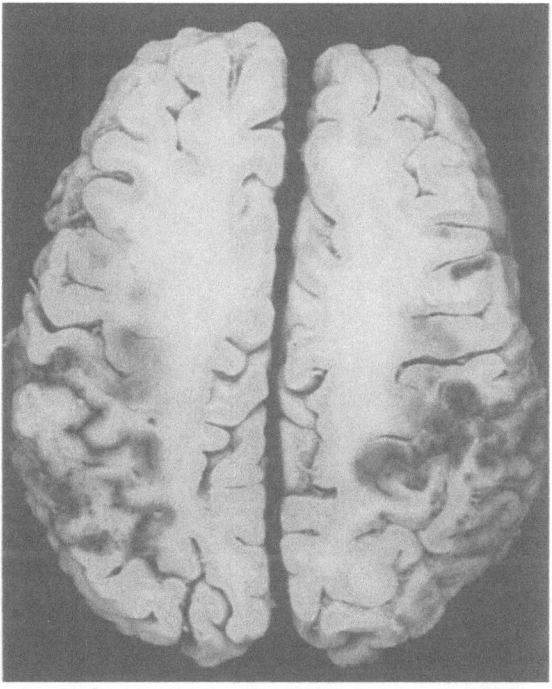

Fig.1: An 87 year-old woman had 12 days before her death a syncopal episode of several minutes. She remained unconscious until death although the blood pressure had returned to normal upon admission to the hospital. Autopsy showed bilateral kinking and ectasia of the internal carotid artery but absence of atheromatous changes or intrinsic occlusion. Extensive hemorrhagic softenings, approximately 12 days old, were seen in the parieto-occipital boundary zones of both cerebral hemispheres (UAB 15,160).

Fig.2: A 63 year-old man who developed gradual onset of dementia and inability to look upward. The patient underwent a sudden deterioration on the day of admission and died 21 days later. An angiomatous malformation was present in the periaqueductal grey matter with evidence of hemorrhage into the upper midbrain. Cortical laminar necrosis is visible in a large section of the frontal-parietal lobes (VA 23-83).

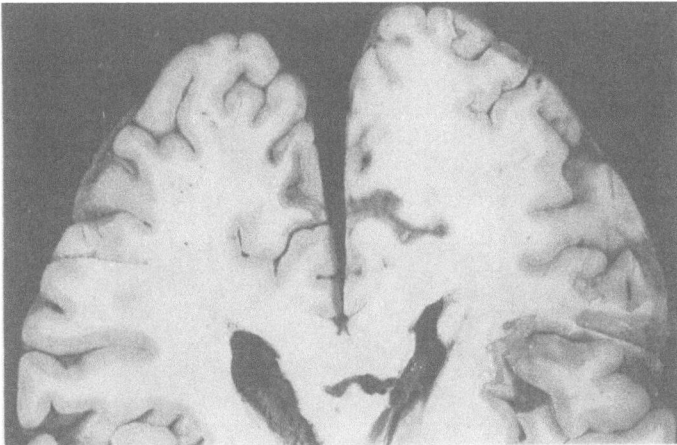

Fig.3: The 40 year-old man had a severe head injury approximately 5 weeks before death when a large epidural hemorrhage was evacuated. The patient remained unconscious during the entire period. The necrosis of the mesial surfaces of the occipital cortex is attributed to brain swelling at the time of injury and to bilateral uncinate gyrus herniation with extrinsic compression of the posterior cerebral arteries (UAB 15,503).

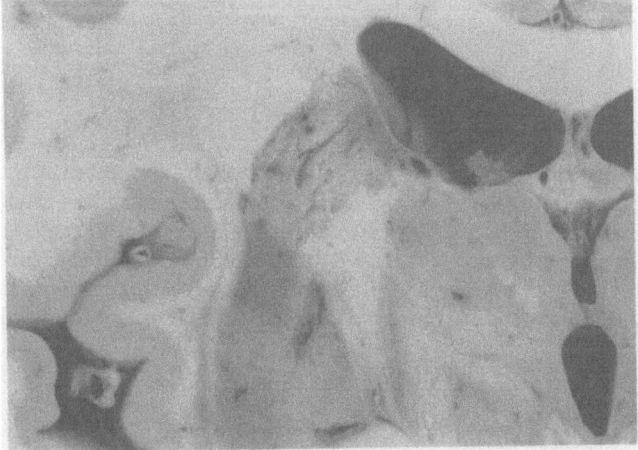

Fig.4: A 62 year-old diabetic and hypertensive man who became hemiplegic about four weeks before death. The close-up photograph of the internal capsule shows subacute infarction in the territory of the lenticulostriate arteriolar branches. Additional small and old lacunes are visible in the putamen (VA 92-82).

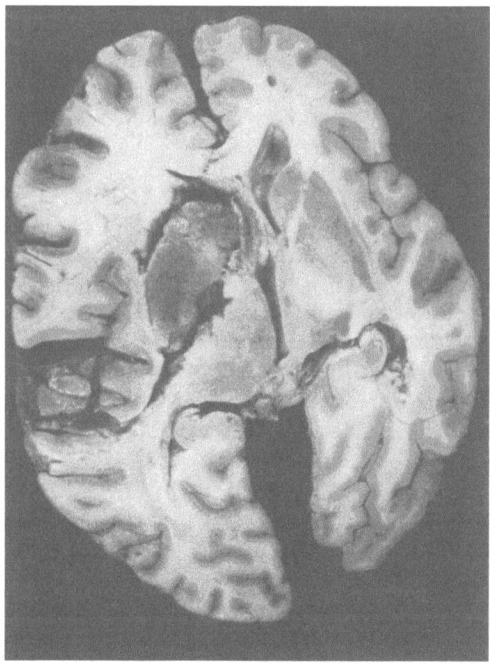

Fig.5: A 72 year-old woman was admitted to the hospital for evaluation of coronary and myocardial function. Upon cardiac catheterization and coronary angiography she abruptly developed right hemiparesis and aphasia. Autopsy disclosed embolic occlusion of the left MCA. The photograph shows a 4-days-old massive infarction in the territory of the occluded artery (UAB 15,499; reproduced with permission from Williams & Wilkins, Baltimore, MD: Davis RL and Robertson DM, eds., Textbook of Neuropathology, pp 590, 1985).

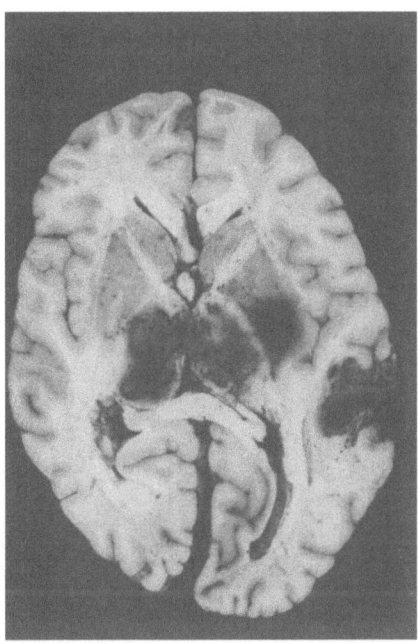

Fig.6: A 26 year-old woman known as diabetic for the previous 6 years. She developed polyuria, polydipsia, intermittent headaches which became persistent and severe when she entered the hospital. An initial CT scan revealed low attenuation lesions in the right thalamus and right basal ganglia suggestive of petechiae. She became unconscious and died on the 7th day in hospital. The cerebral hemispheres show hemorrhagic softening due to venous infarctions of both thalamic nuclei, right putamen, and of a portion of the right temporal lobe. There was thrombotic occlusion of the vein of Galen and of a right temporal cortical vein (UAB 15,518; reproduced with permission from Williams & Wilkins, Baltimore, MD: Davis RL and Robertson, eds., Textbook of Neuropathology, pp 592, 1985).

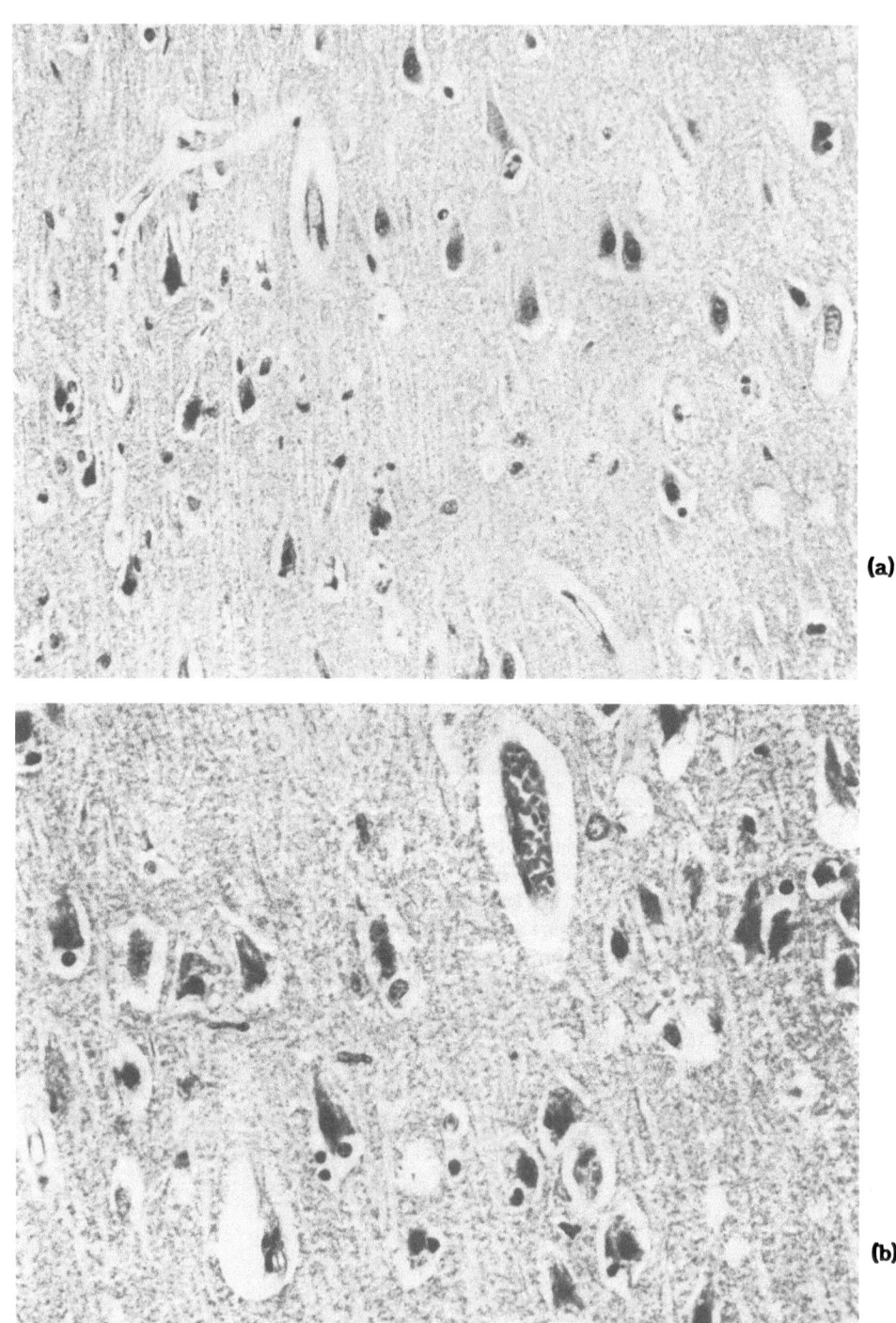

(a)

(b)

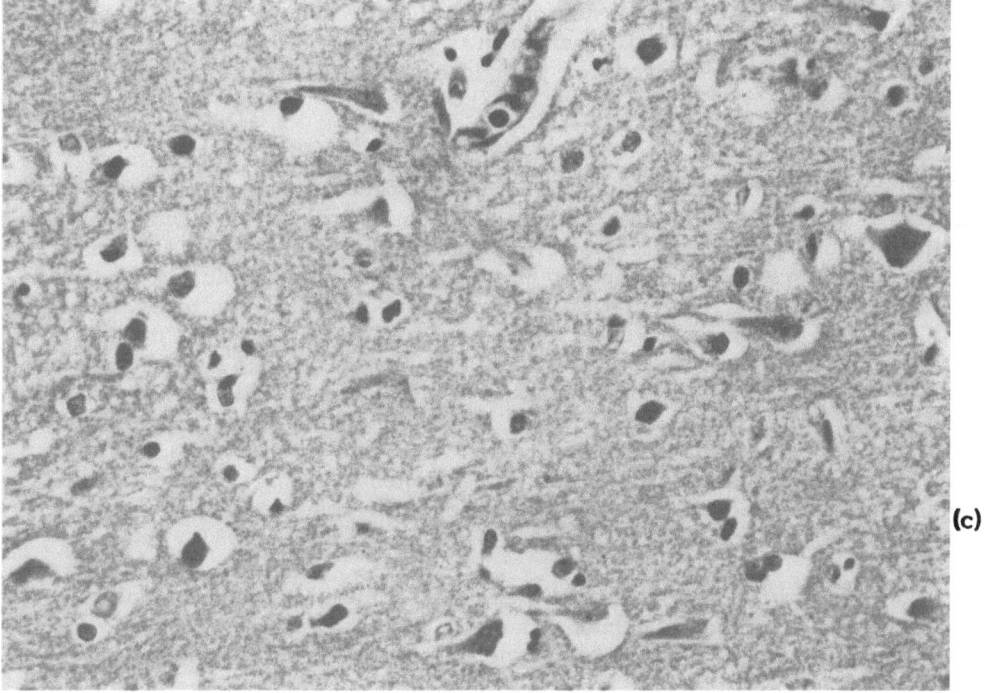

(c)

Fig. 7a,b and c: A 56-year old man suffering from cardiac arrest secondary to myocardial infarction eight days before his death. He
remained unresponsive despite normalization of the blood pressure.
The cerebral cortex shows heterogeneous and multifocal changes
in neuronal stainability and volume. The astrocytic nuclei are
enlarged and the neuropil is spongy. Some frontal neuronal perikarya in **Fig. 7a** are "selectively" injured displaying structural
abnormalities different from others in the same field, or from
those shown in the parietal **(Fig. 7b)** , or temporal cortex (**Fig.
7c** ; orig. magnif.: 63 x; UAB 15,728).

EXPERIMENTAL ISCHEMIA

Three varieties of ischemic brain injury have been experimentally evaluated by several investigators delineating the various steps from onset of ischemia until irreversibility is attained (Garcia, 1984).

(1) Studies of complete and permanent brain ischemia have demonstrated that during circulatory arrest of the brain at room temperature the structural alterations of neuronal, glial and vascular elements are similar and homogeneous. Ultrastructural modifications suggest that viability might be preserved for periods of up to 30-60 min (Arsenio-Nunez et al, 1973; Kalimo et al, 1977).

(2) Ginsberg et al (1978) studied in cats transient (15-30 min) global brain ischemia followed by a 90 min period of normotensive post-ischemic recirculation. After 15 min of ischemia blood flow was restored uniformly though at subnormal levels. Limiting the duration of cerebral ischemia to 15 min was found to prevent focal ischemia during recirculation. After 90 min recirculation energy metabolism largely recovered but did not return to normal, especially in grey matter. After 90 min recirculation, a focal incomplete restitution of the cerebral ATP content and blood flow was noted in animals when ischemia exceeded 15 min. After 30 min of ischemia, there were marked heterogeneities during reperfusion with multiple zones of persistent severe ischemia (Welsh et al, 1978). The homogenous neuronal changes of permanent cerebral ischemia (5-15 min) were modified by 60 min reperfusion into a heterogeneous pattern of selective vulnerability of nerve cells (Jenkins et al, 1981).

(3) Because of its close clinical, pathophysiological and anatomical similarities with the human condition of brain infarction which is the most frequent and important type of stroke cerebral ischemia by single-artery occlusion has been widely studied throughout the world in rodents, cats and subhuman primates. In a recent summary Raichle focussed particular attention on the disturbances in calcium homeostasis, tissue lactacidosis, rise in extracellular K^+ and release of vasoactive substances, e.g. leukotriene C_4 , as mechanisms of the secondary spread of irreversible injury to areas not primarily injured by ischemia (cf. Garcia, 1984). Our analysis of ischemic injury mechanisms has utilized integrated measurements of hemodynamic factors, such as CBF and quantitation of structural abnormalities, such as tissue edema, capillary diameters, capillary densities and numbers of necrotic neurons.

Microscopically, incomplete brain ischemia induced by various methods has common features including varying and unpredictable changes in the volume and stainability of neuronal perikarya including an abundance of "dark" and eosinophilic perikarya once the injury has attained irreversibility, marked sponginess of the neuropil, and in the early stages preservation of astrocytes. The volume and number of these cells increase probably as a result of the rise in extracellular K^+ as suggested by observations in cell cultures (Hertz, 1981). Oligodendrocytes and myelin sheaths occupy an intermediate position between the vulnerability of neurons and the resistance of astroglia (Garcia, Lossinsky, 1978). It has been assumed that the extent of ischemic injury, e.g. reflected by the number of dead neurons could be directly related to the degree of local flow depression. However, this seemingly self-evident premise could not be verified until recently, when a model of regional, incomplete ischemia was established in waking subhuman primates. Local CBF was measured in the territory of the occluded artery by several previously implanted intracerebral hydrogen electrodes. The degree of ischemia was calculated by integrating the time factor with the flow values obtained (Garcia et al, 1983). The extent of structural abnormalities was quantitated in two ways: (1) by differential counts of normal, injured and necrotic neurons, and (2) by point-counting estimates of tissue sponginess, or vacuolization as a measure of edema. In this manner we found an excellent correlation between the degree of local ischemia and the extent of sponginess. Secondly, we found an excellent correlation between the severity of sponginess, i.e. neuropil vacuolization and the number of necrotic neurons. The normal values in non-ischemic tissue were 4-6%, while local sponginess exceeded 30% after ischemia. Then, more than 90% of the neurons were necrotic. Although the observation does not answer the question of whether edema precedes neuronal necrosis, it shows a close spatial and temporal relationship between the two phenomena.

Our observation of capillary dilatation and breakdown as a function of the degree of local ischemia, i.e. percent of local flow decrease x time is intimately related to these alterations. Respective observations were made in transient (30-120 min) regional ischemia followed by reperfusion for 24 hrs. The increased hydrostatic pressure at the capillary level during reperfusion may play a significant role in the structural breakdown of the microvasculature. It remains to be seen whether regional ischemia without reperfusion induces similar changes in the capillary network. Success in understanding stroke, the mechanisms leading to irreversible ischemic injury, and the development of therapeutic regimens depends on the investigator's ability to control the reproducibility of his model. We have characterized several stages of the morphologic evolution of neuronal damage after single artery occlusion (Garcia, 1983). As a next step, we developed an experimental model which allows to maintain a predetermined level of partial ischemia (Conger, 1983). The model uses adult rats with implantation of four platinum electrodes into cerebral cortex several days before the experiment. Ischemia to cerebral cortex is induced by a combination of 3- or 4-vessel ligation (common carotid and vertebral arteries) plus infusion or withdrawal, respectively of aortic blood by an O_2-electrode activated pump (Conger et al, 1985).

Structural alterations similar to those in primates have been found in rats with partial brain ischemia. A marked reduction of cortical blood flow was always achieved, although individual rats required different blood pressures to attain a comparable level of ischemia. We have so far studied tissue samples of three series of animals: (a) controls of either normal rats or animals with implanted electrodes without ischemia, (b)

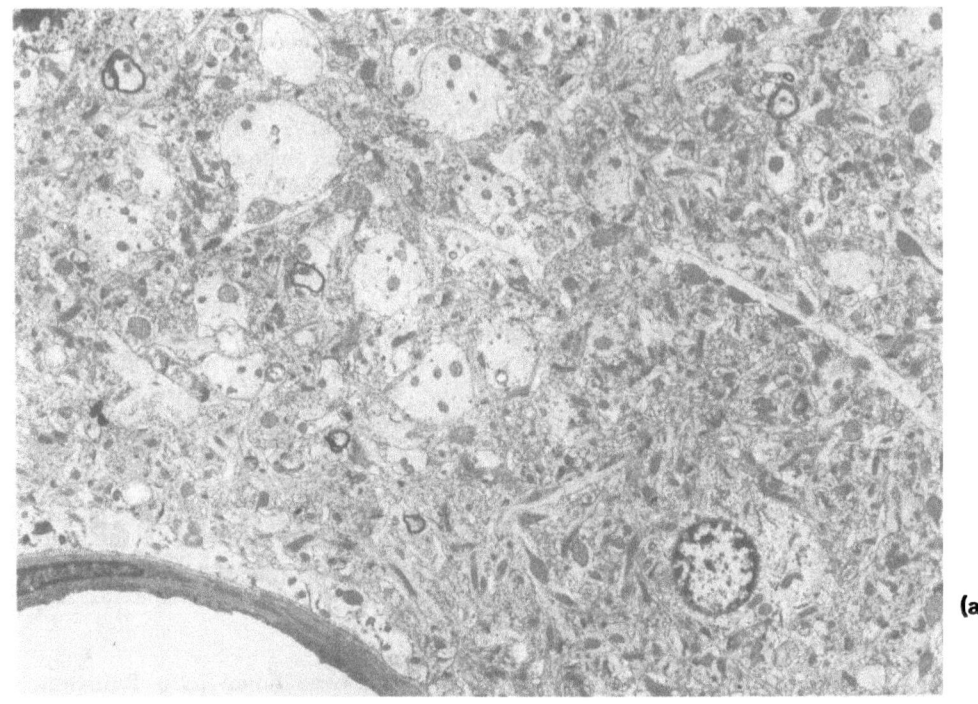

(a)

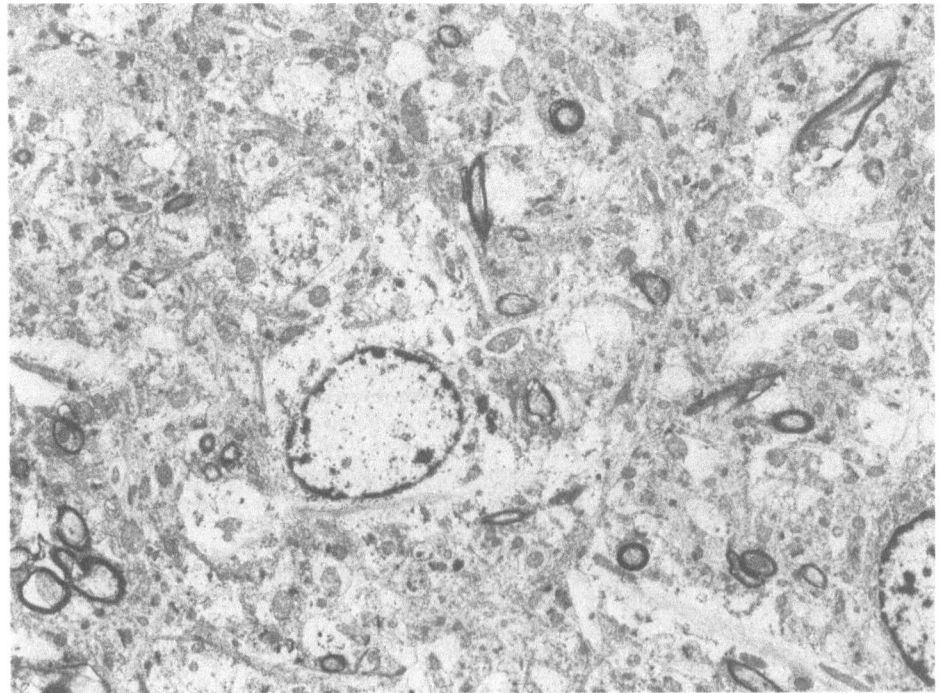

(b)

Fig.8a: Cerebral cortex of rat with "modest" ischemia by 75% reduction of O_2 -availability for 20 min and 24 hrs reperfusion. A minimal enlargement of dendritic processes may be found in the preparation when compared to the control cortex (KAC rat #126; orig. magnif.: 2,700x).

Fig.8b: Cerebral cortex of rat with "moderate" ischemia by 75% reduction of O_2 -availability for 35 min followed by 24 hrs reperfusion. Except for hypertrophic watery astrocytes (cf. center) there are no significant abnormalities (KAC rat #163; orig. magnif.: 2,700x).

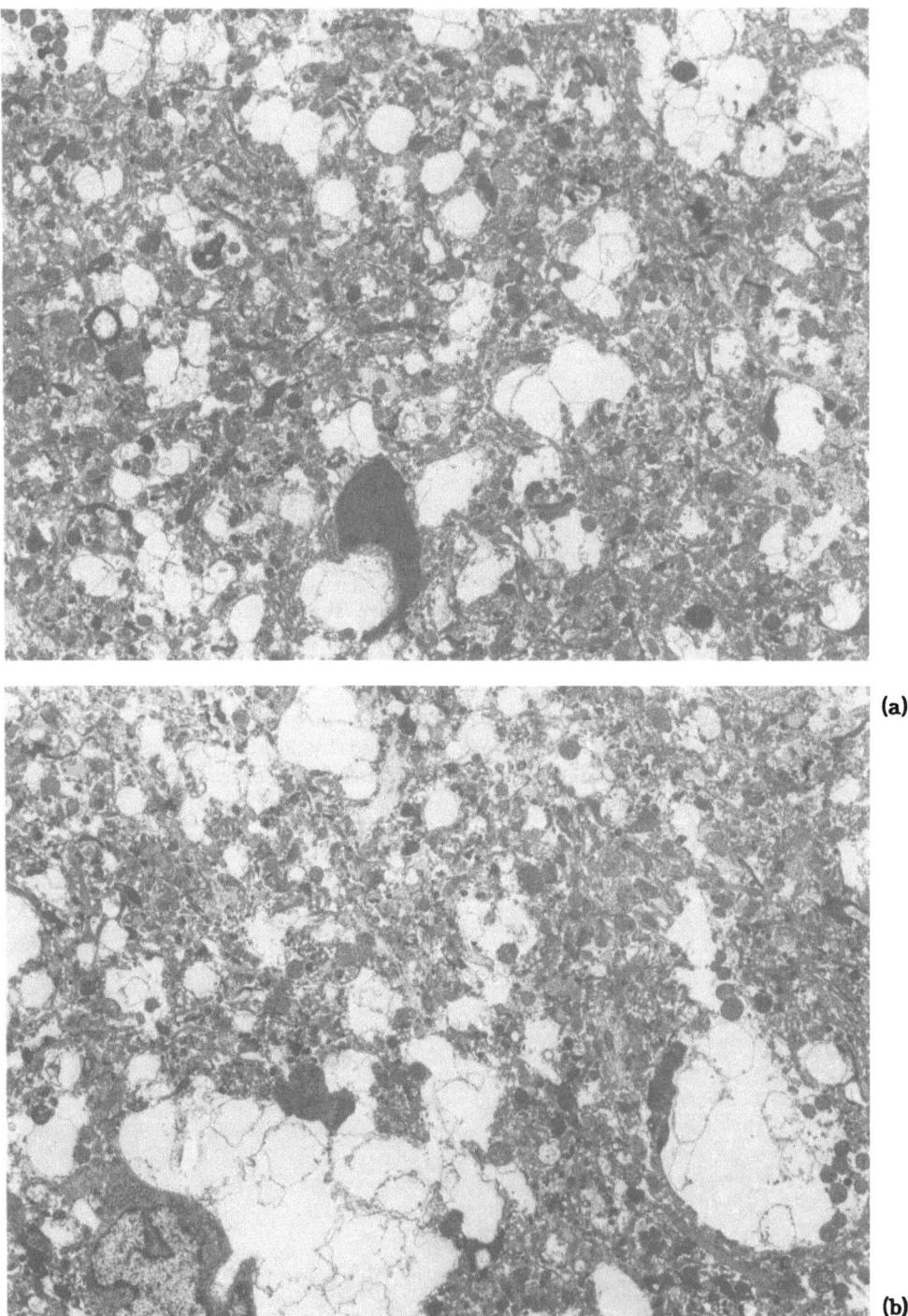

(a)

(b)

Fig. 9a and b: Rat cerebral cortex subjected to "severe" ischemia by 90% reduction of O_2 -availability for 50 min followed by 24 hrs reperfusion. In addition to the numerous necrotic neurons which are not shown, there is marked sponginess mostly attributable to astrocytic swelling (KAC rat #134; orig. magnif.: 2,700x).

rats with moderate injury where oxygen-availability fell by 75% during 20-60 min of incomplete ischemia, and (c) rats with severe injury where oxygen availability fell by 90% for up to 60 min of incomplete ischemia. All animals subjected to ischemia were alive for 24 hrs after the experiment.

Clinical observations. After awaking from anesthesia no noticeable differences in motor activity, physical appearance, and feeding/drinking patterns were observed between rats in series (a) and (b). In contrast, rats of series (c) showed a progressive loss of brain function with longer ischemic exposure.

Structural abnormalities. We found minimal neuronal damage in rats of series (a) and (b) (Fig. 8a,b). However, reduction of oxygen-availability to 10% of normal for 20 to 60 minutes in animals of series (c) produced severe neuronal damage within 40 minutes (Fig. 9a, b). A marked increase in systemic blood pressure was required to maintain a steady oxygen supply under these severe ischemic conditions. This is attributable to the development of an increasing tissue resistance to cerebral blood flow. The computerized feed back control between the regional O_2-pressure and the systemic withdrawal, or reinfusion, respectively of blood is considered essential for a reproducible induction of partial cerebral ischemia permitting comparison of morphological results between different animals.

ACKNOWLEDGEMENTS

We thank very sincerely Sharon Mardis for excellent secreterial support and Ralph Roseman for the high-quality illustrations. Financial support: Grants NS 08802 from the USPHS and from the Department of Pathology, UAB.

REFERENCES

Arsenio-Nunez ML, Hossmann KA, Farkas-Bargeton E, Ultrastructural and histochemical investigations of the cerebral cortex of cat during and after complete ischemia, Acta Neuropathol 26:329 (1973).

Astrup J, Barbiturate protection in focal cerebral ischemia, Scand J Clin Lab Invest 40:201 (1980).

Astrup J, Siesjö B, Symon L, Thresholds in cerebral ischemia: The ischemic penumbra, Stroke 12:723 (1981).

Conger KA, Moraes HP, Strong ER, Hino K, Briggs L, Garcia JH, Halsey JH, Jr, Computer controlled blood pressure changes required for steady state oxygen control of partial ischemia in the rat brain cortex. International Society on Oxygen Transport to Tissue. Louisiana Tech University, Ruston, Louisiana, p 36 (1983).

Conger KA, Moraes HP, Strong, ER, Hino K, Briggs L, Garcia JH,
    Halsey JH Jr, Computer controlled blood pressure changes requi-
    red for steady state oxygen control of partial ischemia in the
    rat brain cortex. in: "Oxygen transport to tissue IV", Bruley D,
    Bicher HI, Reneau D, eds., Plenum Publ. Corp. (1985).
Garcia JH, Kamijyo Y, Cerebral infarction. Evolution of histopathologic
    changes after occlusion of a middle cerebral artery in primates,
    J Neuropathol Exp Neurol 33:409 (1974).
Garcia JH, Lossinsky AS, Kauffman FC, Conger KA, Neuronal ischemic
    injury: Light microscopy, ultrastructure and biochemistry,
    Acta Neuropathol (Berl) 43:85 (1978).
Garcia JH, Ischemic injuries of the brain: Morphologic evolution, Arch
    Path Lab Med 107:157 (1983).
Garcia JH, Mitchem HL, Briggs L, Morawetz R, Hudetz AG, Hazelrig J,
    Halsey JH, Conger KA, Transient focal ischemia in subhuman
    primates: Neuronal injury as a function of local cerebral blood
    flow, J Neuropath Exp Neurol 42:44 (1983).
Garcia JH, Lowry SL, Briggs L, Mitchem HL, Morawetz R, Halsey JH,
    Conger KA, Brain capillaries expand and rupture in areas of
    ischemia and reperfusion. in: "Cerebrovascular disease" Vol 13,
    Reivich M, Hurtig HI, eds., Raven Press, New York (1983).
Garcia JH, Experimental ischemic stroke. A review, Stroke 15:5 (1984).
Ginsberg MD, Budd WW, Welsh FA, Diffuse cerebral ischemia in the cat:
    I. Local blood flow during severe ischemia and recirculation,
    Neurology 3:482 (1978).
Heiss WD, Flow thresholds of functional and morphological damage of
    brain tissue, Stroke 14:329 (1983).
Hertz L, Features of astrocyte function apparently involved in the
    response of the CNS to ischemia-hypoxia, J Cereb Blood Flow
    Metabol 1:143 (1981).
Hudgins WR, Garcia JH, Transorbital approach to the middle cerebral
    artery of the squirrel monkey: A technique for experimental
    cerebral infarction applicable to ultrastructural studies,
    Stroke 1:107 (1970)
Jenkins LW, Povlishock JT, Lewelt W, Miller JD, Becker DP, The role of
    postischemic recirculation in the development of ischemic neuro-
    nal injury following complete cerebral ischemia, Acta Neuro-
    pathol 55:205 (1981).
Kalimo H, Garcia JH, Kamijyo Y, Tanaka H, Trump BF, The ultrastructure
    of "brain death": II. Electron microscopy of feline cortex after
    complete ischemia, Virchows Arch B Cell Pathol 25:297 (1977).
Kamijyo Y, Garcia JH, Cooper J, Temporary MCA occlusion: a model of
    hemorrhagic and subcortical infarction, J Neuropathol Exp
    Neurol 36:338 (1977).
Longstreth WT, Jr., Inui TS, High blood glucose level on hospital admis-
    sion and poor neurological recovery after cardiac arrest, Ann
    Neurol 15:59 (1984).
Mohr JP, Lacunes, Stroke 13:3 (1982).

Nemoto EM, Bleyaert AL, Stezoski SW, Moossy J, Rao GR, Safar P, Global brain ischemia: a reproducible monkey model, Stroke 8:558 (1977).

Rehncrona S, Rosen I, Siesjö BK, Excessive cellular acidosis: an important mechanism of neuronal damage in the brain? Acta Physiol Scand 110:435 (1980).

Rossen R, Karat H, Anderson JP, Acute arrest of cerebral circulation in man, Arch Neurol Psychiat 50:510 (1943).

Sundt TM, Sharborough FW, Piepgras DG, Kearns TP, Messick JM, Jr., O'Fallon WM, Correlation of cerebral blood flow and carotid endarterectomy with results of surgery and hemodynamics of cerebral ischemia, Mayo Clin Proc 56:533 (1981).

Treasure T, Naftel DC, Conger KA, Garcia JH, Kirklin JW, Blackstone EH, The effect of hypothermic circulatory arrest time on cerebral function, morphology, and biochemistry: An experimental study, J Thorac Cardiovasc Surg 86(5):761 (1983).

Welsh FA, Ginsberg MD, Rieder W, Budd WW, Diffuse cerebral ischemia in the cat: II. Regional metabolites during severe ischemia and recirculation, Ann Neurol 3:493 (1978).

Yamaguchi T, Waltz AG, Okazaki H, Hyperemia and ischemia in experimental cerebral infarctions: Correlation of histopathology and regional blood flow. Neurology 21:565 (1971).

SECONDARY CHANGES IN HUMAN AND EXPERIMENTAL BRAIN

INFARCTION WITH PARTICULAR CONSIDERATION OF MICROEMBOLISM

K.J. Zülch

Max-Planck-Institute for Neurological Research
Ostmerheimer Str. 200, 5000 Köln 91, FRG

It may be useful to start the discussion of the subject with a definition of "secondary" changes although this is difficult in cerebral infarction. Therefore, I will try to define secondary changes of another entity, namely in brain tumor. Here, perifocal edema certainly is the most important secondary process which may induce further changes such as an increase of the intracranial pressure, shift of the brain mass and herniation. A final result of these processes may be "edema necrosis" (Jacob, 1940), which is characterized in the white matter by demyelinization. However, even this may not constitute the final consequence, since shrinking may follow secondary to organization of the tissue debris. Obviously, a chain of secondary changes follows the primary process in neoplastic disease of the brain.

MACROINFARCTION

Returning to the primary subject, cerebral edema also develops subsequently to ischemia, which in its first stage is intracellular (cytotoxic) and extracellular (vasogenic) in its second stage (Hossmann, 1982). Both, intra- and extracellular edema allow recognition of an infarct by computertomography as hypodense area. Finally, a rather inconspicuous "perivenous" edema follows at the margin and outside of an infarct. Although present in morphological investigations, it is insignificant because of its small volume. In exceptional cases brain edema secondary to infarction can lead to an extraordinary increase of volume, causing significant shifts of brain mass and herniation. A typical example is hemorrhagic infarction of the middle and anterior cerebral arteries, as described in Figs. 1, 4a,b, 5 (Zülch, 1981). I will not discuss prevention of this form of edema since extensive experimental and clinical experience in this field is available.

MICROINFARCTION

A pattern of cerebral microinfarction easy to study and analyze is cerebral fat embolism (Lazorthes et al., 1965; Ghatak and Zimmerman, 1971; v. Hochstetter and Friede, 1977). Cerebral fat embolism occurs mainly in patients with an open or partly open foramen ovale, or following passage of emboli through the capillary system of the lung. Figure 1 demonstrates the impressive changes, particularly in white matter resulting from microembolization. A case raising my interest in particular was a patient, who after an accident with fracture of a femur died of cerebral fat embolism 6 days later. At necropsy the brain showed the typical macroscopic aspects of "flea-bite" encephalitis (Fig.1).

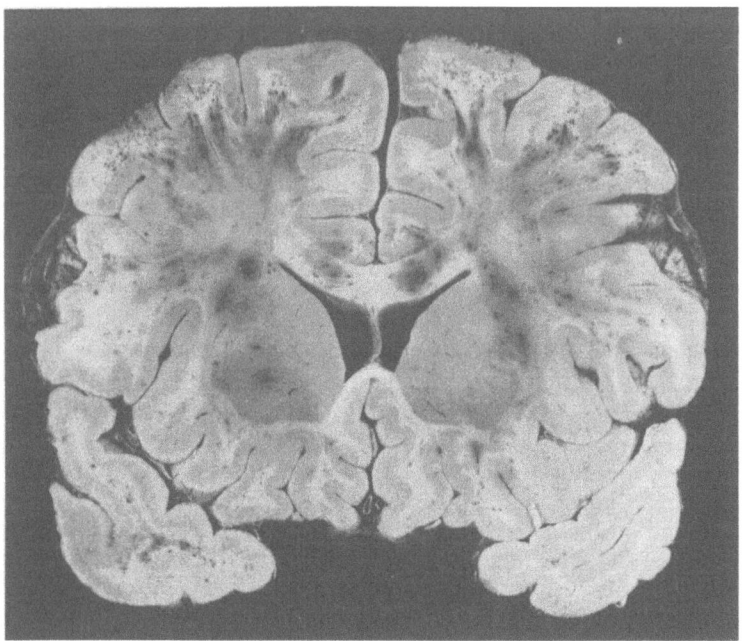

Fig.1: Example of "flea-bite" encephalitis resulting from fat embolism. Note extensive perivenous edema and perivascular hemorrhage.

"Flea-bite" encephalitis is a pattern common in many neuropathological entities, for instance in the so-called "perivenous" encephalitis in association with immunological reactions to vaccinations, drugs (arsphenamide, penicillin), or certain viral diseases (measles, rubeolae).

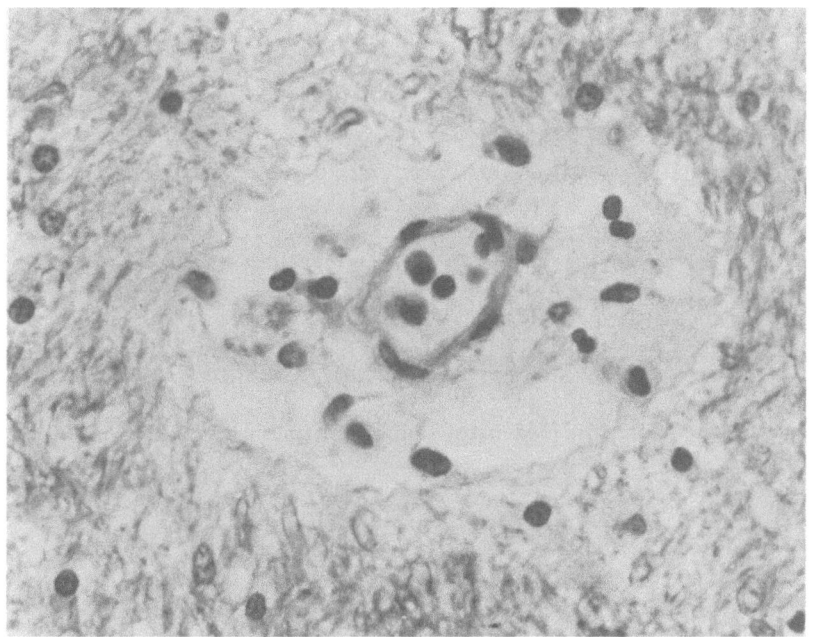

Fig.2: Perivenous edema. Luxol fast blue, 500 x.

The case shown in Figure 1 is impressive because of its numerous fat emboli, which occluded cortical capillaries and which were microscopically still visible on the 6th day. Other secondary changes, such as small erythrodiapedetic hemorrhages around capillaries were minimal. The predominant lesions were found in the white matter, very few fat emboli were seen with major changes around the large veins. Here, a broad mantle of protein-rich edema (Fig.2) had formed. Ring-like hemorrhages were seen outside the zone of edema. The axons were necrotic and the myelin sheaths were distended and undergoing demyelinization (Fig.3), while the necrotic mantle zone became organized, partly by formation of a ring-like granuloma, partly by resorption of debris. However, it is not clear why the actual primary changes were predominantly observed in cortical vessels, whereas the major secondary changes around distant veins. Since a pathogenetic concept could not be obtained from the observations in humans, an animal model was developed.

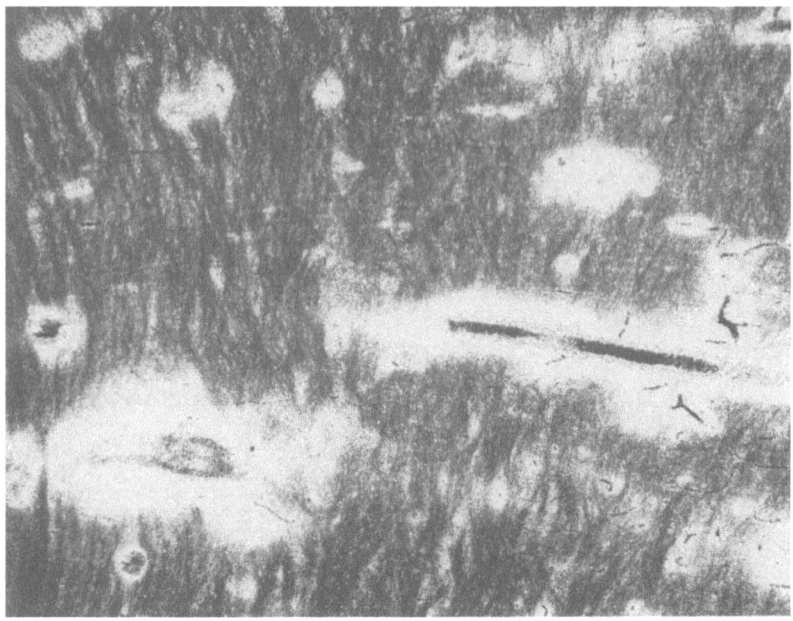

Fig.3: Demyelinization in "flea-bite" encephalitis due to fat embolism
(cf. Fig. 1).

EXPERIMENTAL MICROINFARCTION

    In studies to imitate the process of microembolization we observed
some of the changes described by Swank and Hain (1952; cf. Table I). The
paraffin-lamp-black procedure for embolization, however, was unreliable
because the size of emboli varied between 10 to 60 μ. We tried Hostalen-
spheres (Hoechst, FRG) instead (Zülch and Tzonos, 1964; 1965), which have
a standard size of 37 μ (Table II). The emboli occluded small arterioles
and arteries in the second and third cortical layer of cats, causing local microinfarcts. After 4 hours a protein-rich edema intensively
staining with trichrome Masson formed around the deep veins of the white
matter. Edema began to spread diffusely between the myelin sheaths of
the white matter, also towards the ventricle and down to the ependyma.
There was marked post-edematous demyelinization and marked necrosis of
neurons resulting in liquefaction and cyst formation in the white matter
3 - 4 weeks later. As a final result the white matter became almost
completely destroyed.

TABLE I

Animals Embolic material	Size	Number	Results
Cats Lamp black paraffine oil	10 - 60μ	100,000 ——→ 1.24 mio emboli carotid artery	Edema transuda- tion at veins

There was a close similarity to fat embolism in man, although perivenous hemorrhages around the mantle zone of edema were absent. Because of the difference in size of the occluding emboli and, consequently, of the cortical vasculature sustaining microinfarcts, we continued
our studies by using smaller emboli of the 3-M Company, Minnesota, USA
of 15 ± 5 μ. Together with Pakula and Schuier (Zülch et al., 1976) and
later with Tamura (Tamura and Zülch, 1978), we conducted large series of
experiments on microembolization in cats. Ultimately, the results cor-
responded both with the processes in white matter observed in the former
experiments, and the sequelae of fat embolism in man (Table III).

TABLE II

Animals Embolic material	Size	Number	Survival time	Technique
110 Cats "Hostalen" micro- spheres	37 μ	75,000 - 300,000 pheres	15 min ——→ 4 weeks	Intracarotid in- jection

Cortex	White matter	Acute sequelae	End result	Correlation to the human
Microinfarcts after occlusion of arterioles (3. - 4. layer) Scarring	4[h] heavy perivenous edema, diffuse streaming	Gross demyelination, "edema necrosis"	Cyst formation white matter. Hydrocephalus. Relatively minor damage in cortex	Emboli too large! Yet strong corre- lation to human peri- venous edema

TABLE III

Animals Embolic material	Size	Number	Survival Time	Technique
28 cats M-3 Minnesota Comp. microspheres	$15 \pm 5 \mu$	4 mio	1 3/4 - 4 hours ⟶ 8 weeks	Previous contralateral decompression, carotid injection via lingual artery; 20% losses
Cortex	White matter	Final results	Pathophysiology	Correlation
89% microspheres in 3.-4. layer; some ring hemorrhages. Pericapillary and perivenous edema at some vessels 1 3/4^h ⟶ 2/3 day no spreading	11% microspheres rare infarcts. Heavy pericapillary and perivenous edema, sometimes hemorrh. spreading ! Diminishing after 5 days; disappeared after 1 week	Hemisphere swollen - herniations - post-edematous softening, cystic degenerations near ventricles	EEG: depression during embolization; 1-3 sec isoelectric, gradual recovery. Low grade changes contralateral with early recovery	Similar to fat embolism in the human: cortex mild if any damages, heavy edema white matter, even edema lakes.

TABLE IV

Animals Embolic material	Size	Number	Survival time	Technique
46 cats M-3 Minnesota Comp. microspheres	$15 \pm 5 \mu$	10,5 mio	4 hours	Carotid injection via lingual artery
Cortex/Basal ganglia	White matter	Result	Final gross aspects	Correlation to the human
87% emboli. Multifocal BBB damage 2.5 min, edematous extravasations. Maximum 30 min. Recovery 1 - 4 ^{h}. Hyperemia.	12.1% a) Minimal injury b) Edema spread from cortex? c) Genuine perivenous edema.	Swelling neuroglia a.axons. Degenerative foci "Edema predominantly around large vessels" (veins)	Hemispheric swelling. Evans blue staining	Only 4^h survival. Death rate 100%. Early BBB damage with recovery after 1-2^h. Clearance of cortical edema, seeping into white matter. Transudation around veins ?

In the previous discussion particular attention was payed to the
changes in white matter. However, other studies of our laboratory (Vise
et al., 1977; Schuier et al., 1978) were also concerned with changes in
the cortex induced by injection of high amounts of 15 μ-emboli. Only the
first four hours were of particular interest. It should be emphasized
that most of the animals then already died from high intracranial pressure due to edema (Table IV). Examinations with double tracer techniques
showed rapid leakage of the blood-brain barrier in the cortex with extravasation of an albumin-rich edema from the capillaries. The maximum was
reached already within 30 min (!), while the process came to an end one
hour later. To our surprise the blood-brain barrier began to regain its
integrity at that time. The barrier had completely recovered after four
hours. Contrary to the cold injury model, spread of edema into the white
matter could not be observed. At the same time edema from perivenous
areas began to flow into the white matter (Vise et al., 1977; Schuier et
al., 1978). A disadvantage of this model was the short survival of only
little longer than four hours. The cause of death was the marked increase
of intracranial pressure due to edema in the cortex.

In earlier series, considerable amounts of perivenous edema formed
only after 4 hours. Furthermore, although damage to the blood-brain barrier became quickly established, recovery, i.e. cessation of edema production in the cortex started already one to two hours later. The intensity of the process and high mortality rate probably resulted from the
enormous number of microspheres used. As to the pathogenesis of "secondary" edema in the white matter, two possibilities were considered:
(1) local perivenous edema, or alternatively (2) secondary spread of the
primary edema from cortex into white matter. However, the latter was
not impressive during the first four hours. Yet, Vise and Schuier (1977,
1978) made several observations supporting my concept of perivenous
edema primarily spreading from veins of the white matter as a result of
embolized cortical capillaries.

Since the pathogenesis of the perivenous changes in the deep tissue
structures remained unclear, another experimental series ensued (Tamura
and Zülch, 1978), where only 4×10^6 microspheres of 15 ± 5 μ (3-M Company, Minnesota) were injected. However, this had still a mortality of
60% due to edema. We employed therefore surgical decompression according to Cushing by a large osteoplastic flap on the contralateral side.
This decreased the acute loss of animals to a tolerable level of 20%. The
primary changes were similar to those of the former series. 89% of the
microspheres lodged in the third and fourth layer of the homolateral cortex, the rest in the homolateral white matter. In some cases the microspheres were also located in areas supplied by the anterior cerebral
artery of the contralateral side. Deep perivenous edema started after 1h
and 45 min. The cortical edema was not nearly as extensive as in the
studies with large numbers of microspheres. In addition, according to
the tracer studies there was no downward flow of edema in any of the
sections. To the contrary, the marked perivenous edema originated apparently from white matter veins and spread then diffusely through the

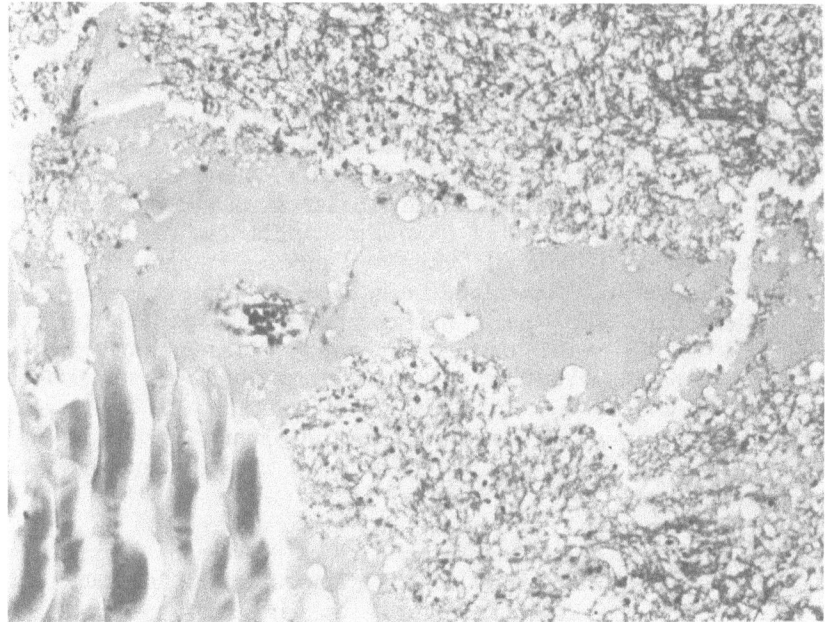

Fig.4: Accumulation of edema in experimental microembolism. A central vein can be recognized. Masson stain, 105 x.

subcortex of the hemisphere. It diminished after five days and disappeared completely after one week. Resolution of edema fluid was associated with post-edematous demyelinization, subsequent softening, necrosis, and finally cystic degeneration through the white matter down to the ventricular wall as already found in the model using 37 μ spheres. The pathogenesis of the experimental findings described herewith appears to be similar to that of cerebral fat embolism in man (Zülch, 1967). Yet, hemorrhages in the perivenous mantle were absent, while damage of the white matter was considerably more extensive. Due to the larger dimensions of human brain, single patches of perivenous edema fluid may be unable to accumulate and form edema lakes as seen in the experimental model (Fig. 4).

PATHOGENESIS

In a first study (Zülch and Tzonos, 1965) the concept was developed that after obstruction of the capillaries or arterioles by microemboli the territory was supported by diffusion from non-affected vessels in the neighbourhood. Patent neighbouring capillaries would deliver more oxygen than normal which should result in a decrease of the arterio-venous O_2-difference. Thus, the veins would be perfused by hypoxic blood (Fig. 5).

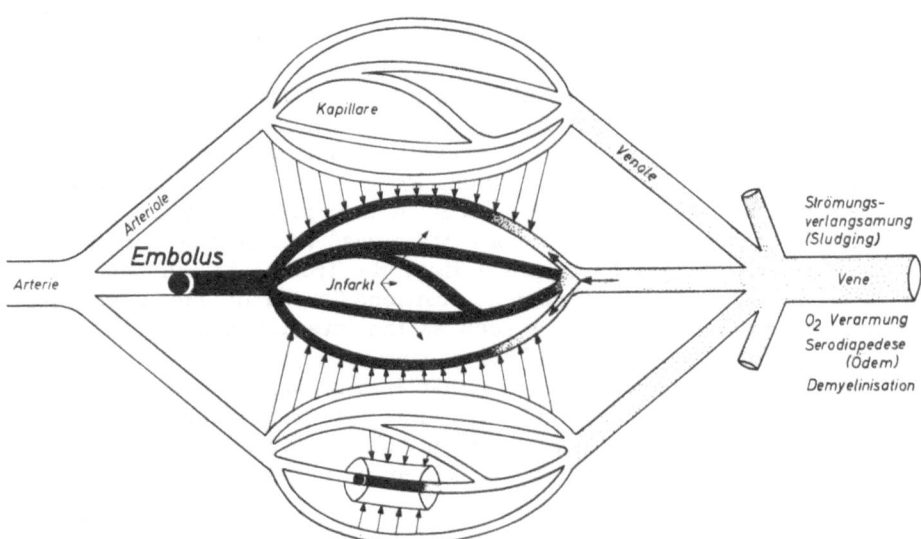

Fig.5: Schematic drawing of the pathomechanism of "perivenous" brain edema (cf. text; "Arterie": artery; "Arteriole": arteriole; "Kapilla-
re": capillary; "Venole": venule; "Infarkt": infarct; "Strömungs-
verlangsamung": flow retardation; "Vene": vein; "O_2 -Verarmung":
O_2 -depletion; "Serodiapedese": protein extravasation).

Since it is understood that the venous endothelium is supplied with
nutrients from blood, formation of endothelial lesions might occur more
easily under the above mentioned circumstances. The venous vessel wall
damaged hereby could become more permeable and allow for extravasation
of perivenous edema, or even erythrodiapedesis. The latter occurs in human in so-called "flea-bite" encephalitis (cf. above).

The above concept could be tested in the light of available data.
The experiment of Vise (1977) provided important biochemical results. The
arterio-venous O_2 -difference fell from 10.5 to 5.3 ml/100ml while the
arterio-venous glucose concentration difference from 11.7 to 2.6 mg/100
ml. These data seem to contradict my concept. However, I must emphasize
that these observations were made in a model with rapid formation of edema in the cortex and increase of the intracerebral pressure (cf. above).
On the other side, the observations suggest that hyperbaric oxygen could
be effective as in experimental air embolism. Hyperbaric oxygen has been
successfully used in experimental allergic encephalitis (Gosset et al.,
1969; Warren et al., 1978; Hansbrough et al., 1980) and provides transient improvement in multiple sclerosis (cf. Fisher et al., 1983).

CONCLUSIONS

As mentioned above it may be concluded that perivenous edema induced by microembolism is a secondary phenomenon. However, prevention of this process remains unclear because its pathogenesis is hardly understood. Brain edema is a well established aspect of microembolism both in man and experimental animals. Nevertheless, its enormous quantity is surprising. Its seeps and flows through the hemisphere down to the ventricle (Fig. 1; cf. Tamura and Zülch, 1978). The question is whether such quantities of edema may also occur in man in other neuropathological processes. We have observed "streams of edema" in infected cerebral wounds. Here, the edema bulk spread from the frontal to the occipital lobe with the patient in a recumbent position. Spread of edema changed direction only, where fiber tracts such as the internal capsule, the optic tract etc. prevented further fluid expansion (Fig. 9c; cf. Zülch, 1943). At that time our concept was that resolution of edema is brought about by: (a) resorption from the extra- into the intracellular space causing secondary cell swelling, (b) by spreading towards the ventricles, where the edema fluid is drained via CSF-pathways, or (c) by capillary resorption.

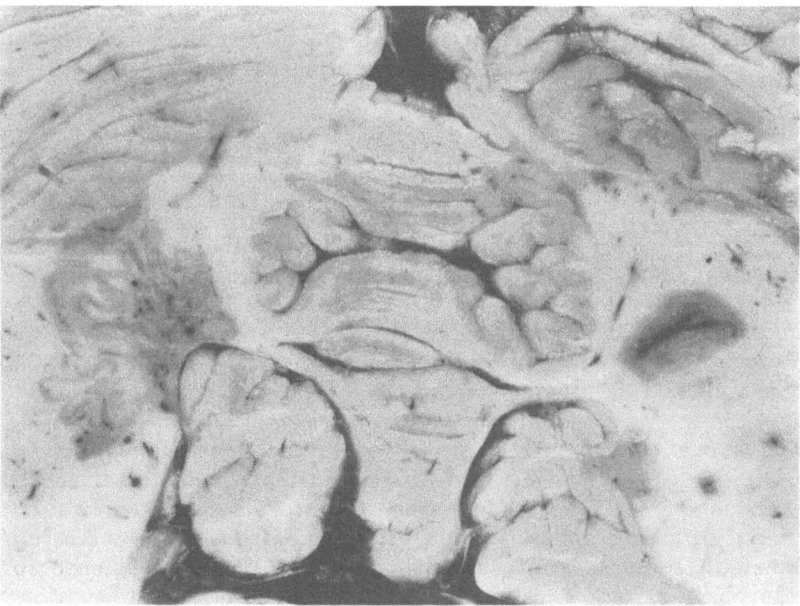

Fig.6: Marked plaque around a large vein in the cerebellar white matter.

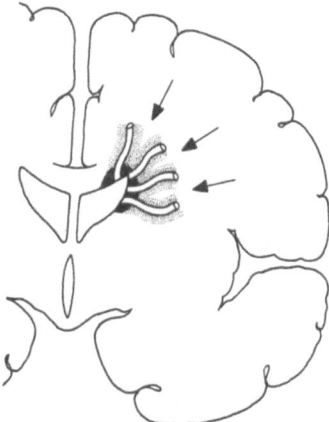

Fig.7: Concept of "Steiner's Wetterwinkel" in multiple sclerosis by perivenous demyelinization.

As to the therapeutical prevention of edema and cerebral fat embolism, a better understanding of its pathogenesis is required. Hyperbaric
oxygen may be considered to prevent the secondary changes by ensuring O_2
-supply to the venous endothelium, which is particularly prone to sustain
damage under these conditions. Since brain edema in microembolization is
extracellular (vasogenic), steroid treatment should be tried as in tumors
and other space-occupying processes. Yet, Siegel et al. (1972) have shown
steroids to be ineffective, if large microemboli (80 ± 20 μ) were used.
It is important to uncover the mechanism of perivenous edema formation,
since it is a phenomenon common in many neuropathological processes. In
multiple sclerosis, Rindfleisch (1863) and Borst (1897) have emphasized
the perivenous pattern (Fig. 6) of demyelinization found at the beginning. Based on this phenomenon Courville (1959) developed a concept of
the pathogenesis of multiple sclerosis. Figure 7 shows my concept of a
"perivenous syndrome" with demyelinization in multiple sclerosis in order
to explain "Steiner's Wetterwinkel". It is hoped that a better understanding of the "perivenous syndrome" provides information to prevent
or treat this destructive process more effectively.

REFERENCES

Borst M, Zur pathologischen Anatomie und Pathogenese der multiplen
    Sklerose des Gehirns und Rückenmarks, Beitr pathol Anat 21:308
    (1897).
Courville CB, Multiple sclerosis as an incidental complication of a dis-
    order of lipid metabolism, Bull Los Angeles Neurol Soc 24:60
    (1959).
Fisher BH, Marks M, Reich T, Hyperbaric-oxygen treatment of multiple
    sclerosis, N Engl J Med 308:181 (1983).
Ghatak NR, Zimmerman HM, Cerebral bone marrow embolism,
    Arch Pathol 92:112 (1971).
Gosset A, Gambarelli-Dubois D, Riche D, Fructus X, Naquet R, Tentatives
    therapeutiques par l'hyperbarie de l'embolie gazeuse caroti-
    dienne chez le babouin, Rev Neurol 121:531 (1969).
Hansbrough JF, Piacentine JG, Eiseman B, Immunosuppression by hyper-
    baric oxygen, Surgery 87:662 (1980).
v.Hochstetter AR, Friede RL, Residual lesions of cerebral fat embolism,
    J Neurol 216:227 (1977).
Hossmann KA, Pathophysiology of vasogenic and cytotoxic brain edema,
    in: "Treatment of cerebral edema", Hartmann A, Brock M, eds.,
    Springer, Berlin (1982).
Jacob H, Über die diffuse Markdestruktion im Gefolge eines Hirnödems
    (Diffuse Ödemnekrose des Hemisphärenmarkes), Z Ges Neurol
    Psychiatr 168:382 (1940).
Lazorthes G, Espagno J, Arbus L, Van Hong N, Les embolies grasseuses
    en pratique neuro-chirurgicale, Neuro-chirurgie 11:253 (1940).
Reulen HJ, Schürmann K, eds., "Steroids and brain edema", Springer, Ber-
    lin (1972).
Rindfleisch E, Histologisches Detail zu der grauen Degeneration von Ge-
    hirn und Rückenmark, Arch Pathol Anat Physiol 26:474 (1863).
Schuier FJ, Vise WM, Hossmann KA, Zülch KJ, Cerebral microembolization.
    II. Morphological studies, Arch Neurol 35:264 (1978).
Siegel BA, Meidinger R, Elliot AJ, Studer R, Curtis C, Morgan J, Potchen
    EJ, Experimental cerebral microembolism: Multiple tracer as-
    sessment of brain edema. Arch Neurol 26:73 (1972).
Swank RL, Hain F, The effect of different size emboli on the vascular
    system and parenchyma of the brain, J Neuropathol Exp Neurol
    11:280 (1952).
Tamura M, Zülch KJ, Experimental microembolism of the brain,
    Neurosurg Rev 1:111 (1978).
Vise WM, Schuier FJ, Hossmann KA, Takagi S, Zülch KJ, Cerebral micro-
    embolization, I. Pathophysiological studies, Arch Neurol 34:660
    (1977).
Warren J, Sacksteder MR, Thunin CA, Oxygen immunosuppression: modi-
    fication of experimental allergic encephalomyelitis in rodents,
    J Immunol 121:315 (1978).
Zülch KJ, Hirnödem und Hirnschwellung, Virchows Arch (Pathol Anat)
    310:1 (1943).

Zülch KJ, Neuropathological aspects and histological criteria of brain edema and brain swelling, in: "Brain edema", Klatzo I, Seitelberger F, eds., Springer, New York (1967).

Zülch KJ, Cerebrovascular pathology and pathogenesis as a basis of neuroradiological diagnosis, in: "Handbuch der medizinischen Radiologie", Vol. XIV, Part 1A, Diethelm L, Wende S, eds., Springer, Berlin (1981).

Zülch KJ, Tzonos T, Transsudationsphänomene an den tiefen Hirnvenen nach Blockade von Arteriolen oder Kapillaren der Rinde durch Mikroembolien, Naturwissenschaften 51:539 (1964).

Zülch KJ, Tzonos T, Transudation phenomena at the deep veins after blockage of arterioles and capillaries by micro-emboli, Bibl Anat 7:279 (1965).

Zülch KJ, Pakula H, Schuier F, Distant perivenous demyelination after microembolization of the brain, Neuropathol Appl Neurobiol 2:163 (1976).

ISCHEMIC BRAIN DAMAGE IN THE RAT IN A LONG TERM
RECOVERY MODEL

Roland N. Auer, M.-L. Smith and B.K. Siesjö

Laboratory for Experimental Brain Research
E-Blocket, University Hospital
S-22185 Lund, Sweden

INTRODUCTION

Since the pioneering experiments of Grenell (1), who demonstrated neuronal necrosis in dogs after periods of ischemia from only two to less than ten minutes, there has been controversy as to the minimum time necessary to produce permanent brain damage after ischemia. The maximum duration of ischemia which the animal can survive indefinitely ("revival time") has also been the subject of some dispute, with times as long as one hour having been reported for recovery of at least basic neurophysiologic functions (2,3).

More recently, there has been a reversal of the trend of accepting such long ischemic periods as survivable, due to the advent of long term recovery models, which demonstrate neuronal necrosis after periods as short as 3 to 5 minutes (4,5). As such models demonstrate progressive neuronal disease and then death in longitudinal studies through time, and correlate better with clinical experience of revival times, the importance of such long term study of the post-ischemic period cannot be overemphasized.

As these results were obtained in the gerbil, and there has been doubt expressed as to the general applicability of these observations due to an idiosyncratic epileptic predisposition in the gerbil species (6), pathologic examination of another species after such short ischemic periods seems warranted. Already, results in rats have provided evidence supporting a similar "maturation" or "evolution" of damage in the post-ischemic period (7,7a). However, the minimum time necessary to produce neuronal necrosis in the rat has not been clearly established, and a model producing dense cerebral ischemia is required for this. Such observations are critical for the interpretation of results obtained in

99

the gerbil, and for generalization of the observed phenomena to a wider range of species.

The present article describes brain damage in the rat after ischemic periods of two to ten minutes, with an emphasis on secondary mechanisms. Brain damage is considered here as acidophilic neuronal necrosis, or total tissue necrosis (infarction).

MATERIAL AND METHODS

The pathologic material reported here derives from original publications concerning a long term recovery model for forebrain ischemia in the rat (8,9).

Male Wistar rats were anaesthetized with halothane and intubated. Tail artery and vein catheters were inserted. After exposure of the carotid arteries, ligatures were loosely placed around the vessels and the animals allowed to stabilize on a 2:1 mixture of N_2O : O_2. The carotid arteries were held by the ligatures, and clasps were applied. The blood pressure was lowered to 50 mm Hg through infusion of Arfonad, and controlled exsanguination. After the desired period of ischemia (two to ten minutes) the clasps were released, the exsanguinated blood was re-infused, and the animals extubated.

After survival of one week, the rats were sacrificed by trans-cardiac perfusion with 4% buffered formaldehyde. The brains were embedded in paraffin, sub-serially sectioned at 8μ, and stained with acid fuchsin/cresyl violet.

RESULTS

Results pertaining to the animal model have been reported previously (8). Briefly, rats with shorter than eight minutes ischemia were normal on handling. However, a number of rats with ten minutes ischemia, which proved to have severe brain damage on subsequent histologic examination, walked on extended limbs, and were hyperexcitable on handling (8). This was felt to be due to severe damage to components of the motor system (see below). These animals demonstrating the most severe clinical and pathologic damage had plasma glucose levels above 12 mM/l.

Hippocampus

This was the first structure affected, already after two min ischemia. A subgroup of CA4 neurons seemed to be affected, since with increasing grades of damage, all of CA4 was never recruited. After four minutes ischemia, CA1 was regularly affected, with necrosis of pyramidal neurons in CA1 and the subiculum. After six to eight minutes, some CA3 pyramids were necrotic, and after ten minutes the granule cells of the dentate gyrus, the most resistant cells, were sometimes affected. Counts of acidophilic neurons at the seven layers indicated (Fig. 1) showed an increase in the numbers of these neurons substantiating the above order of recruitment of the hippocampal neuronal cell types.

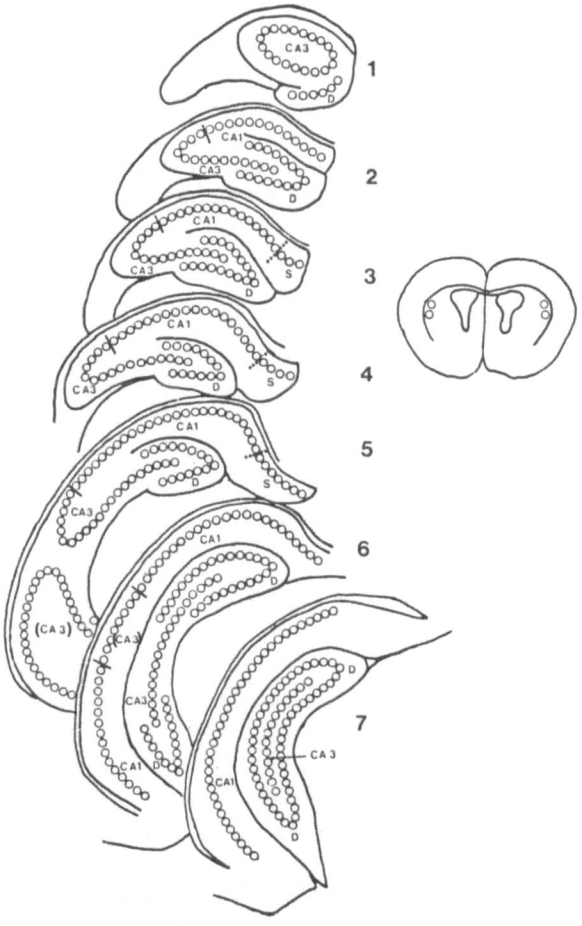

Fig.1: Schematic diagram to indicate the levels where the hippocampus and caudate nucleus were examined and quantitated. The location of the counted fields in the caudate nucleus, used to generate the data in Fig.4, is given on the right

Cerebral Cortex

Here we distinguish selective neuronal necrosis from infarction, since infarction either occurred or did not occur in an all or none fashion, and these two conditions represent vastly different grades of tissue injury. We define selective neuronal necrosis as necrosis of neurons with preservation of astrocytes, and infarction as cavitation of tissue due to necrosis of astrocytes, oligodendrocytes, pericytes, etc.; in short, all cells of the tissue.

Selective neuronal necrosis was seen in the cerebral cortex after four minutes ischemia, and tended to affect the middle laminae of the cortex. Bilaminar affectation was specifically looked for, but was unusual.

With regard to the geographic distribution of the damage over the hemisphere, two patterns were seen: Arterial and venous. The first area of the cerebral hemisphere which was consistently affected was the superolateral convexity, in the zone between the territories of supply of the anterior and middle cerebral arteries. Selective neuronal necrosis was present here in the mildly affected animals, and infarction developed in an identical location in the more severely affected animals (Fig. 2).

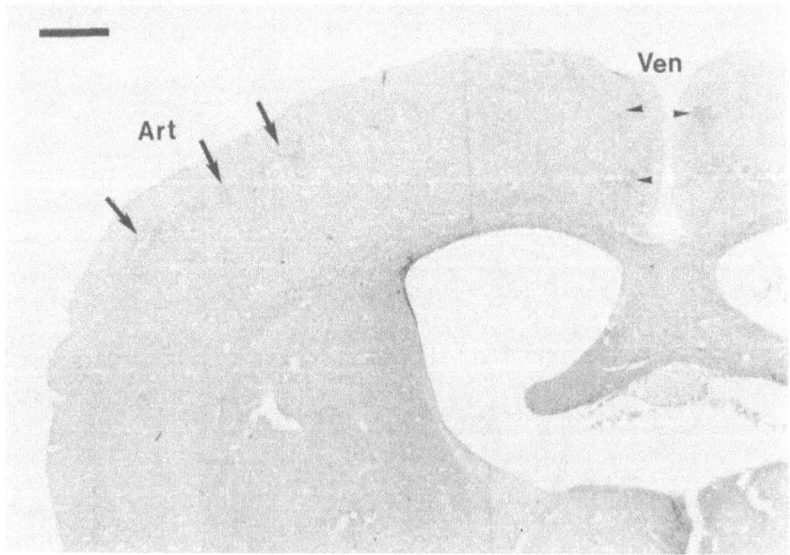

Fig.2: Low power coronal section of the hemisphere to show the arterial (Art) and venous (Ven) patterns. The superolateral convexity shows patchy infarction (arrows), with selective neuronal necrosis intervening, in the borderzone between the anterior and middle cerebral arteries. The venous pattern is seen as patchy parasaggital microinfarcts (arrowheads). Bar = 500 μ

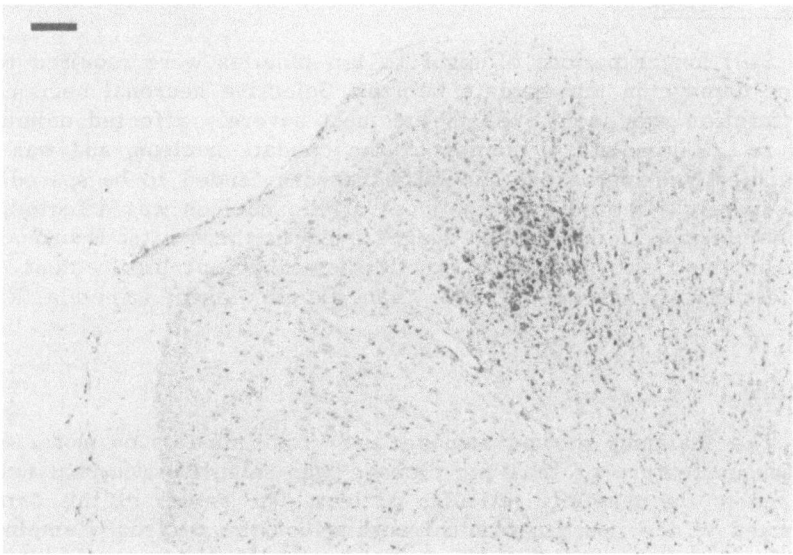

Fig.3: The venous infarcts were also seen over the superior surface of the brain, but always near the midline, remote and distinct from the more lateral areas of borderzone damage. They were present from the frontal to the occipital cortex. Bar = 100 μ

Infarction could alternate with intervening selective neuronal necrosis (Fig. 2). The cortex adjacent to an infarct showed selective neuronal necrosis in the middle layers, similar to that seen in a less severely affected brain. In other words, infarction developed in an augmentation of the pattern of selective neuronal necrosis, but the borders between infarcted tissue and tissue with selective neuronal necrosis were sharply demarcated: Infarction with cavitation either did or did not occur.

The second pattern of hemispheric cortical damage was venous, located parasagittally in patches near the median cerebral fissure. These microinfarcts alternated from side to side as one progressed from anterior to posterior through the brain sections, and were often related to superficial cerebral veins draining medially. They were present either on the mesial surface of the hemisphere (Fig. 2) (cingulate gyrus), or on the convexity near the midline (Fig. 3). They were present in the occipital lobes above the midbrain, an area not subjected to as dense ischemia as the anterior forebrain. Because of the anatomy of these parasagittal infarcts, they were regarded as venous, in spite of the lack of hemorrhage associated with them.

Caudate Nucleus

     Still longer periods of eight to ten minutes were required to
produce damage in the caudate nucleus. Selective neuronal necrosis, but
not infarction was seen, even in the most severely affected animals. This
began in the dorsolateral portion of the caudate nucleus, and was
geographically demarcated. The large neurons tended to be spared. In the
most severely affected animals, most of the nucleus was affected, but a
portion inferiorly tended to be spared even in these rats. Neuronal
counts in the fields indicated (Fig. 1) showed abrupt involvement of
caudate neurons only after longer than six minutes of ischemia (Fig. 4).

Thalamus

     The thalamus showed damage after four minutes or more. A
conspicuous feature of thalamic damage was selective neuronal necrosis
localized to the thalamic reticular nucleus. The border of this damage
conformed to the neuroanatomical outlines of the nucleus. Remaining
thalamic nuclei were affected in a patchy and less regular fashion.

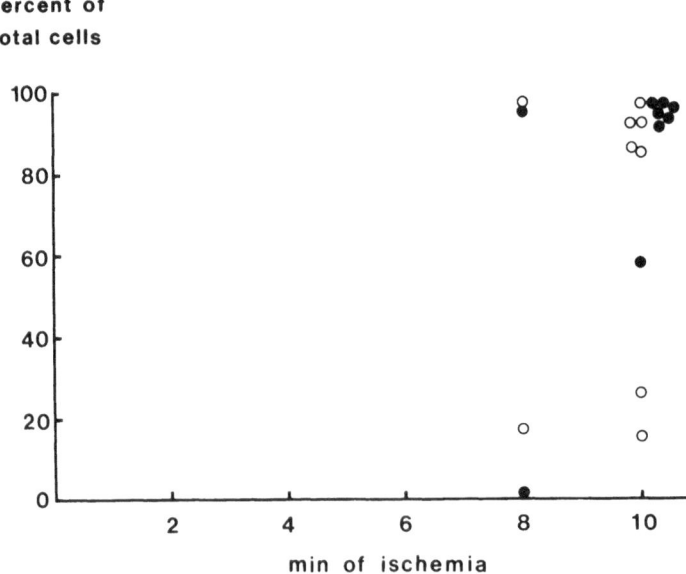

Fig.4: Per cent cells affected in the lateral caudate nucleus as a
        function of the number of minutes ischemia. Data for left (open
        circles) and right (filled circles) are shown. No damage is seen
        after six min ischemia or less. In all groups, n = 6, except the
        ten minutes group, where n = 8

Posterior Fossa Structures

Damage was occasionally seen in infratentorial structures, somewhat surprising at first in view of the forebrain ischemia produced by this model. The substantia nigra, pars reticularis, showed occasional small infarcts in the geographic center of the nucleus. The cerebellum showed damage in the vermis, appearing first as selective neuronal necrosis of Purkinje cells. In the most severely affected rats, frank infarction of the vermis was seen.

DISCUSSION

The present results have demonstrated degrees of brain damage from selective neuronal necrosis after only two minutes to frank infarction, especially in high glucose animals after ten minutes. Ten minutes was the maximum "revival time". Infarction occurred in arterial and venous patterns. These findings will be discussed in order, in relation to the literature on ischemia.

"Revival Time" and Maturation of Cell Damage in Ischemia

Revival time refers to the longest period of ischemia after which an organism can be revived. Data in dogs (1) have suggested ten minutes as a tolerable revival time, when survival times were several weeks. Data in rabbits (10) showed similar revival times and histologic damage, seen also after survival times of days to weeks. More recently, prolonged periods of anoxia or ischemia have been considered to produce little brain damage (3,11,12,13), and the discrepancies between these two extremes have not yet been satisfactorily resolved. It is noteworthy, however, that survival times in the later experiments were shorter, on the order of minutes to hours, than in the earlier experiments, where survival was days to weeks, and survivability of the organism was demonstrated beyond doubt. Hence, survival time may have an influence on the apparent revival time , with seemingly longer doses of insult becoming tolerable if only short recovery is desired in the experiments as designed. The resolution of this difficulty may lie in the recently documented phenomenon of "maturation" of ischemic brain damage (4,5,7,14,15) where cell and tissue damage is prolonged into the recovery period. This concept of maturation of ischemic brain damage underscores the dynamic and evolving nature of the process over hours to days following a square wave insult. It may well be responsible for the discrepancy of short revival times seen at long survival times, and apparently long revival times if sufficiently short survival are used.

Infarction

Infarction occurred in the rats with the highest pre-ischemic plasma glucose values, over 12 mM/l. The detrimental effect of high glucose levels on ischemic brain damage is well known (16,17,18,19,20,21,22) probably through the mechanism of increased lactic acidosis and concomitant tissue edema (23,24,25).

Edema in the Recovery Period and Parasagittal Damage

The phenomenon of post-ischemic edema is well known, the edema reaching a maximum several hours to several days after an ischemic insult (26,27,28,29,30,31). It is the most likely explanation for the venous pattern of damage seen in the present model, seen only in the most severely affected animals. Parasagittal accentuation of ischemic damage has been seen before, in the cat (32,33). The extension of these infarcts into the occipital poles, areas not subject to dense ischemia in this model, combined with the anatomical pattern of these infarcts, support venous obstruction due to post-ischemic edema as the etiology of these infarcts.

Attempts to carry out more than ten minutes ischemia with the present model results in rats dying in status epilepticus, and the brains show massive edema, similar to previous reports (17).

Clinical Correlation

The clinical deficits seen in the most severely affected animals consisted of a peculiar posture while walking. The rats walked with their belly off the floor, due to hyperextension of all limbs. The abnormal posture and movement on walking is most likely related to the severe damage to basal ganglia and motor cortex in these animals. The borderzone damage in the motor cortex is similar to that seen in human neuropathology in cardiac arrest or carotid artery occlusion.

The present results demonstrate that short periods of dense ischemia, of the order of ten minutes or less, can give rise to substantial brain damage and even infarction, and point toward post-ischemic edema as an important secondary mechanism of brain damage.

ACKNOWLEDGEMENTS

The authors appreciate the technical expertise of Lille-Mor Lindeström in preparation of the sections. Birgit Olsson created the illustrations. This research was supported by the Swedish Medical Research Council (projects 12X-03020, 14X-263) and the National Institutes of Health of the United States Public Health Service (grant number 5 R01 NS07838). Dr. Auer is the recipient of a Medical Research Council of Canada Fellowship.

REFERENCES

1. Grenell RG, Central nervous system resistance. I. The effects of temporary arrest of cerebral circulation for periods of two to ten minutes, J Neuropathol Exp Neurol 5:131 (1946).
2. Hossmann KA, Kleihues P, Reversibility of ischemic cell damage, Arch Neurol 29:375 (1973).
3. Hossmann KA, Zimmermann V, Resuscitation of the monkey brain after 1 h complete ischemia. I., Brain Res 81:59 (1974).
4. Kirino T, Delayed neuronal death in the gerbil hippocampus following ischemia, Brain Res 239:57 (1982).
5. Kirino T, Sano K, Selective vulnerability in the gerbil hippocampus following transient ischemia, Acta Neuropathol (Berl.) 62:201 (1984).
6. Brown AW, Levy DE, Kublik M, Harrow J, Plum F, Brierley JB, Selective chromatolysis of neurons in the gerbil brain. Ann Neurol 5:127 (1979).
7. Pulsinelli WA, Brierley JB, Plum F, Temporal profile of neuronal damage in a model of transient forebrain ischemia, Ann Neurol 11:491 (1982).
7a. Kirino T, Tamura A, Sano K, Delayed neuronal death in the rat hippocampus following transient forebrain ischemia, Acta Neuropathol (Berl.) 64 :139 (1984).
8. Smith ML, Bendek G, Dahlgren N, Rosen I, Wieloch T, Siesjö BK, Models for studying long-term recovery following forebrain ischemia in the rat. 2., Acta Neurol Scand 69:385 (1984).
9. Smith ML, Auer RN, Siesjö BK, The distribution of ischemic brain damage in the rat after two to ten minutes forebrain ischemia, Acta Neuropathol (Berl.) 64:319 (1984).
10. Hirsch H, Müller HA, Funktionelle und histologische Veränderungen des Kaninchengehirns nach kompletter Gehirnischämie, Pflügers Archiv 275:277 (1962).
11. Neely WA, Youmans JR, Anoxia of canine brain without damage, JAMA 183:1085 (1963).
12. Hossmann KA, Olsson Y, Supression and recovery of neuronal function in transient cerebral ischemia, Brain Res 22:313 (1970).
13. Miller JR, Myers RE, Neurological effects of systemic circulatory arrest in the monkey, Neurology 20:715 (1970).
14. Ito U, Spatz M, Walker JT Jr, Klatzo I, Experimental cerebral ischemia in Mongolian gerbils. I. Light microscopic observations, Acta Neuropathol (Berl.) 32:209 (1975).
15. Mrsulja BB, Mrsulja BJ, Ito U, Walker JT Jr, Spatz M, Klatzo I, Experimental cerebral ischemia in Mongolian gerbils. II. Changes in carbohydrates, Acta Neuropathol (Berl.) 33:91 (1975).
16. Ginsberg MD, Welsh FA, Budd WW, Deleterious effect of glucose pretreatment on recovery from diffuse cerebral ischemia in the cat. I. Local cerebral blood flow and glucose utilization, Stroke 11:347 (1980).
17. Myers RE, Yamaguchi M, Effects of serum glucose concentration on

brain response to circulatory arrest, J Neuropathol Exp Neurol
35: 301 (1976).

18. Pulsinelli WA, Waldman S, Rawlinson D, Plum F, Moderate hyper-
    glycemia augments ischemic brain damage: A neuropathologic
    study in the rat, Neurology 32:1239 (1982).

19. Siemkowicz E, Hansen AJ, Clinical restitution following cerebral
    ischemia in hypo-, normo-, and hyperglycemic rats,
    Acta Neurol Scand 58:1 (1978).

20. Siemkowicz E, Gjedde A, Post-ischemic coma in rat: effect of
    different pre-ischemic blood glucose levels on cerebral meta-
    bolic recovery after ischemia, Acta Physiol Scand 110:225
    (1980).

21. Diemer NH, Siemkowicz E, Regional neuronal damage after cerebral
    ischemia in the normo- and hypoglycemic rat, Neuropathol Appl
    Neurobiol 7:217 (1981).

22. Welsh FA, Ginsberg MD, Rieder W, Budd WW, Deleterious effect of
    glucose pretreatment on recovery from diffuse cerebral ischemia
    in the cat. II. Regional metabolite levels, Stroke 11:355
    (1980).

23. Myers RE, Lactic acid accumulation as a cause of brain edema and
    cerebral necrosis resulting from oxygen deprivation, in: "Advan-
    ces in perinatal neurology", Korobkin R, Guillemineault G, eds.,
    Spectrum Publishers, New York (1979).

24. Rehncrona S, Rosen I, Siesjö BK, Brain lactic acidosis and ischemic
    cell damage: 1. Biochemistry and neurophysiology, J Cerebr Blood
    Flow Metabol 1:297 (1981).

25. Kalimo H, Rehncrona S, Söderfeldt B, Olsson Y, Siesjö B, Brain lactic
    acidosis and ischemic cell damage. 2. Histopathology, J Cerebr
    Blood Flow Metabol 1:313 (1981).

26. Ng LK, Nimmannitya I, Massive cerebral infarction with severe brain
    swelling, Stroke 1:158 (1970).

27. O'Brien M, Jordan MM, Waltz AG, Ischemic cerebral edema and the
    blood-brain barrier. Distributions of pertechnetate, albumin,
    sodium, and antipyrine in brains of cats after occlusion of the
    middle cerebral artery, Arch Neurol 30:461 (1974).

28. O'Brien MD, Waltz AG, Intracranial pressure gradients caused by
    experimental cerebral ischemia and edema, Stroke 4:694 (1973).

29. Shaw CM, Alvord EC, Berry RG, Swelling of the brain following isch-
    emic infarction with arterial occlusion, Arch Neurol 1:161
    (1959).

30. Schuier FJ, Hossmann KA, Experimental brain infarcts in cats. II.
    Ischemic brain edema, Stroke 11:593 (1980).

31. White OB, Norris JW, Hachinski VC, Lewis A, Death in early stroke,
    causes and mechanisms, Stroke 10:743 (1979).

32. Welsh FA, Ginsberg MD, Rieder W, Budd WW, Diffuse cerebral ischemia
    in the cat. II. Regional metabolites during severe ischemia and
    recirculation, Ann Neurol 3:493 (1978).

33. Ginsberg MD, Graham DI, Welsh FA, Budd WW, Diffuse cerebral isch-
    emia in the cat: III. Neuropathologic sequelae of severe isch-
    emia, Ann Neurol 5:350 (1979).

MICROTHROMBOSIS: A CONTRIBUTING FACTOR TO THE

PROGRESSION OF CEREBRAL INFARCTION IN MAN

J. Cervos-Navarro, J. Figols*, G. Ebhardt

Institute of Neuropathology, Klinikum Steglitz, Free University of Berlin, Hindenburgdamm 30, 1000 Berlin 45, FRG

INTRODUCTION

Eighty percent of strokes are the consequence of cerebral infarcts resulting from occlusion of a large vessel (Cervos-Navarro, 1980). Furthermore, non-occlusive mural thrombi can be the source of emboli and of embolic cerebral infarcts in the territories of major cerebral arteries. Various diseases affecting the whole organism can lead to microthrombosis of brain vessels, such as polycythemia vera, leukaemia, consumption coagulopathy, etc. Relatively little attention has been paid to the role of microthrombosis in progressive stroke. A significance of microthrombosis has only been considered in watershed infarcts (Romanul and Abramowic, 1964). We have focussed our attention on the role of microthrombosis as a possible factor of further progression of brain infarcts.

MATERIAL AND METHODS

A retrospective study was performed on brains of 100 patients who suffered from hemispheric ischemic lesions. Samples of the infarct proper and of its border were cut and stained with hematoxylin-eosine (HE), cresyl-violet (Nissl), elastica Van Gieson (EvG), Mallory, Heidenheim (myelin) and phosphotungstic acid hematoxylin (PTAH) after paraffin embedding. In many cases sections stained with HE had intraluminal aggregates of red blood cells which could be mistaken for true microthrombi. To avoid such a misinterpretation PTAH staining was performed in all sections because of its specificity for fibrin. The brain infarcts were classified into three stages according to Spatz (1939). Stage I represents the phase of acute necrosis, with a duration of up to 72 hours after the onset of ischemia. Stage II is the phase of resorption, when the disintegrated tissue becomes digested by macrophages. After 12-14 days the characteristic cellular reactions of stage II can be found in all areas of the infarct. This time coincides with the beginning of stage III, the terminal cystic defect.

According to our observation an exact classification is not feasible in a high percentage (23%) of cases. These cases of inhomogeneous infarcts showing areas of stage I-II had enough distinctive features to warrant establishing a new category comprising a transition between stages I and II. The pre-stage described at the ultrastructural level (Cervos-Navarro, 1980) cannot be detected by light microscopy. Consequently we followed the classical histological criteria of Spatz (1939). Attention was focussed upon arterioles, capillaries and venules from the periphery of the infarct and of the meninges covering the infarcted area. Distinguishing the peripheral zone from the stage I infarct area was rather difficult. We have therefore examined an area of 1.5 cm width around a zone of characteristic necrotic changes. The frequency of microthrombi was scored from 0-3.

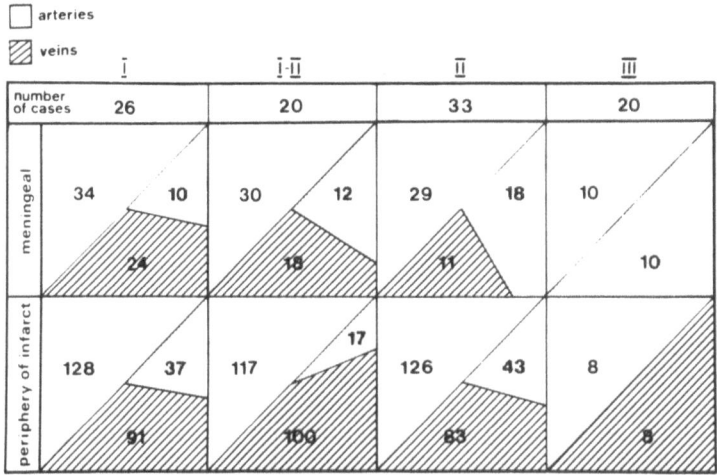

Fig.1: The roman numerals above each field represent scores (0-3) which correlate directly with the frequency of microthrombi. The total number of microthrombi of each stage are given in the white field.

RESULTS

The results (scores) obtained from all cases of the same stage are summarized in Figure 1. Separate data of intracerebral and meningeal vessels were also collected. A higher frequency of microthrombosis was observed in intracerebral vessels as compared to meningeal vessels. However, this may have resulted from the smaller number of meningeal vessels present in a tissue sample. Another important feature was an extensive involvement of microvessels in stage I, I-II and II as compared to stage III, where microthrombi were rarely found. The frequency of microthrombi in arterioles and venules are presented in Figure 2. Microthrombi in both types of vessels were separately studied taking into account their conspicuous histological characteristics. We did not count microthrombi in capillaries, because it was rather difficult to distinguish by light microscopy between postcapillary venules and capillaries. Mirothrombi were more frequently found in venules than in arterioles with a proportion of about 2:1.

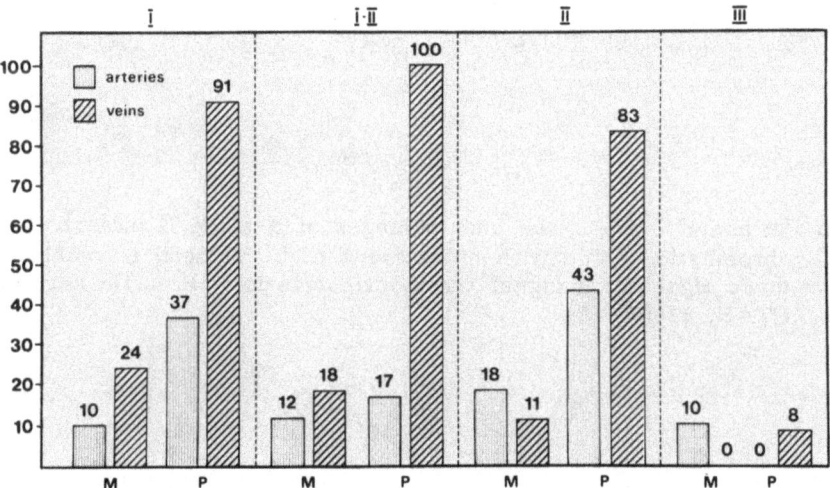

Fig.2: Frequency of microthrombi in arterioles and venules found in different stages of infarction. M: meningeal vessels, P: vessels of brain tissue at the periphery of a brain infarct.

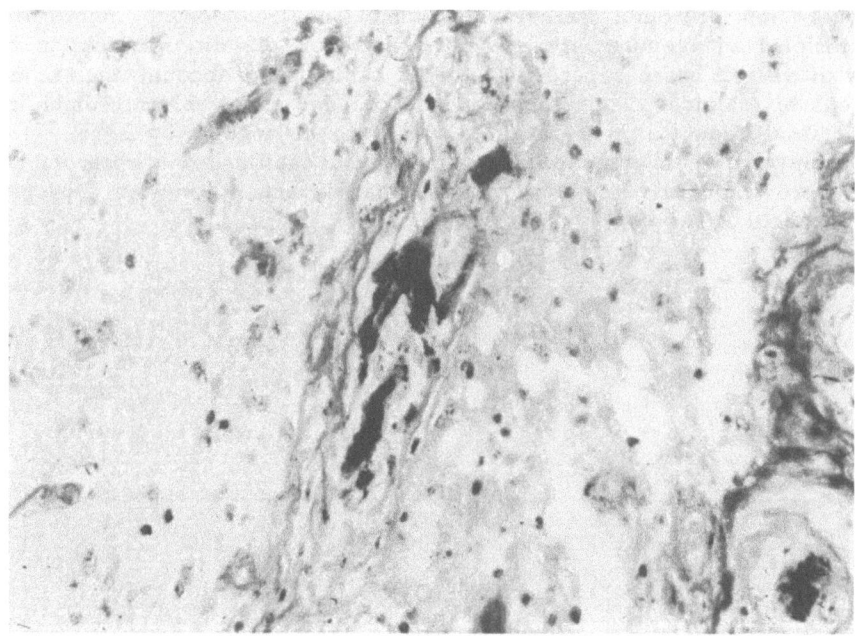

Fig.3a: Peripheral brain tissue and meninges of a stage II infarct. A thrombotic venule with entrapment of blood cells is visible. Note down right a meningeal thrombotic arteriole. Paraffin section; PTAH; x300

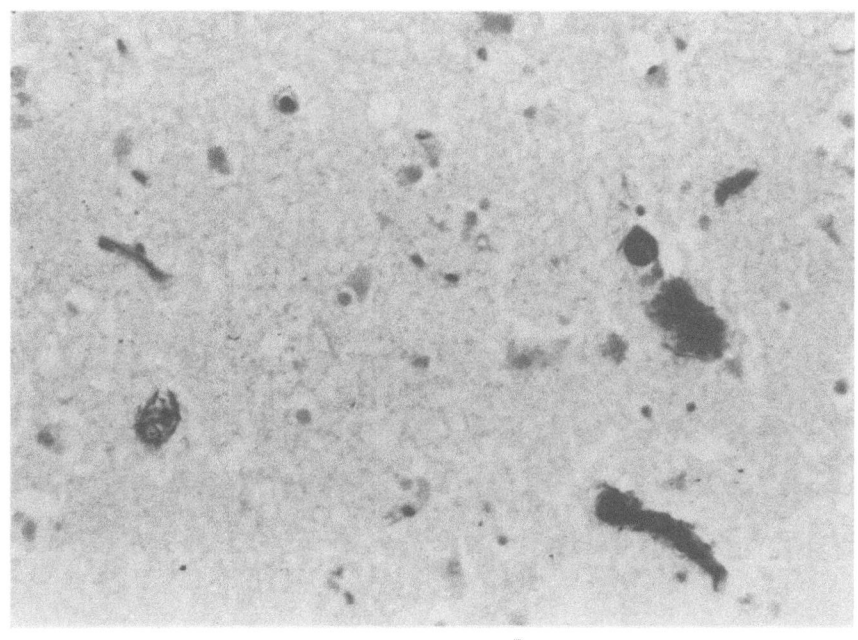

Fig.3b: Multiple thrombotic microvessels (fibrin thrombi) from the periphery of a stage I-II infarct. Paraffin section; PTAH; x300

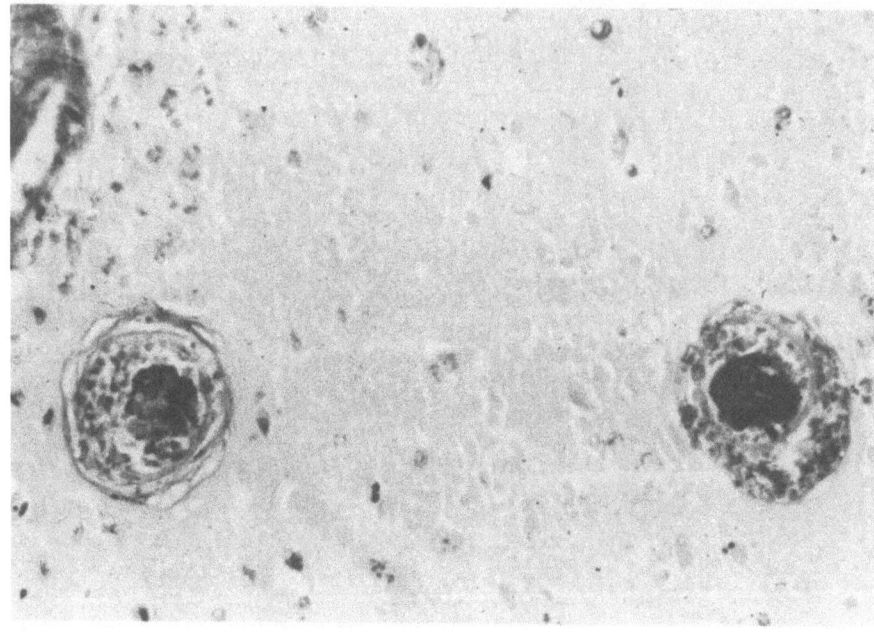

Fig.4a: Thrombotic arterioles with hemorrhagic extravasation at the periphery of a stage II infarct. Paraffin section; PTAH; x300

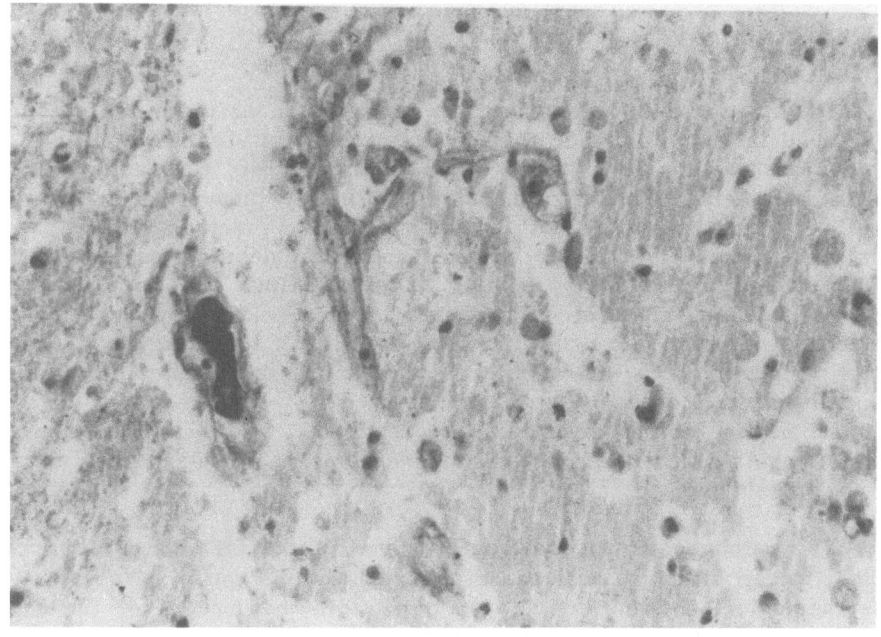

Fig.4b: Periphery of a stage II infarct with a thrombotic venule.
Paraffin section; PTAH; x300

The microthrombi were composed of fibrin - staining strongly positive deep blue with PTAH - and platelets with occasional entrapment of
red and white blood cells (Figure 3a, 4b). In other cases the vascular
lumina were full of dense fibrin (Figure 3b). Occasionally, thrombotic
arterioles and venules of the cortex were surrounded by perivascular
hemorrhage (Figure 4a). A small number of capillaries at the periphery
of stage I and I-II infarcts had "shock bodies". In later stages (III) a
moderate number of arterioles had greatly thickened walls and pinpoint
lumina. Many of them were opened up again by recanalization of old
thrombi.

DISCUSSION

Although it is surprising that a greater number of microthrombi
were found in venules rather than in arterioles, this feature is relatively easy to explain. It is known since Virchow (1846) that veins rather
than arteries tend to undergo thrombosis. Already then, Virchow drew
attention to the importance of (a) damage of the vascular wall, (b) retardation of blood flow and (c) an increased coagulability, so-called
Virchow's triad. These pathophysiological mechanisms hold for larger
veins, particularly for those of the limbs. Nevertheless, applying this
rule to the cerebral microcirculation is not simple, because blood flow
at the periphery of an infarct (i.e. "luxury perfusion"; Lassen, 1966) is
extremely rapid. Moreover, we have observed a higher frequency of
microthrombi in venules which may be attributed to a higher portion of
these segments in comparison to arterioles present in sections of normal
brain (Rickenbacher, 1972).

The pathogenetic mechanisms involved in the formation of microthrombi during brain infarction are summarized in Figure 5. Vascular and
hemorrheological changes are the major factors responsible for the formation of microthrombi under certain pathological conditions. Crater and
balloon-like defects were observed on the endothelial surface of carotid
as well as of meningeal arteries and arterioles in experimental ischemia
(Nelson, 1975). Similar lesions may occur in intracerebral arterioles.
The role of endothelial injury in the pathogenesis of ischemic microthrombosis is, first, to increase platelet adhesiveness (Meyer, 1958),
second, to activate coagulation causing deposition of fibrin (Marchesi,
1977), and third, to induce a breakdown of the blood-brain barrier
leading to exudation of water and ions into the surrounding tissue.

In border zones of ischemic lesions, hyperemia resulting from increased levels of acid metabolites in the hypoxic tissue has been frequently demonstrated (Kuschinsky, 1972; Leniger-Follert, 1978). In certain cases hyperemia exceeds normal blood flow by a factor of two or
more. An extremely rapid blood flow in areas with "luxury perfusion"
(Lassen, 1966) is a prerequisite for the adhesion of platelets causing
formation of platelet aggregates and subsequent thrombosis (Baumgartner,
1973).

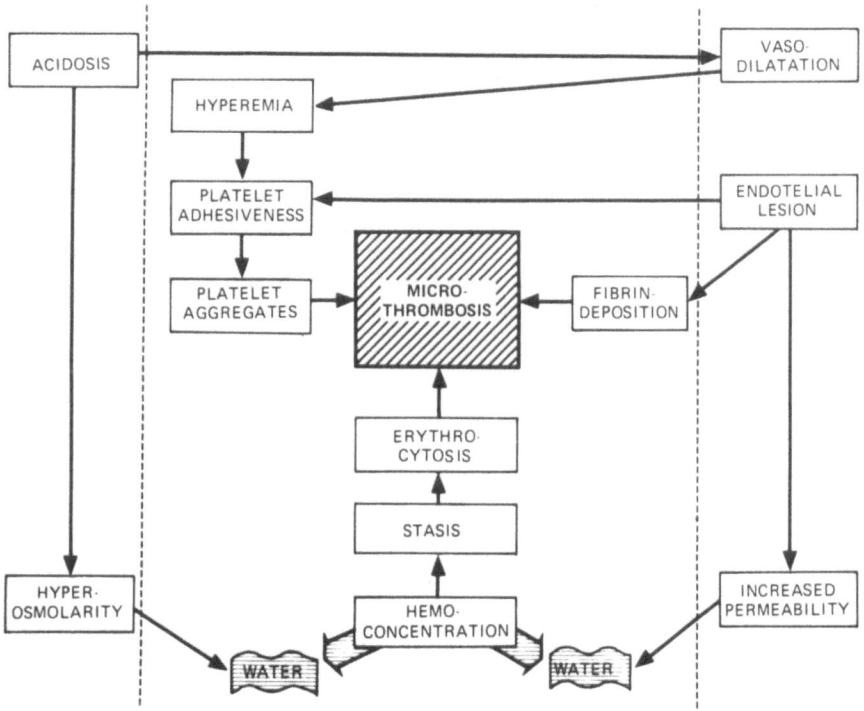

Fig.5: Pathogenetic mechanisms involved in the formation of microthrombi during brain infarct.

A release of serotonin from platelets secondary to ischemia and anoxia, is an early factor in the production of enhanced platelet adhesiveness. Moreover, ischemic cerebral tissue can release catecholamines, arachidonic acid and thromboxane B_2 , activate platelets and increase local tissue acidosis (Zervas, 1975; Wolfe, 1976, Marcus, 1983). Platelet aggregation may represent a hematologic response to tissue injury during human cerebral ischemia contributing to the propagation of ischemic brain damage (Petito, 1979). Our results support that effective antithrombotic and antiaggregation treatment may prevent further expansion of an evolving infarct.

ABSTRACT

We examined 100 brains of patients who suffered from a cerebral hemispheric infarct, focussing attention on the thrombotic obstruction of small vessels. Survival times ranged from a few hours (I) to three months (III). Special attention was given to the border of large ischemic necroses. In many cases dying at stages I and II, large numbers of arterioles, venules and capillaries, showed thrombotic occlusion. At later stages (III), a small number of arterioles (less than 10%) had greatly thickened walls and pinpoint lumina as a consequence of organization of microthrombi and recanalization. In a small percentage of infarcts at stages I and II, foci of cell loss around the thrombotic vessels were found. We postulate that microthrombosis is a relevant factor which contributes to the progression of cerebral infarcts.

* Fellow of the Alexander von Humboldt Foundation.

REFERENCES

Baumgartner HR, The role of blood flow in platelet adhesion, fibrin deposition and formation of mural thrombi, Microvasc Res 5:167 (1973).

Cervos-Navarro J, Hirninfarkt, in: "Spezielle pathologische Anatomie", Bd. 13/I, Doerr W, Seifert G, eds., Springer-Verlag, Berlin (1980).

Goldstein GW, Metabolism of brain capillaries in relation to active ion transport, in: Advances in neurology, Vol. 20: "Pathology of cerebrospinal microcirculation", Cervos-Navarro J, Betz E, Ebhardt G, Ferszt R, Wüllenweber R, eds., Raven Press, New York (1978).

Kuschinsky W, Wahl M, Bosse O, Thurau K, Perivascular potassium and pH as determinants of focal pial arterial diameter in cats: A microapplication study, Circ Res 31:240 (1972).

Lassen NA, Luxury perfussion syndrome and its possible relation to acute metabolic acidosis within the brain, Lancet 2:1113 (1966).

Leniger-Follert E, Urbanics R, Lübbers DW, Behavior of extracellular H^+ - and K^+ -activities during functional hyperemia of microcirculation in the brain cortex, in: Advances in neurology, Vol. 20: "Pathology of cerebrospinal microcirculation", Cervos-Navarro J, Betz E, Ebhardt G, Ferszt R, Wüllenweber R, eds., Raven Press, New York (1978).

Marcus AJ, Recent progress in the role of platelets in occlusive vascular disease, Stroke 14:475 (1983).

Meyer JS, Importance of ischemic damage to small vessels in experimental cerebral infarction, J Neuropath Exp Neurol 17:571 (1958).

Nelson E, Kawamura J, Sunaga T, Endothelial ischemia. Scanning (SEM) and transmission (TEM) electron microscope studies in rabbit. in: "Cerebral circulation and metabolism", Langfitt TW,

McHenry LC, Reivich M, Wollman H, eds., Springer-Verlag, New York (1975).

Petito CK, Platelet thrombi in experimental cerebral infarction, Stroke 10:192 (1979).

Rickenbacher J, Normale und pathologische Anatomie des Hirngefäßsystems, in: "Der Hirnkreislauf", Gänshirt H, ed., Georg Thieme Verlag, Stuttgart (1972).

Spatz H, Pathologische Anatomie der Kreislaufstörungen des Gehirns, Z Neurol Psychiat 167:301 (1939).

Wolfe LS, Rostworosky K, Marion J, Endogenous formation of the prostaglandin endoperoxide metabolite thromboxane B_2 by brain tissue, Biochem Biophis Res Commun 70:907 (1976).

Zervas NT, Lavyne MH, Negoro M, Neurotransmitters and the normal and ischemic cerebral circulation, N Engl J Med 293:812 (1975).

THE CT NORMAL BUT LOW FLOW INFARCT PERIPHERY:

SELECTIVE NEURONAL NECROSIS?

Jens Astrup, M. Nedergaard* and L. Klinken*

Department of Neurosurgery, Rigshospitalet, Copenhagen and Institute of Neuropathology*, University of Copenhagen, Denmark

INTRODUCTION

Large infarcts in the MCA territory usually appear sharply demarcated from tissue of normal attenuation on CT. On 133-Xenon inhalation blood flow tomography, on the other hand, infarcts appear surrounded by a wide borderzone of low blood flow (1). Such low flow in the CT normal infarct periphery can be explained as low function and metabolism either due to deafferentiation and disconnection of neurons in the infarct surroundings, or as so-called incomplete infarction with CT invisible microfoci of infarction and wider zones of selective loss of neurons (2,3,4). This study explores the latter possibility.

MATERIAL AND METHODS

Six patients who had suffered a major stroke in the MCA territory more than two months prior to death from other causes were studied. Coronal slices of the formalin fixated brain were divided into blocks, and 7 um sections were Klüver-Barrera stained. Neuron and glial cell densities were measured at predetermined distances moving away from the infarct and following cortex on the surface and in the sulci of the hemisphere. Sections at mirror positions in the controlateral hemisphere served as control. Tissue columns 0.90 mm wide were counted for neurons and glial cells, their densities calculated, and the cortex thickness measured.

NEURON DENSITY

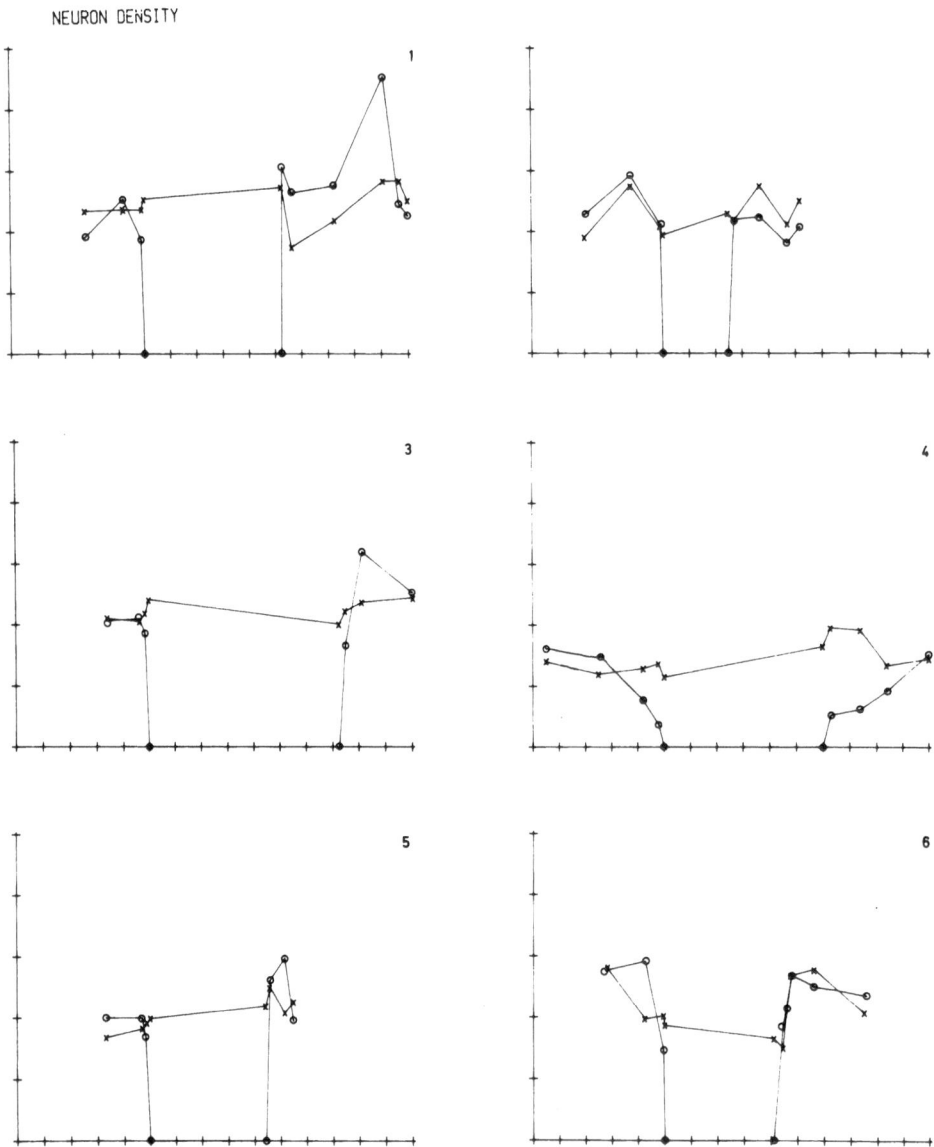

Fig.1a: The infarct is indicated by absence of nerve- and glial cells and
absent cortex. It is marked on the x-axis by its borders to the
anterior (left side) and posterior (right side) watersheds. o indicates the infarcted hemisphere, x the contralateral hemisphere.
Neuron density: x-axis: intervals in cm; y-axis: intervals in 100
neurons per 0.81 mm^2 .

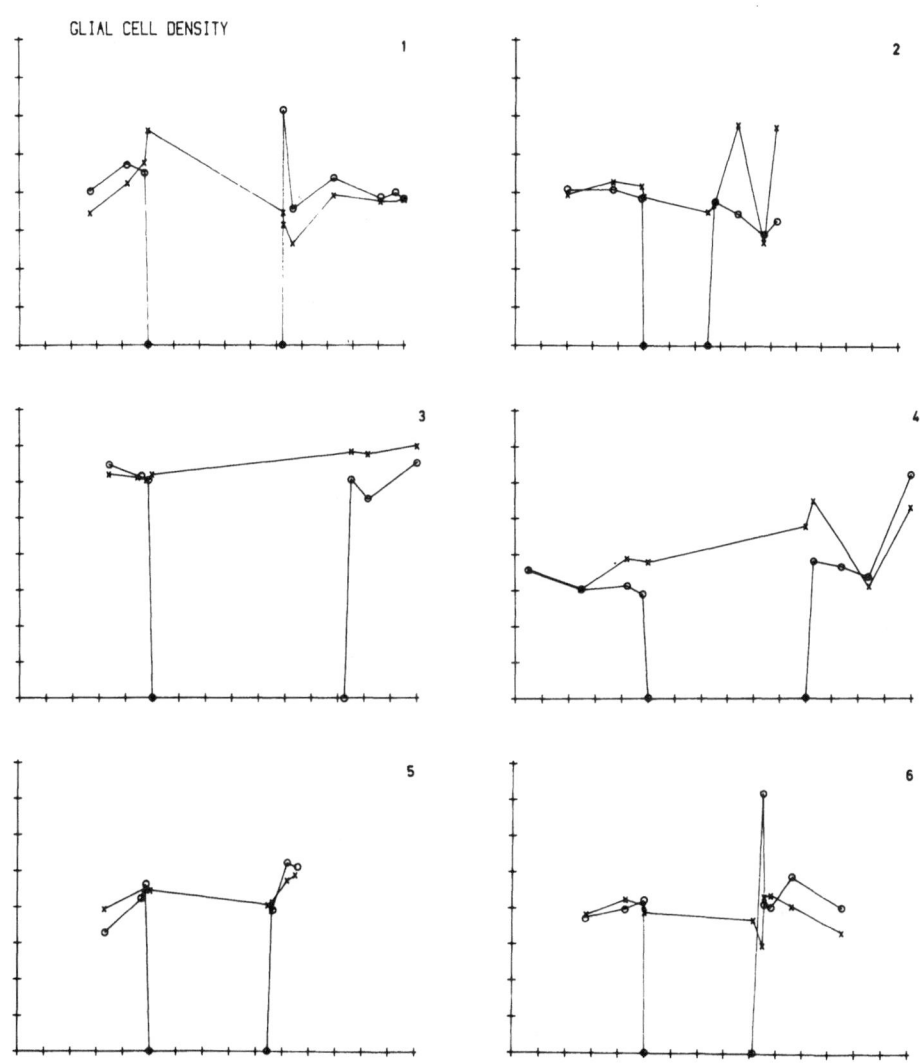

Fig.1b: <u>Glial cell density:</u> x-axis: intervals in cm; y-axis: intervals in 100 glial cell per 0.81 mm^2

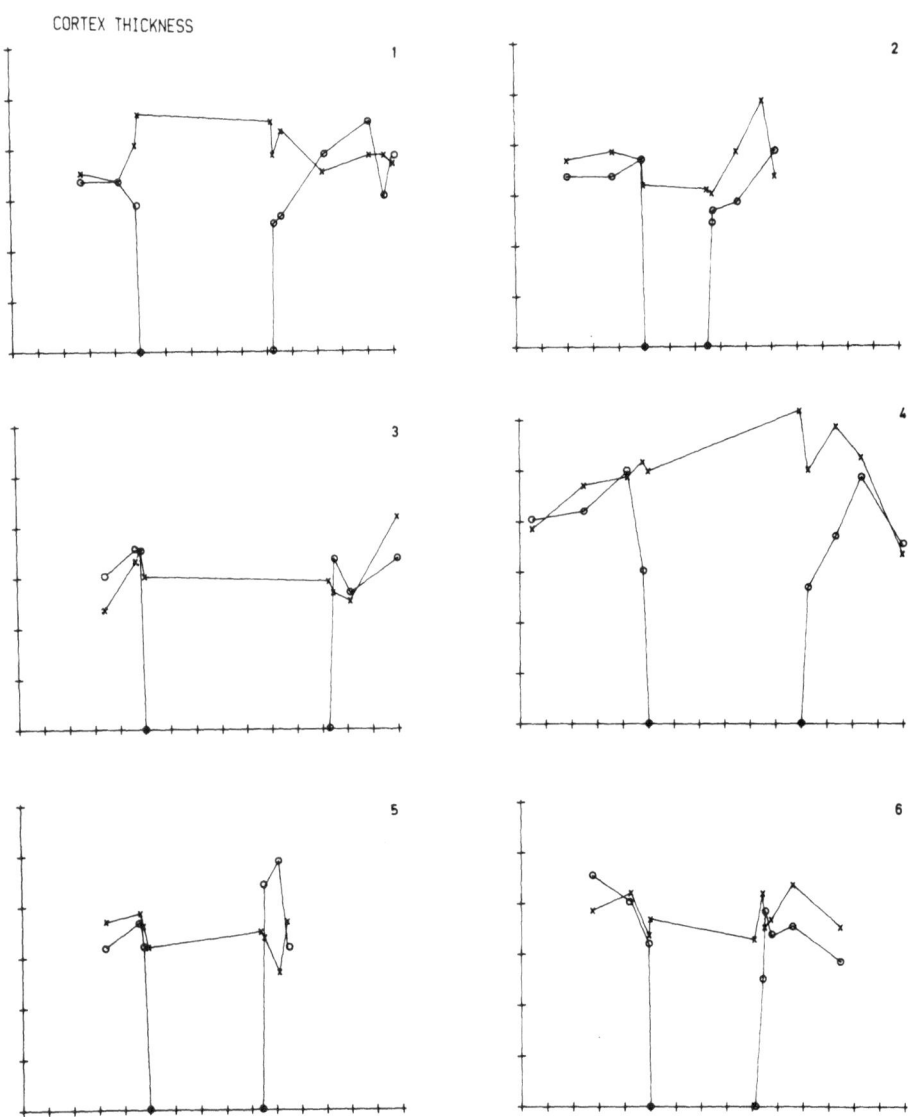

Fig.1c: <u>Cortex thickness:</u> x-axis: intervals in cm; y-axis: cortex
thickness in mm

RESULTS AND DISCUSSION

In light microscopy the infarct appeared sharply demarcated. The zones of transition from infarcted to normal tissue were less than 2 mm, and in half the counting points less than 0.5 mm. This finding is in accordance with the generally held view among neuropathologists of a fairly abrupt transition from infarcted to normal tissue (5,6). This narrow zone of transition had rims and small islands of infarcted tissue interchanging with normal tissue indicating a bulging irregular infarct surface.

In the infarct surroundings we found normal neuron and glial cell densities and normal cortex thickness at all counting points in five out of the six patients, Fig. 1. One patient had a selective neuron loss in a few cm wide zone observed in the anterior and particularly in the posterior watershed. In this zone encrusted calcified ganglion cells were observed, corresponding to the clinical information in this patient of a superimposed recent infarct with aggravation of the hemiparesis within 4 days before death.

In conclusion, this study cannot support the concept of incomplete infarction with microfoci of infarcted tissue and selective loss of neurons in the surroundings of complete larger MCA infarcts. Our findings of normal neuron and glial cell densities and normal cortex thickness around the infarcts favor disconnection of neurons with deafferentiation in the infarct surroundings as explanation of the low function, metabolism (7) and blood flow in this zone.

REFERENCES

1. Vorstrup S, Henrikson L, Paulson OB, Lindewald H, Haase J, Lassen NA, Regional cerebral blood flow by Xenon-133 inhalation and emission tomography before and after extracranial-intracranial by-pass operation in ischemic cerebral disease: Scandinavian meeting on cerebrovascular disease, Acta Neurol Scand 66, Suppl. 91: 33 (1982).
2. Strong AJ, Tomlinson BE, Venables GS, Gibson G, Herdy JA, The cortical ischemic penumbra associated with occlusion of the middle cerebral artery in the cat: 2. Studies of histopathology, water content, and in vitro neurotransmitter uptake, J Cereb Blood Flow Metabol 3:97 (1983).
3. Mies G, Auer LM, Ebhardt G, Traupe H, Heiss WD, Flow and neuronal density in tissue surrounding chronic infarction, Stroke 14:22 (1983).
4. Lassen NA, Skyhoj Olsen T, Hojgaard K, Skriver E, Incomplete infarction: A CT-negative irreversible ischemic brain lesion, J Cereb Blood Flow Metabol 3, Suppl. 1:S602 (1983).
5. Blackwood W, Corsellis J, "Greenfield's Neuropathology", 3rd edition, Arnold, London (1976).

6. Symon L, Brierley F, Morphological changes in cerebral blood vessels in chronic ischemic infarction: Flow correlation obtained by the hydrogen clearence method, in: "The cerebral vessel wall", Cervos-Navarro J, Betz E, Matakas F, Wüllenweber R, eds., Raven Press, New York (1976).

7. Kuhl DE, Phelps ME, Kowell AP, Metter EJ, Selin C, Winter J, Effects of stroke on local cerebral metabolism and perfusion: mapping by emission computed tomography of 18-FDG and 13-NH$_3$ Ann Neurol 8:47 (1980).

DISTURBANCES OF EXTRACELLULAR HOMEOSTASIS AFTER A PRIMARY INSULT AS A MECHANISM IN SECONDARY BRAIN DAMAGE

K.G. Go

Department of Neurosurgery, University of Groningen
The Netherlands

In the analysis of the multitude of lesions causing damage to the brain, the concept of secondary brain damage has evolved as a form of damage that is not inherent to the primary insult, although it is the consequence of a detrimental chain of events that may follow. In a follow-up study on craniocerebral trauma, it appeared that among patients with severe primary injury as indicated by the low scores on the Glasgow Coma Scale, there was a group with an initial higher score, which subsequently deteriorated and ended fatally (Minderhoud, 1977). It is tempting to assume, that the latter group represented those patients, who might have suffered from secondary brain damage in addition to their severe primary injury.

Brain edema may be considered to occupy a significant place among the factors involved in secondary brain damage. The vasogenic type is of particular significance. It is inherent to disruption of the blood-brain barrier in many focal lesions of the brain, and characterized by the exudation of plasma constituents into the extracellular space. The exudate migrates into white matter structures where it accumulates. The driving force of the exudation seems to be the arterial blood pressure (Klatzo et al., 1967), which is modulated by vasomotor activity on its way to the capillary bed (Go et al., 1974). In the presence of vasomotor paralysis following the injury, the pressure head may quite unimpeded be transmitted into the capillary bed. The accumulation of exudate along the pathway of its migration elevates the local tissue pressure, which is higher near the source of exudation constituting a downstream pressure gradient (Reulen et al., 1977). Moreover, the accumulation of exudate in the white matter by its mass effect eventually raises global intracranial pressure. The consequent decrease of regional and global cerebral perfusion to the extent of actual ischemia (Marmarou et al., 1976) clearly provides one mechanism by which cerebral edema may induce secondary damage. Finally, massive brain edema may provoke shifts of brain structures, resulting in lethal herniation.

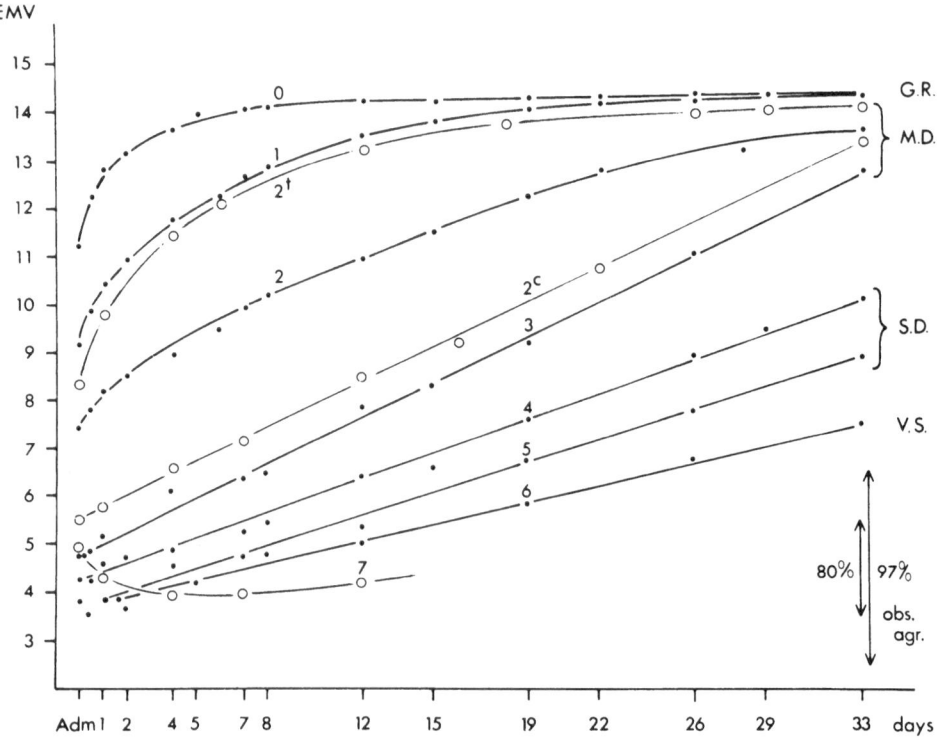

Fig. 1: Outcome after coma in head injured patients, as evaluated 6
months later: GR = good recovery, MD = moderate disability, SD =
severe disability, VS = very severe disability. Obs.agr. = extent
of observer's agreement. Mass lesions are excluded. Patients of
group 7 starting with the same score as group 3 subsequently
deteriorated and died (courtesy by Prof. J.M. Minderhoud; re-
printed with permission from Excerpta Medica).

Many studies on vasogenic brain edema have been conducted on
the model first introduced by Clasen (1953), in which a freezing injury
was inflicted to the cerebral cortex of experimental animals. It enables
a separate analysis of a stage of exudation taking place in the cortex
and a stage of migration into the white matter. The exudation may be
studied in various animals including rodents, particularly by electronmicroscopy using protein tracers, such as horseradish peroxidase (HRP),
or endogenous antibodies (IgG) raised against HRP and visualized immuno-
cytochemically (Houthoff et al., 1981). The migration of exudate into the
white matter should be studied in animal species with large white matter
compartments, such as carnivores or primates. By a modification of the
freezing injury to the cat brain it is possible to obtain edema fluid in

amounts that can be subjected to chemical analysis (Patberg et al., 1977). Previous estimates of edema fluid constituents had to be derived from data pertaining to whole brain tissue and their distribution volumes in the tissue.

TABLE

	Edema Fluid	CSF	Plasma
Sodium (mEq/l)	142.54 ± 0.83	159.20 ± 5.10	144.58 ± 0.96
Potassium (mEq/l)	7.05 ± 2.16	3.34 ± 0.22	4.24 ± 0.10
Calcium (mEq/l)	2.35 ± 0.18	4.08 ± 0.88	4.65 ± 0.41
Colloid Osmotic Pressure (mm Hg)	12.89 ± 3.28	0.55 ± 0.20	19.08 ± 0.74

Determination of Na^+ in isolated edema fluid (Table 1) yielded a value (142.5 mEq/l), which substantiated the vasogenic origin of the exudate, as it more resembled the Na^+-level in plasma than in CSF. CSF is considered to be in extensive diffusional exchange with brain tissue extracellular fluid. The edema fluid K^+-concentration was elevated up to 9 mEq/l, in particular in the early stages after the freezing injury. This illustrates a new element as the source of edema fluid constituents, next to the leakage through a disrupted blood-brain barrier. The elevated level of K^+ in edema fluid is clearly the result of the freezing damage to cellular elements and its subsequent release into the extracellular space. Later, the K^+ level in edema fluid showed a rapid decline, which indicated interference by the glial cells. The glial cells respond to the elevation of extracellular K^+ by its uptake together with water (Somjen et al., 1976). Therefore, the glial swelling observed in vasogenic brain edema may be explained to some extent by this mechanism, which serves to maintain ionic homeostasis in the extracellular space while preventing loss of K^+ from the tissue. However, it may be concluded that the K^+-retrieval mechanism may not be entirely successful in brain edema. Loss of K^+ probably occurs through the disrupted blood-brain barrier or by way of the CSF, which showed a significantly raised K^+-level (3.87 mEq/l) with respect to the control (3.34 mEq/l).

Apart from control by the glia, there are indications that the $Na^+ K^+$-ATPase abounding in cerebral capillary endothelium may subserve the regulation of extracellular K^+ by exchange against Na^+ (Goldstein, 1979). Moreover, fluctuations of extracellular electrolytes will ultimately dissipate by diffusional exchange with the CSF. The CSF can adequately fulfill this homeostatic function by its large volume, which is

continuously replenished and by its relatively constant composition
determined during secretion by the choroid plexus (Ames et al., 1964).
The cellular uptake of K^+ is probably associated with an inward shift of
Cl^-. The distribution of Cl^- has been assumed to obey Donnan forces, but
in glial cells a translocation mechanism for its exchange against HCO_3^-
has been postulated (Van Gelder, 1983). The calcium content of edema
fluid was low compared to CSF and plasma. This resembles the findings
in cerebral ischemia, where a decrease of extracellular Ca^{++} has been
observed with Ca^{++}-sensitive electrodes (Harris et al., 1981). On the
other hand, whole tissue calcium was increased in ischemic infarction,
as well as in freezing lesions in rats (Korf et al., 1984). The findings
imply an entrance of Ca^{++} into the intracellular compartment. In the case
of the cold injury, damage to the Ca^{++}-extrusion mechanisms in the
plasma membrane affecting the Ca^{++} dependent ATP-ase, and the Ca^{++}/
Na^+ antiport system must be assumed. Notably, in the model of vasogenic
brain edema, where following intraventricular injection of collagenase
only the blood-brain barrier was disrupted and the parenchyma otherwise
undamaged, no changes of tissue calcium content could be observed (Korf
et al., 1984). Changes of extracellular calcium concentration will not
only affect neuronal function, but they may also change vasomotor reactivity. Calcium seems to be required for the maintenance of contractility
of isolated cerebral arteries (Brandt et al., 1981).

The characteristic feature of vasogenic brain edema is exudation
of plasma proteins due to the breakdown of the blood-brain barrier. It is
reflected by the rise of edema fluid colloid osmotic pressure, which may
attain values of 18 mm Hg (Gazendam et al., 1979), with 1 g/100 ml of
plasma corresponding to 3.4 mm Hg. The validity of the Starling hypothesis of transcapillary fluid exchange may be questioned for the cerebral
vasculature in the light of the exceedingly high content of Na^+ K^+ ATPase
of cerebral capillary endothelium allegedly involved in the active transport of fluid across the endothelium (Eisenberg & Suddith, 1979). Nevertheless, the dependence of blood-to-brain passage of Na^+ on arterial
blood pressure as observed by ourselves (Go & Pratt, 1975) indicates the
significance of hydrostatic forces in the net movement of fluid across
the cerebral capillary wall. This corresponds to the convectional flow of
water across the capillary wall. The exchange diffusional flow of water
by contrast is determined by blood flow as demonstrated during single capillary transit (Go et al., 1981). On account of its high protein content
and elevated colloid osmotic pressure, vasogenic edema fluid is presumably retained in the tissue for relatively long periods of time, contrary
to osmotic edema fluid. Osmotic brain edema, which is the consequence
of hypoosmotic hydration, appears to resolve readily and for the greater
part by way of the CSF greatly increasing CSF formation rate (DiMattio
et al., 1975). On the few occasions that edema fluid could be obtained
in osmotic brain edema, a low colloid osmotic pressure corresponding to
a low protein content has been found indeed (Patberg et al., 1977).

Among the proteins that enter brain tissue secondary to disruption
of the blood-brain barrier, the kininogens should be mentioned. There are

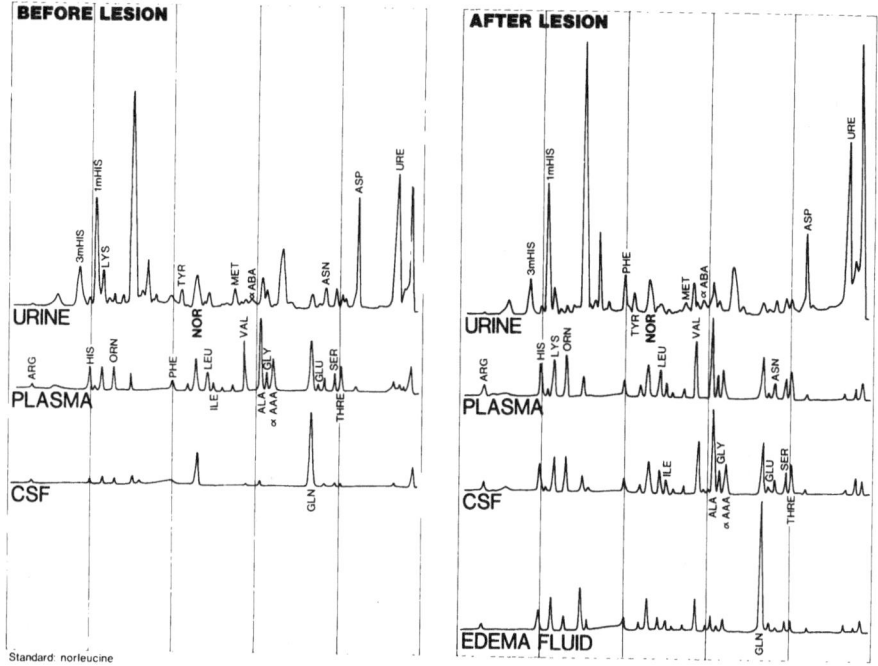

Fig. 2: Ion exchange aminoacid chromatogram of cat plasma, urine and
cerebrospinal fluid before and after freezing injury and of edema
fluid.

indications that kininogens, which are constituents of the plasma α_2
-globulin fraction are converted into kinins in a freezing lesion and in
perifocal edematous areas (Unterberg et al., 1984). The kinins may be regarded as potential mediators of brain edema, since superfusion of cerebral cortex with bradykinin has been shown to cause extravasation of
sodium fluorescein. Ventriculo-cisternal perfusion of brain with plasma
allegedly containing kininogens resulted in brain edema of the periventricular tissue concomitant with a decrease of kininogens in the perfusate indicative of kinin formation (Oettinger et al., 1976). The exudation of kininogens is illustrative of the instance, in which a substance
that enters through the impaired barrier, is consumed by interaction with
the tissue, before it can spread extensively with the edema fluid.

The freezing injury model of brain edema has also proved suitable
for the study of the formation and spreading of other edema mediators,
notably glutamate and free fatty acids. Ventricular perfusion with
glutamate has been reported to induce brain edema (Baethmann et al.,
1979). Topical application of glutamate to cerebral tissue provoked cellular swelling and inward shift of Na^+ from the extracellular space (Van

Harreveld & Fifkova, 1971). Glutamate has been demonstrated in the edema fluid from cats with a freezing lesion in concentrations which amounted to 15-20 times of those in normal CSF (Maier-Hauff et al., 1984). In normal brain tissue glutamate is abundant (10 mM/kg fresh weight). It is endogenously produced by amination of α-ketoglutarate with metabolic ammonia, which is catalyzed by glutamic acid dehydrogenase (Van den Berg, 1978). Since excessive amounts of glutamate in the extracellular space have deleterious effects, extracellular glutamate is maintained at a low level by high-affinity uptake mechanisms of the glia. Only glial cells have glutamine synthetase to convert glutamate into glutamine by the addition of ammonia. The excess of glutamine is readily released into the extracellular space, where it is retrieved by the neuronal elements (Van Gelder, 1983). An amino-acid chromatogram of edema fluid (Fig. 2) not only shows an increased amount of glutamate, but also an increased level of glutamine as compared to CSF. Apparently the mechanisms for the clearance of extracellular glutamate are operative along the pathway of edema fluid migration without, however, attaining a normalization within the time span of the experiment.

    Free fatty acids, especially the polyunsaturated fatty acids cause cellular swelling with increase of Na^+ and decrease of K^+ when applied to brain slices (Chan & Fishman, 1978). Among the polyunsaturated fatty acids, notably arachidonic acid (C 20:4) provoked blood-brain barrier disruption, apart from an increase of brain water when injected intracerebrally (Chan et al., 1983). It was also shown to induce changes of membrane integrity of neuronal and glial elements in culture (Chan & Fishman, 1982). Furthermore, polyunsaturated fatty acids were released in a freezing lesion (Bazan et al., 1984), presumably by the degradation of membrane phospholipids, as there was a local decrease of tissue phospholipid content concomitant to the increase of the fatty acids (Chan et al., 1983). The release of arachidonic acid in the lesion could not be inhibited by dexamethasone, or indomethacin (Pappius & Wolfe, 1984). Apart from direct effects on the tissue, arachidonic acid may conceivably operate by its conversion into prostaglandins, thromboxanes and other products of the arachidonic acid cascade (Wolfe, 1982). In the lesion an increase of prostaglandin $F_{2}\alpha$ has been demonstrated, which could be inhibited by indomethacin. This treatment also exerted a beneficial influence on the disturbance of cellular metabolism as reflected by the local glucose utilization in the hemisphere with the freezing lesion. Therefore, a role of prostaglandins has been suggested in the functional disturbances, possibly through the modulation of ionic channels and neurotransmitter release (Pappius & Wolfe, 1984). Free fatty acids could also be demonstrated in edema fluid from cats with a freezing lesion. However, only arachidonic acid was present in a concentration exceeding that of plasma implying formation in the lesion, and subsequent release into the edema fluid (Maier-Hauff, 1984).

    The cellular swelling frequently observed in vasogenic edema may conceivably be ascribed to the effects of these mediators, operating separately or in concert. With regard to those mediators which are capable of disrupting the blood-brain barrier, a mechanism of extension of brain edema may be envisaged, in which the disruption of the barrier is propagated by the spreading of the mediators from the site of the lesion. The freezing injury model with its features of exudation in the cortical lesion and migration of exudate into the white matter seems particularly suitable to assess the significance of such a mechanism. It has been demonstrated by Steinwall & Klatzo (1966) that after the intravenous administration of two different tracers at various times following the freezing injury, both tracers entered the brain at the site of the lesion only, and not in the edematous white matter. This indicated that in spite of the presence of edema in the white matter, it was not the barrier impairment, but only the exudate that has spread. More recently this has been confirmed again by studies of Blasberg et al. (1979). These findings have to be borne in mind when considering the relevance of mediators that disrupt the blood-brain barrier. Obviously the in situ demonstration

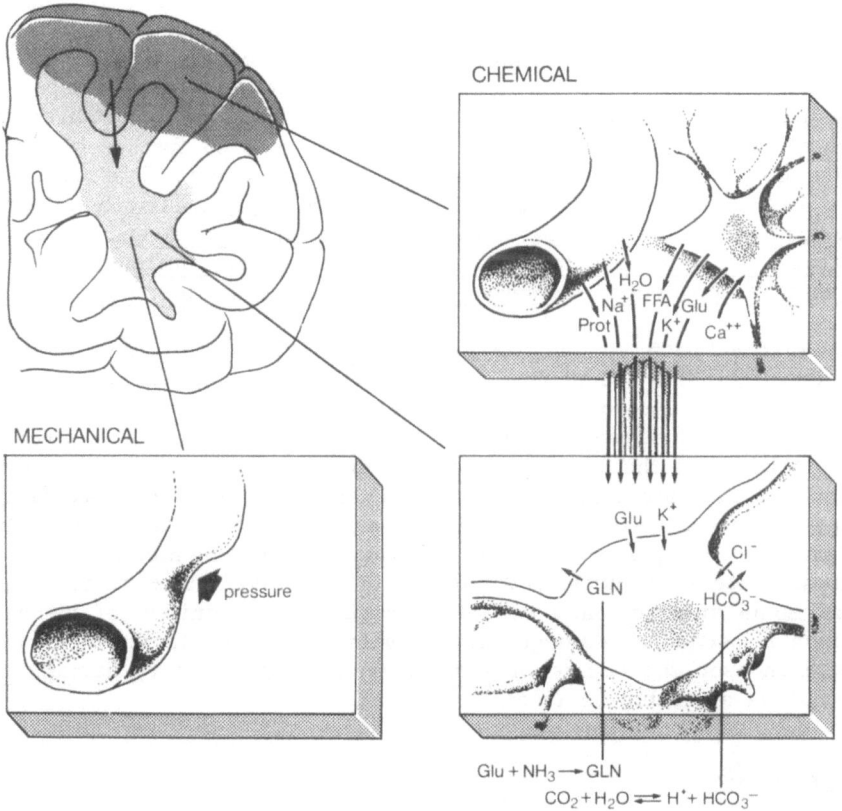

Fig. 3: Schematic diagram of interactions of edema fluid and its constituents with brain tissue.

of such mediators does not necessarily imply that the effect is actually being exerted. It is conceivable that through the disrupted barrier plasma constituents, such as albumin which e.g. binds free fatty acids, capable of antagonizing the mediators have entered the tissue. Alternatively, highly reactive substances with short biological half-lives, such as kinins or free radicals may have been degraded and inactivated by interaction with adjoining perifocal tissue before it can spread by the edema fluid.

Analysis of edema fluid has also enabled the study of changes that take place in the extracellular space during hypoxia or ischemia. These consist of a pronounced elevation of K^+, and a sustained accumulation of enzymes and proteins in the edema fluid (Go et al., 1979). The changes indicate increased release and concentration of K^+ and of high-molecular substances in the extracellular space, as fluid was shifted into the cells by inadequate functioning of the cellular Na^+ / K^+ -pump during hypoxia. A similar situation may arise during ischemia, when local or global perfusion fails as a consequence of massive edema and elevation of intracranial pressure. Under these circumstances, a 100-fold increase of glutamate with respect to normal CSF values was found in edema fluid (Maier-Hauff et al., 1984). This indicates a failure of the glutamate-uptake mechanism which depends on the integrity of active pumps. Although arachidonic acid has been reported to be released in ischemic cerebral tissue (Hirashima et al., 1984), there was no further elevation of the arachidonic acid content in the vasogenic edema fluid with additional ischemia (Maier-Hauff, 1984). This may imply a strictly intracellular localization of the process. The findings in hypoxia and ischemia illustrate the additional deleterious effects of energy deficiency resulting in a depression of energy dependent homeostatic mechanisms.

In conclusion, among the processes associated with primary lesions and responsible for secondary brain damage, the development of vasogenic brain edema should be considered as a significant event, in which mechanical and biochemical effects may be envisaged (Fig. 3). By the influx of fluid into the brain parenchyma and its subsequent accumulation, not only mass effects may result that eventually interfere with vascular perfusion. Severe disturbances of the ionic environment are carried by the exudate to unperturbed areas together with biochemically active substances. These substances may partake in extensive interactions with the tissue, some of which may be detrimental. The recognition of mechanisms involved in secondary damage may provide the means to impose a beneficial turn upon the course of events, even though primary damage per se may not be amenable to treatment.

REFERENCES

Ames III A, Sakanoue M, Endo S. Na, K, Ca, Mg and Cl concentrations in choroid plexus fluid and cisternal fluid compared with plasma ultra-filtrate, J Neurophysiol 27:672 (1964).

Baethmann A, Oettinger W, Rothenfußer W, Kempski O, Unterberg A, Geiger R, Brain edema factors: current state with particular reference to plasma constituents and glutamate, in: "Brain edema; pathology, diagnosis and therapy", Cervos-Navarro J, Ferszt R, eds., Raven Press, New York, Adv Neurol 28:171 (1979).

Bazan NG, Politi E, Rodriguez de Turco EB, Endogenous pools of arachidonic acid-enriched membrane lipids in cryogenic brain edema, in: "Recent progress in the study and therapy of brain edema", Go KG, Baethmann A, eds., Plenum, New York (1984).

Blasberg RG, Gazendam J, Patlak CS, Fenstermacher JD, Quantitative autoradiographic studies of brain edema and a comparison of multi-isotope autoradiographic techniques, in: "Brain edema; pathology, diagnosis and therapy", Cervos-Navarro J, Ferszt R, eds., Raven Press, New York, Adv Neurol 28:255 (1979).

Brandt L, Andersson KE, Edvinsson L, Ljunggren B, Effects of extracellular calcium and of calcium antagonists on the contractile responses of isolated human pial and mesenteric arteries, J Cereb Blood Flow Metabol 1:339 (1981).

Chan PH, Longar S, Fishman RA, Phospholipid degradation and edema development in cold-injured rat brain, Brain Res 277:329 (1983).

Chan PH, Fishman RA, Brain edema: induction in cortical slices by polyunsaturated fatty acids, Science 201:358 (1978).

Chan PH, Fishman RA, Alteration of membrane integrity and cellular constituents by arachidonic acid in neuroblastoma and glioma cells, Brain Res 248:152 (1982).

Clasen RA, Brown DVL, Leavitt S, Hass GM, The production by liquid nitrogen of acute closed cerebral lesions, Surg Gynecol Obstet 96:605 (1953).

DiMattio J, Hochwald GM, Malhan C, Wald A, Effects of changes in serum osmolarity on bulk flow of fluid into cerebral ventricles and on brain water content, Pflügers Arch 359:253 (1975).

Eisenberg HM, Suddith RL, Cerebral vessels have the capacity to transport sodium and potassium, Science 206:1083 (1979).

Gazendam J, Go KG, Van Zanten AK, Composition of isolated edema fluid in cold-induced brain edema, J Neurosurg 51:70 (1979).

Go KG, Lammertsma AA, Paans AMJ, Vaalburg W, Woldring MG, Extraction of water labeled with oxygen-15 during single-capillary transit, Arch Neurol 38:581 (1981).

Go KG, Pratt JJ, The dependence of the blood to brain passage of radioactive sodium on blood pressure and temperature, Brain Res 93:329 (1975).

Go KG, Gazendam J, Van Zanten AK, The influence of hypoxia on the composition of isolated edema fluid in cold-induced brain edema, J Neurosurg 51:78 (1979).

Go KG, Zijlstra WG, Flanderijn H, Zuiderveen F, Circulatory factors influencing exudation in cold-induced cerebral edema, Exp Neurol 22:332 (1974).

Goldstein GW, Relation of potassium transport to oxidative metabolism in isolated brain capillaries, J Physiol 286:185 (1979).

Harris RJ, Symon L, Branston NM, Bayhan M, Changes in extracellular
    calcium activity in cerebral ischaemia, J Cereb Blood Flow
    Metabol 1:203 (1981).
Hirashima Y, Koshu K, Kamiyama K, Nishijima M, Endo S, Takaku A, The
    activities of phospholipase A_1, A_2, lysophospholipase and acyl
    CoA: lysophospholipid acyltransferase in ischemic dog brain,
    in: "Recent progress in the study and therapy of brain edema",
    Go KG, Baethmann A, eds., Plenum, New York (1984).
Houthoff HJ, Go KG, Huitema S, The permeability of cerebral capillary
    endothelium in cold injury. Comparison of an endogenous and
    exogenous protein tracer, in: "Cerebral microcirculation and
    metabolism", Cervos-Navarro J, Fritschka E, eds.,
    Adv Neurol 28:331 (1981).
Klatzo I, Wisniewski H, Steinwall I, Streicher E, Dynamics of cold injury
    edema, in: "Brain Edema", Klatzo I, Seitelberger F, eds.,
    Springer, Berlin (1967).
Korf J, Gramsbergen JBP, Prenen GHM, Go KG, Cation shifts and excito-
    toxins in Alzheimer's and Huntington's disease and experimental
    brain damage, Progr Brain Res (in press).
Maier-Hauff K, Lange M, Schürer L, Guggenbichler C, Vogt W, Jacob K,
    Baethmann A, Glutamate and free fatty acid concentrations in
    extracellular vasogenic edema fluid, in: "Recent progress in the
    study and therapy of brain edema", Go KG, Baethmann A, eds.,
    Plenum, New York (1984).
Marmarou A, Shulman K, Shapiro K, Pöll W, The time course of brain
    tissue pressure and local CBF in vasogenic edema, in: "Dyna-
    mics of brain edema", Pappius HM, Feindel W, eds., Springer,
    Berlin (1976).
Minderhoud JM, Outcome after coma due to head injury. Discussion in
    the 11th World Congress of Neurology, Amsterdam Sept. 1977,
    Excerpta Medica: 255 (1978).
Oettinger W, Baethmann A, Rothenfußer W, Geiger R, Mann K, Tissue
    and plasma factors in cerebral edema, in: "Dynamics of brain
    edema", Pappius HM, Feindel W, eds., Springer, Berlin (1976).
Pappius HM, Wolfe LS, Effect of drugs on local cerebral glucose utiliza-
    tion in traumatized brain: mechanisms of action of steroids re-
    visited, in: "Recent progress in the study and therapy of brain
    edema", Go KG, Baethmann A, eds., Plenum, New York (1984).
Patberg WR, Go KG, Teelken AW, Isolation of edema fluid in cold-induced
    cerebral edema for the study of colloid osmotic pressure,
    lactate dehydrogenase activity and electrolytes, Exp Neurol
    54:141 (1977).
Reulen HJ, Graham R, Spatz M, Klatzo I, Role of pressure gradients and
    bulk flow in dynamics of vasogenic brain edema, J Neurosurg
    46:24 (1977).
Somjen GG, Rosenthal M, Cordingley G, LaManna J, Rothman E, Potassium,
    neuroglia, and oxidative metabolism in central grey matter,
    Fed Proc 35:1266 (1976).

Steinwall O, Klatzo I, Double tracer methods in studies on blood-brain
    barrier dysfunction and brain edema, Acta Neurol Scand 41,
    Suppl. 13:591 (1965).
Unterberg A, Maier-Hauff K, Wahl M, Lange M, Baethmann A, Cerebral
    uptake and consumption of plasma-kininogens in vasogenic brain
    edema: recent findings of kinin-mechanisms, in: "Recent progress
    in the study and therapy of brain edema", Go KG, Baethmann A,
    eds., Plenum, New York (1984).
Van den Berg CJ, Matheson DF, Nijenmanting WC, Compartmentation of
    amino acids in brain: the gaba-glutamine-glutamate cycle, in:
    "Amino acids as chemical transmitters", Fonnum F, ed., Plenum,
    New York (1978).
Van Gelder NM, Metabolic interactions between neurons and astroglia:
    glutamine synthetase, carbonic anhydrase and water balance,
    in: "Basic mechanisms of neuronal hyperexcitability", Jasper HH,
    Van Gelder NM, eds., Liss, New York (1983).
Van Harreveld A, Fifkova E, Light- and electronmicroscopic changes in
    central nervous tissue after electrophoretic injection of gluta-
    mate, Exper Molec Pathol 15:61 (1971).
Wolfe LS, Eicosanoids: prostaglandins, thromboxanes, leukotrienes, and
    other derivatives of carbon-20 unsaturated fatty acids,
    J Neurochem 38:1 (1982).

ROLE OF MEDIATOR COMPOUNDS IN SECONDARY BRAIN DAMAGE - CURRENT EVIDENCE[*]

Andreas Unterberg, K. Maier-Hauff, C. Dautermann,
U. Hack, L. Schürer, A. Baethmann

Institute for Surgical Research, Klinikum Großhadern, Ludwig-Maximilians-University, 8000 Munich 70, W.-Germany

Formation or release of noxious factors in damaged brain tissue requires an ever increasing interest among the various mechanisms studied in the development of secondary brain damage (Baethmann, 1978; Baethmann et al., 1979, 1980; Siesjö, 1981). This review attempts to evaluate the current state of knowledge of this interesting subject. Mediator compounds are believed to form, or to become released in primarily damaged brain tissue areas, e.g. the focal necrosis, or to enter the brain parenchyma from the intravascular compartment. These factors spread then from the site of formation, or entrance into perifocal, edematous tissue. Toxic effects of mediator compounds are likely to occur not only in the boundary around a focal lesion, but also in perifocal tissue.

Definition of "Secondary Brain Damage" Caused by Mediators

Various reversible and irreversible pathophysiological mechanisms may be attributed to mediator compounds. Mediators might enhance damage to the blood-brain barrier thereby promoting influx of vasogenic edema. They might induce circulatory disturbances, such as loss of the cerebral autoregulation and vasocongestion resulting in secondary ischemia and increase of intracranial pressure. Secondary cell swelling, i.e. cytotoxic edema and, finally cell necrosis might also be attributed to mediator compounds. Finally, these substances could impair nervous function and induce metabolic depression of the brain.

Requirements for Identification of Mediator Compounds of Secondary Brain Damage

Since necrotic brain tissue and the intravascular compartment are

139

the source of origin of mediator compounds, a formidable number of compounds must be taken into consideration. This makes necessary to establish certain requirements a mediator candidate has to fulfill. The following guidelines were adpoted in our laboratory for the identification of mediator compounds (Baethmann et al., 1980):

1. Cerebral administration of a mediator should induce brain damage, such as tissue necrosis, cell swelling, or vasogenic edema (cf: Definition of "Secondary Brain Damage").
2. The compound should be released, or formed in effective concentrations under conditions associated with cerebral damage, and the degree of formation, or release should correlate with the degree of brain damage.
3. Methods which specifically inhibit release, or the pathological functions of a mediator should prevent or attenuate secondary brain damage.

TABLE I

Mediator Compounds of Secondary Brain Damage

Proteolytic enzymes

Kallikrein-Kinin-System

Biogenic amines
    o Serotonin
    o Histamin

Neurotransmitters
    o Glutamate

Fatty acids
    o Arachidonic acid and
      metabolites, e.g.
        - Prostaglandins
        - Leukotrienes

Free Radicals

acc. to: Baethmann, 1979

Table I gives a survey on mediator candidates which were subjected to respective investigations. These are lysosomal enzymes, biogenic amines, such as serotonin and histamine, glutamate, unsaturated fatty acids, e.g. arachidonic acid and its metabolites, free radicals and the kallikrein-kinin-system (Bingham et al., 1969; Bulle, 1957; Chan and Fishman, 1978; Chan et al., 1982; Clendendon et al., 1978; Fenske et al., 1976; Gross, 1982; Joo, 1979; Kempski, 1982; Khattab, 1967; Kobrine and Doyle, 1976; Kontos et al., 1980; Osterholm et al., 1969; Westergaard, 1975). The following discussion focusses on the kallikrein-kinin-system, glutamate and arachidonic acid, since these substances appear particularly promising.

Mediator Role of the Kallikrein-Kinin-System

The active peptides of the kallikrein-kinin-system, bradykinin and kallidin are cleaved from kininogens by the enzyme kallikrein. All components of the system are natural constituents of plasma, but inhibited under normal conditions. In the periphery, the sytem is activated by tissue damage, coagulation, inflammatory processes, etc. (Erdös, 1976, 1979; Frey et al, 1968). The components of the kallikrein-kinin-system are also found in brain tissue, though at markedly lower levels (Shikimi et al., 1983). The damaging potential of bradykinin in brain tissue represented first evidence for a mediator function of the system in secondary brain damage. Exposure of cerebral parenchyma to bradykinin, the active peptide of the KK-system, by ventriculo-cisternal perfusion in dogs was shown to cause a marked increase of the water content in cerebral cortex, basal ganglia, and periventricular white matter (Unterberg and Baethmann, 1984). Recent studies on ventriculo-cisternal perfusion of the brain with bradykinin using measurements of the electrical brain tissue impedance demonstrated an extracellular fluid accumulation (Unterberg et al., 1985). In another series of experiments bradykinin was found to open the blood-brain barrier which would support formation of vasogenic brain edema (Unterberg et al., 1984). Fluorescencemicroscopic investigations of cerebral vessels revealed opening of the blood-brain barrier by bradykinin to be restricted to small molecules, such as Na^+-fluorescein. Larger molecules, such as FITC-labeled dextran (MW: 20,000 - 70,000) did not penetrate the barrier under these conditions. Moreover, bradykinin dilated cerebral arteries and constricted, albeit moderately, cerebral veins when applied from the extravascular side (Unterberg et al., 1984; Wahl et al., 1983). Recent observations are available in addition that bradykinin depresses nerve cell activity and enhances cerebral glucose uptake (Reiser and Hamprecht, 1982; Unterberg et al., 1983; Unterberg et al., 1985).

It has also been studied whether kinins are actually released in cerebral damage. Respective investigations were conducted in cold-induced brain edema analysing samples of focal necrotic and perifocal edematous brain tissue (Baethmann et al., 1981; Maier-Hauff et al., 1984a). For that purpose a method was developed to quantify the amount of kinins formed in the damaged brain. Kinins are elusive, short-lived substances, which are instantaneously degraded upon formation by active and ubiquitous present kininases. In these experiments, the release of kinins was determined at 3, 5 or 7 hours after induction of cold injury. In animals with a moderate rise of intracranial pressure only, significant formation of kinins was found in focal necrotic tissue, whereas it was less prominent in perifocal edematous brain. Since intracranial pressure rose markedly in approximately 50% of the experiments, we studied the effect of secondary cerebral ischemia on formation of kinins in the brain in addition. In these experiments perfusion pressure fell below 40 mmHg, eventually resulting in a flat EEG for at least 30 minutes in the final stage. With additional ischemia formation of kinins was not only found in the primary necrotic lesion, but also in the perifocal edematous tissue indicative of a relationship between release of kinins and the severity of tissue damage. The amount of kinins formed under these circumstances was 10^{-5} - 10^{-6} M, which may suffice to increase blood-brain barrier permeability. Concentrations of 10^{-7} M were found to to open the barrier for intravascular Na^+-fluorescein (Unterberg et al., 1984).

Final support to identify kinins as mediator substances was obtained in studies on inhibition of kinin-formation employing again the cold-injury model. Aprotinin was used as inhibitor of kallikrein, the enzyme catalyzing formation of kinins. Aprotinin was intravenously infused in a dose of 150 mg/kg b.w. starting 15 minutes after induction of trauma. The resulting weight increase of the traumatized cerebral hemispheres 24 hours after cold injury was utilized as a measure to quantify brain swelling. The hemispheric water content was determined in addition. Aprotinin reduced significantly brain swelling from 13% in controls to 10% in treated animals. The cerebral tissue water content of treated animals was decreased both in the traumatized and contralateral hemisphere (Unterberg et al., 1986). Although it was not studied then whether aprotinin actually prevented formation of kinins, the findings complete currently available evidence for a pathophysiological function of the kallikrein-kinin-system in secondary brain damage.

Role of Glutamate in Secondary Brain Damage

The aminoacid glutamate is another interesting candidate as mediator of secondary brain damage. Glutamate occurs within cells in concentrations of ca. 15-18 mM, while it is almost negligible in the cerebral extracellular space. The intra- to extracellular glutamate concentration ratio is ca. 1000:1 (Perry and Jones, 1961; Perry et al., 1975; Harvey and McIlwain, 1968). Increase of glutamate in the extracellular fluid raises Na^+-permeability of cell membranes (Ames et al., 1967; Harvey and McIlwain, 1968). This may constitute a mechanism of the glutamate-dependent cell swelling. A noxious potential of glutamate on cerebral tissue has been reported for the first time by Van Harreveld and Fifkova. Concentrations of 8 mM were found to cause cell necrosis in brain tissue (Van Harreveld and Fifkova, 1971b; Van Harreveld, 1972).

Based on these and our own findings we have proposed release of glutamate from cells of primarily damaged brain tissue as a mechanism of secondary cell swelling and necrosis in cerebral ischemia or trauma (Baethmann, 1980). Jorgensen and Diemer (1982) suggest glutamate to mediate
the selective loss of neurones in hippocampal areas sustaining brief episodes of ischemia. Extracellular concentrations of only 100-300 µM induce
cell swelling in the chicken retina in vitro (Van Harreveld and Fifkova,
1971a,b). Studies of Kempski and Gross revealed the cell volume of C6-
glioma cells to increase immediately by administration of glutamate (cf:
Kempski, this volume). Cell swelling amounted to 10-15 %. With the
experiment progressing a tendency of cell volume normalization became apparent, however (Kempski et al., 1982). Cell swelling in the brain secondary to glutamate administration is also supported by in vivo studies.
Ventriculo-cisternal perfusion of rat brain with glutamate causes brain
edema (Kempski, 1982; Rothenfusser, 1982), impedance studies confirm its
cytotoxic nature (Kempski, 1982).

Evidence is available not only that glutamate causes cerebral tissue damage, but also that it is released in conditions associated with
cerebral injury. Respective experiments were conducted in the model of
cold injury edema. Maier-Hauff and coworkers (1984b) analyzed glutamate
concentrations in extracellular edema fluid obtained by interstitial
fluid drainage according to the method of Gazendam and coworkers (1979).
In normal CSF serving as reference medium glutamate concentrations were
only about 20 µM. In edema fluid, however, glutamate concentrations were
300 µM, or higher. Accumulation of glutamate in the edema fluid could
not be attributed to a mere uptake from the intravascular space into the
edematous parenchyma, since glutamate concentrations in plasma were only
100 µM. Major portions of glutamate found in edema fluid must have been
released from damaged brain tissue itself. In experiments where cold injury led to a marked increase of intracranial pressure and, consequently,
cerebral ischemia the glutamate concentrations in edema fluid were up to
1000 µM. This was about 100 times the level found in normal CSF. The
observations demonstrate therefore again a relationship between the
severity of the insult and extent of mediator release. Extracellular glutamate concentrations accumulating under these conditions were certainly
high enough to induce secondary changes. E.g., 100-300 µM were reported
to induce cell swelling in cerebral tissue (Van Harreveld and Fifkova,
1971b). Contrary to the kallikrein-kinin-system, evidence on a therapeutically successful inhibition of the aminoacid to attenuate brain tissue
damage is as yet scant. However, preliminary results on delayed neuronal
necrosis of ischemia in the hippocampus appear promising (Jorgensen and
Diemer, 1982; Wieloch et al., 1985).

Mediator Function of Arachidonic Acid and its Metabolites in Secondary Brain Damage

Another most promising mediator candidate of secondary brain damage is arachidonic acid and its metabolites. The neurotoxic potential of

arachidonic acid has been established in many studies. Chan and Fishman
(1978) have shown that incubation of cortical slices with arachidonic
acid results in swelling. The fatty acid inhibits Na^+/K^+-ATPase and un-
couples oxidative phosphorylation (Chan and Fishman, 1982; Chan et al.,
1983). The many active arachidonic acid metabolites, such as prostaglandins, leukotrienes, and oxygen-derived free radicals have also a
potential as mediator substances (Chan and Fishman, 1980; Wolfe, 1982).
Cytotoxic properties of arachidonic acid were observed in studies on cell
volume of C6 glioma cells in vitro. Administration of arachidonic acid in
concentrations of 4-12 µg/l resulted in a rapid and dose-dependent increase in cell volume of up to 13 % above normal (cf: Kempski, this
volume).

Aside from cytotoxic mechanisms in the cerebral parenchyma,
structural damage by arachidonic acid to the blood-brain barrier must
be considered as well. Injury of endothelial cells has been reported
by Kontos et al. (1980) who studied pial arterioles after topical administration of arachidonic acid, or of its metabolites. Prostaglandin
G_2 and arachidonic acid produced domelike protrusions and craters in the
cerebrovascular endothelium. We found that superfusion of cerebral cortex
with arachidonic acid induces gross opening of the blood-brain barrier
(Unterberg, et al., in preparation). Bazan and others have extensively
studied whether arachidonic acid and other fatty acids are spontaneously
released in various forms of cerebral damage (Agardh and Siesjö, 1981;
Bazan 1970, 1971, 1976; Bazan et al., 1980; Gardiner et al., 1981;
Maier-Hauff et al., 1984b; Rehncrona et al., 1982). Liberation of arachidonic and of other free fatty acids was found during cerebral ischemia,
seizures, hypoglycemia and cold injury to the brain. The process is probably initiated by an activation of phospholipase A_2 (Siesjö, 1981;
Wolfe, 1982).

Concentrations of arachidonic acid sampled in interstitial edema
fluid after induction of cold injury to the brain amounted to 10^{-4} M.
This was clearly above the concentrations found in plasma. Therefore, occurrence of arachidonic acid in edema fluid is attributable to a release
from damaged brain rather than to transport from the intravascular compartment (Maier-Hauff, et al., 1984b). The concentrations detected in the
interstitial edema fluid were clearly above the threshold to induce cytotoxic damage, or opening of the blood-brain barrier, respectively.

Pharmacological inhibition of release and metabolization of arachidonic acid has frequently been attempted (cf: Pickard, 1981; Wolfe, 1982;
Pappius and Wolfe, 1983, 1984; Pappius, this volume). Kontos and coworkers (1980) report, e.g. that pretreatment with indomethacin prevents
histopathological abnormalities produced by arachidonic acid in the cerebral vasculature. Based on these and subsequent studies, they suggest a
release of oxygen-derived free radicals to be responsible, which is associated with the conversion of arachidonic acid to prostaglandin G_2 and to
prostaglandin E_2 (cf: Becker, this volume). Other drugs which prevent
liberation or metabolization of arachidonic acid are corticosteroids,

barbiturates, mannitol, and α-tocopherol. Although many studies have demonstrated attenuation of lipolysis or lipid peroxidation by these substances, their specificity is questionable at the moment. It remains to be seen, whether these drugs interfere specifically with the formation of arachidonic acid and of free radicals, or whether secondary effects, e.g. on cerebral blood flow, oxygen consumption, or other factors are involved. Most important, inhibition of lipolysis and lipid peroxidation should benefit the outcome from head injury or cerebral ischemia to become clinically relevant.

TABLE II

Evidence for Mediators of Secondary Brain Damage

	Damage	Release	Therapeutic Inhibition
Kinins	+	+	+
Glutamate	+	+	./.
Arachidonic Acid	+	+	(+)
Prostaglandins	(+/?)	+	–
Free Radicals	+	+	?

+ positive evidence, – negative evidence, ? questionable, ./. not tested

Interactions of Mediator Compounds

Until now the role of mediator compounds has been discussed as of substances which are independent from each other. This is certainly a simplistic view of the complex processes evolving from cerebral ischemia or traumatic injury. Although little is currently known about interactions between this interesting group of compounds, evidence has been presented, e.g. which functionally associates the kallikrein-kinin system with the arachidonic acid cascade. Such an interaction of both systems would represent an enormous potential for amplification and enhancement. Kinins were shown to stimulate liberation of fatty acids and prostaglandins, most likely by activation of phospholipase A_2 (Nasjletti and Malik, 1979; Vargaftig and Dao Hai, 1972). Conversely, Vio et al. (1982) reported arachidonic acid to activate kallikrein, a reaction where prostaglandins might be involved. Thus, activation of one system triggers the other. In addition, there are functional interrelationships between the KK-system and coagulation, the complement- and fibrinolytic system, as well as with various blood elements (Movat, 1979).

SUMMARY AND CONCLUSIONS

The current evidence on pathophysiological functions of mediator compounds is summarized in Table II considering different levels of significance. Induction of brain damage by a mediator compound is a minimal requirement, while additional demonstration of its release during cerebral injury and of specific therapeutical inhibition providing cerebral protection should suffice for its final identification. The kallikrein-kinin system meets all the necessary requirements. Kinins induce brain damage, such as brain edema, they are released after trauma, and their inhibition was shown to reduce secondary brain damage. The evidence available is less complete in the case of arachidonic acid. Currently, induction of cerebral damage and its release has been ascertained, while to the best of our knowledge the specificity of its therapeutic inhibition has not been clearly established yet. This is comparable with glutamic acid, where the evidence on a specific therapeutic inhibition is as yet preliminary. Although prostaglandins and free radicals were demonstrated to induce damage to cerebral tissue and to become released under respective conditions, the causal relationship between inhibition and therapeutical attenuation of tissue damage is questionable at the moment. Table II makes obvious the significance of further experiments to provide missing evidence. Experimental and, finally, clinical results are required on specific methods of inhibition of mediator compounds resulting in prevention of secondary damage. The intriguing role of mediator compounds in secondary brain damage can not comprehensively be discussed in this context. The list of candidates dealt with in this report is incomplete by all means. Further, it would be naive to believe one single mediator compound being responsible for the many and complex pathophysiological processes evolving in traumatized or ischemic brain tissue. It is safe to assume that progress in treatment can be expected from methods which efficiently influence as many as possible mediator compounds.

ACKNOWLEDGEMENTS

* Supported by Deutsche Forschungsgemeinschaft: Ba 452/6-3. The technical and secretarial assistance of Ruth Demmer, Isolde Juna and Ulrike Goerke is gratefully acknowledged.

REFERENCES

Agardh CD, Siesjö BK, Hypoglycemic brain injury: Phospholipids, free fatty acids and cyclic nucleotides in the cerebellum of the rat after 30 and 60 minutes of severe insulin-induced hypoglycemia, J Cereb Blood Flow Metabol 1:267 (1981)

Ames A, Tsukada Y, Nesbett FB, Intracellular Cl^-, Na^+, K^+, Ca^{++}, Mg^{++} and P in nervous tissue; response to glutamate and to changes in extracellular calcium, J Neurochem 14:245 (1967)

Baethmann A, Pathophysiological and pathochemical aspects of cerebral edema. Neurosurg Rev 1:85 (1978)

Baethmann A, Oettinger W, Rothenfußer W, Geiger R, Biochemical aspects of cerebral edema, in: "Pathophysiology of cerebral energy metabolism", Mrsulja BB, Rakic LM, Klatzo I, Spatz M, eds., Plenum Press, New York (1979)

Baethmann A, Oettinger W, Rothenfußer W, Kempski O, Unterberg A, Geiger R, Brain edema factors: Current state with particular reference to plasma constituents and glutamate, in: "Brain edema", Cervos-Navarro J, Ferszt R, eds., Raven Press, New York, Adv Neurol 28:171 (1980)

Baethmann A, Maier-Hauff K, Kempski O, Activation of the kallikrein-kinin system in brain tissue secondary to cerebral injury and ischemia, J Cereb Blood Flow Metabol 1 (Suppl 1):S218 (1981)

Bazan NG, Effects of ischemia and electroconvulsive shock on free fatty acid pool in the brain, Biochim Biophys Acta 218:1 (1970)

Bazan NG, Changes in free fatty acids of brain by drug-induced convulsions, electroshock, and anesthesia, J Neurochem 18:1379 (1971)

Bazan NG, Free arachidonic acid and other lipids in the nervous system during early ischemia and after electroshock, Adv Exp Med Biol 72:317 (1976)

Bazan NG, Rodriguez de Turco EB, Membrane lipids in the pathogenesis of brain edema. Phospholipids and arachidonic acid, the earliest membrane components changed at the onset of ischemia, in: "Brain edema", Cervos-Navarro J, Ferszt R, eds., Raven Press, New York, Adv Neurol 28:197 (1980)

Bingham WG, Paul SG, Sastry KSS, Effects of cold injury of six enzymes in rat brain, Neurology 21:649 (1969)

Bulle HP, Effects of reserpine and chlorpromazin in prevention of cerebral edema and reversible cell damage, Proc Soc Exp Biol Med 94:553 (1957)

Chan PH, Fishman RA, Brain edema: Induction in cortical slices by polyunsaturated fatty acids, Science 201:358 (1978)

Chan PH, Fishman RA, Transient formation of superoxide radicals in polyunsaturated fatty acid-induced brain swelling, J Neurochem 35:1004 (1980)

Chan PH, Fishman RA, Alterations of membrane integrity and cellular constituents by arachidonic acid in neuroblastoma and glioma cells, Brain Res 248:151 (1982)

Chan PH, Yurko M, Fishman RA, Phospholipid degradation and cellular

edema induced by free radicals in brain cortical slices,
J Neurochem 38:525 (1982)

Chan PH, Kerlan R, Fishman RA, Reduction of γ-aminobutyric acid and
glutamate uptake and (Na⁺-K⁺)-ATPase activity in brain slices
and synaptosomes by arachidonic acid, J Neurochem 40:309 (1983)

Clendendon NR, Allen N, Ito T, Gordon MT, Yashan D, Response of lysosomal hydrolases of dog spinal cord and cerebrospinal fluid to
experimental trauma, Neurology 28:78 (1978)

Erdös EG, The kinins. A status report, Biochem Pharmacol 25:1563 (1976)

Erdös EG, ed., "Bradykinin, kallidin and kallikrein", Handbook of Experimental Pharmacology, Vol XXV (Suppl), Springer, Berlin, Heidelberg, New York (1979)

Fenske A, Sinterhauf K, Reulen HJ, The role of monoamines in the de-
velopment of cold induced edema, in: "Dynamics of brain edema",
Pappius HM, Feindel W, eds., Springer, Berlin, Heidelberg,
New York (1976)

Frey EK, Kraut H, Werle E, Vogel R, Zickgraf G, Trautschold I, Das
Kallikrein-Kinin-System und seine Inhibitoren, F. Enke, Stuttgart
(1968)

Gardiner M, Nilsson B, Rehncrona S, Siesjö BK, Free fatty acids in the
rat brain in moderate and severe hypoxia, J Neurochem 36:1500
(1981)

Gazendam J, Go KG, Van Zanten AK, Composition of isolated edema
fluid in cold-induced brain edema, J Neurosurg 51:70 (1979)

Gross PM, Cerebral histamine: Indications for neuronal and vascular regulation, J Cereb Blood Flow Metabol 2:3 (1982)

Harvey JA, McIlwain H, Excitatory acidic amino acids and the cation
content and sodium ion flux of isolated tissues from the brain,
J Biochem 108:269 (1968)

Joo F, Significance of adenylate cyclase in the regulation of the permeability of brain capillaries, in: "Pathophysiology of cerebral
energy metabolism", Mrsulja BB, Rakic CM, Klatzo I, Spatz M,
eds., Plenum Press, New York (1979)

Jorgensen MB, Diemer NH, Selective neuron loss after cerebral ischemia
in the rat: Possible role of transmitter glutamate, Acta Neurol
Scandinav 66:536 (1982)

Kempski O, Die Lokalisation des Glutamat-induzierten Hirnödems,
Inauguraldissertation, München (1982)

Kempski O, Gross U, Baethmann A, An in-vitro model of cytotoxic brain
edema: Cell volume and metabolism of cultivated glial and nerve
cells, in: "Advances in neurosurgery, Vol 10", Driesen W, Brock M,
Klinger M, eds., Springer, Berlin, Heidelberg (1982)

Khattab FI, Alterations in acid phosphatase bodies (lysosomes) in cat
motoneurons after asphyxiation of the spinal cord, Exp Neurol
18:133 (1967)

Kobrine AI, Doyle T, Role of histamine in posttraumatic spinal cord
hyperemia and the luxury perfusion syndrome, J Neurosurg 44:16
(1976)

Kontos HA, Wei EP, Povlishock JT, Dietrich WD, Magiera CJ, Ellis EF,
Cerebral arteriolar damage by arachidonic acid and prostaglandin

G2, Science 209:1242 (1980)

Maier-Hauff K, Baethmann A, Lange M, Schürer L, Unterberg A, The kallikrein-kinin system as mediator in vasogenic brain edema. Part 2: Studies on kinin formation in focal and perifocal brain tissue, J Neurosurg 61:97 (1984a)

Maier-Hauff K, Lange M, Schürer L, Guggenbichler C, Vogt W, Jacob K, Baethmann A, Glutamate and free fatty acid concentrations in extracellular vasogenic edema fluid, in: "Recent progress in the study and therapy of brain edema", Go KG, Baethmann A, eds., Plenum Press, New York (1984b)

Movat HZ, The plasma kallikrein-kinin system and its interrelationship with other components of blood, in: "Bradykinin, kallidin and kallikrein", Handbook of Experimental Pharmacology, Vol XXV (Suppl), Erdös EG, ed., Springer, Berlin, Heidelberg, New York (1979)

Nasjletti A, Malik KU, Relationship between the kallikrein-kinin and prostaglandin systems, Life Sci 25:99 (1979)

Osterholm J, Bell J, Meyer R, Peyenson J, Experimental effects of free serotonin on the brain and its relation to injury. Part I-III, J Neurosurgery 31:408 (1969)

Pappius HM, Wolfe LS, Functional disturbances in brain following injury: search for underlying mechanisms, Neurochem Res 8:63 (1983)

Pappius HM, Wolfe LS, Effect of drugs on local cerebral glucose utilization in traumatized brain. Mechanisms of action of steroids revisited, in: "Recent progress in the study and therapy of brain edema", Go KG, Baethmann A, eds., Plenum Press, New York (1984)

Perry TL, Jones RT, The aminoacid content of human cerebrospinal fluid in normal individuals and mental defectives, J Clin Invest 40:1363 (1961)

Perry TL, Hansen S, Kennedy J, CSF amino acids and plasma - CSF amino acid ratios in adults, J Neurochem 24:587 (1975)

Pickard JD, Role of prostaglandins and arachidonic acid derivatives in the coupling of cerebral blood flow to cerebral metabolism, J Cereb Blood Flow Metabol 1:361 (1981)

Rehncrona S, Westerberg E, Akesson B, Siesjö BK, Brain cortical fatty acids and phospholipids during and following complete and severe incomplete ischemia, J Neurochem 38:84 (1982)

Reiser G, Hamprecht B, Bradykinin induces hyperpolarizations in rat glioma cells and in neuroblastoma x glioma hybrid cells, Brain Res 239:191 (1982)

Rothenfußer W, Die Bedeutung von Glutamat als Hirnödemfaktor, Inauguraldissertation, München (1982)

Shikimi T, Kema R, Matsumoto M, Yamahata Y, Miyata S, Studies on kinin-like substances in brain, Biochem Pharmacol 22:567 (1983)

Siesjö BK, Cell damage in the brain: a speculative synthesis, J Cereb Blood Flow Metabol 1:155 (1981)

Unterberg A, Baethmann A, The kallikrein-kinin-system as mediator in vasogenic brain edema: I. Cerebral exposure to bradykinin and plasma, J Neurosurg 61:87 (1984)

Unterberg A, Hack U, Baethmann A, Cerebral blood flow and metabolism during bradykinin exposure, in: "Cerebral blood flow, metabolism

and epilepsy", Baldy-Moulinier M, Ingvar DH, Meldrum BS, eds.,
    John Libbey, London, Paris (1983)
Unterberg A, Wahl M, Baethmann A, Effects of bradykinin on permeability
    and diameter of pial vessels in-vivo, J Cereb Blood Flow Metabol
    4:574 (1984)
Unterberg A, Hack U, Baethmann A, Blood flow, metabolism and function
    of the brain during cerebral administration of bradykinin, in:
    "Extra-intracranial vascular anastomoses. Microsurgery at the edge
    of the tentorium", Advances in Neurosurgery, Vol. 13, Dietz H,
    Brock M, Klinger K, eds., Springer, Berlin, Heidelberg, New York
    (1985)
Unterberg, A, Dautermann C, Baethmann A, Müller-Esterl W, The
    kallikrein-kinin system as mediator in vasogenic brain edema.
    III: Inhibition of the kallikrein-kinin system in traumatic brain
    swelling, J Neurosurg 64:269 (1986)
Van Harreveld A, The extracellular space in the vertebrate central ner-
    vous system, in: "The structure and function of nervous tissue,
    IV", Bourne HG, ed., Academic Press, New York, London (1972)
Van Harreveld A, Fifkova E, Effects of glutamate and other amino acids
    on the retina, J Neurochem 18:2145 (1971a)
Van Harreveld A, Fifkova E, Light- and electronmicroscopic changes in
    central nervous tissue after electrophoretic injection of glutama-
    te, Exp Molec Pathol 15:61 (1971b)
Vargaftig RJ, Dao Hai N, Selective inhibition by mepacrine of the release
    of "rabbit aorta contracting substance" evoked by the administra-
    tion of bradykinin, J Pharm Pharmacol 24:159 (1972)
Vio CP, Churchill L, Terragno A, Mc Giff JC, Terragno NA, Arachidonic
    acid stimulated renal kallikrein release in isolated rat kidney,
    Clin Sci 63:235 (1982)
Wahl M, Young AR, Edvinsson L, Wagner F, Effects of bradykinin on pial
    arteries and arterioles in vitro and in situ, J Cereb Blood Flow
    Metabol 3:231 (1983)
Westergaard E, Enhanced vesicular transport of exogenous peroxidase ac-
    ross cerebral vessels, induced by serotonin, Acta Neuropath 32:27
    (1975)
Wieloch T, Lindvall O, Blomquist P, Gage FH, Evidence for amelioration
    of ischaemic neuronal damage in the hippocampal formation by
    lesions of the perforant path, Neurol Res 7:24 (1985)
Wolfe LS, Eicosanoids: Prostaglandins, thromboxanes, leukotrienes,
    and other derivates of carbon-20 unsaturated fatty acids,
    J Neurochem 38:1 (1982)

NEUROCHEMICAL SEQUELAE OF BRAIN DAMAGE AND

THEIR ROLE IN FUNCTIONAL DISTURBANCES

Hanna M. Pappius and L.S. Wolfe

The Goad Unit of The Donner Laboratory of Experimental
Neurochemistry, Montreal Neurological Institute and Hospital
and the Department of Neurology and Neurosurgery, McGill
University, Montreal, Quebec, Canada

INTRODUCTION

Various types of injury have been shown to produce a variety of
chemical changes in brain which are thought to underly functional disturbances resulting from the particular insult. Some of the systems known
to be affected in injured brain are summarized in Figure 1, but the list
is by no means exhaustive. The exact mechanisms by which the chemical
perturbations are related to functional disturbances are not understood
in most cases and it is often difficult to distinguish between the cause
and the effect in this relationship. One problem has been the difficulty
in assessing neurological function in animals since normalization of neurochemical disturbances in brain cannot always be equated with normally
functioning central nervous system and the available methods are not necessarily applicable in the particular experimental situation.

In the work to be described, effects of a focal cortical freezing
lesion in the rat, as an example of traumatic injury, were studied. The
functional state of the injured brain was assessed by measurement of local cerebral glucose utilization (LCGU) with the deoxyglucose technique
(Sokoloff et al., 1977), a method widely used for mapping of cerebral
functional activity in awake animals (Sokoloff, 1981). The neurochemical
correlates examined included the arachidonic acid release and the pro-
staglandins/eicosanoid cascade, and the serotonin and catecholamine neurotransmitter systems. Manipulation of each of these resulted in a modification of the LCGU response to injury. The studies to-date have allowed
development of a working hypothesis according to which prostaglandins,
serotonin and catecholamines, released as a consequence of injury, probably in an interrelated fashion, induce activation of calcium channels,
causing functional depression.

151

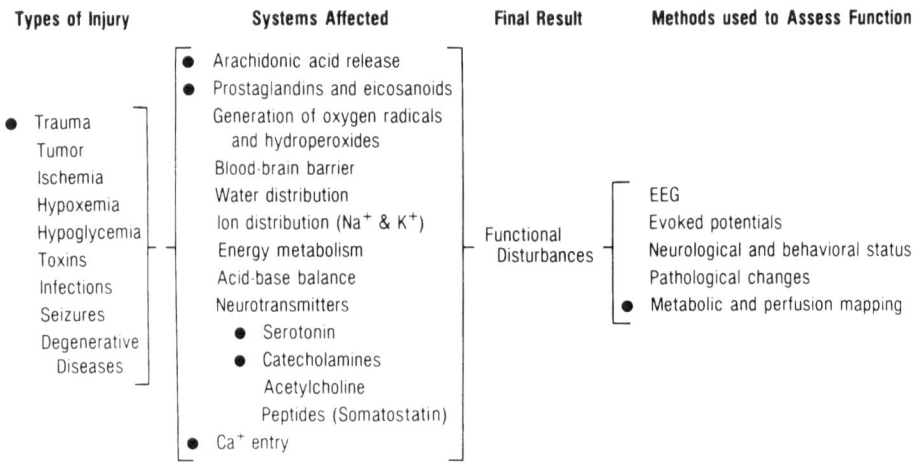

Fig.1: Studies on consequences of brain injury. Summary of types of injury, neurochemical systems affected and methods for assessment of resulting functional disturbances.

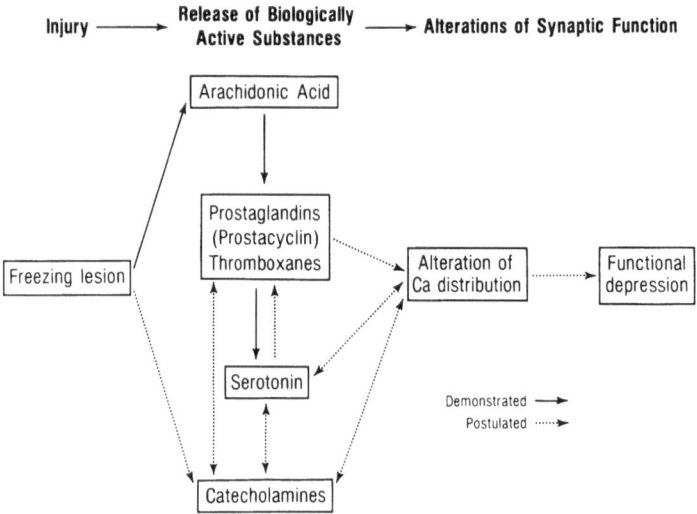

Fig.2: Schematic representation of the working hypothesis on mechanisms underlying functional disturbances in traumatized brain. See text for discussion.

MATERIALS AND METHODS

Except as noted below, all methods and drug doses and regimen have been described in detail (Pappius, 1981, 1982; Pappius and Wolfe, 1983a,b,c).

General Procedure

Briefly, the general procedure was as follows: Freezing lesions standardized to produce superficial focal cortical injury in the rat were made in the left parietal region of halothane anesthetized Sprague-Dawley male rats (280-320 g) by applying a freezing probe (-50°C) to the dura for 5 seconds through a 4x4 mm opening in the skull. After the lesion was made the wound was sutured and the animals allowed to awaken.

^{14}C-deoxyglucose and occasionally ^{14}C-iodoantipyrine studies were carried out as originally described (Sokoloff et al., 1977; Sakurada et al., 1978) in awake rats usually 3 days after the lesion was made. Evans blue dye (2%; 0.35 ml per rat) was injected intravenously before the start of either study so that the state of the blood-brain barrier at the particular time period following the lesion could be determined. As reported previously (Pappius, 1981) blue staining was seen in the area of the lesion in animals killed 24 hours after injury but not 3 days later and the drugs investigated in the present experiments had no effect on this. Animals without lesion untreated or given the drug under investigation according to the same regimen as in the lesioned rats served as "normal" or "no lesion" controls. The physiological state of the awake animals was assessed by monitoring arterial blood pressure and rectal temperature and by measurement of the hematocrit (Ht), serum glucose and blood gases prior to each study.

For determination of arachidonic acid, prostaglandins (PG), serotonin (5-HT) and 5-hydroxy-indoleacetic acid (5-HIAA) the animals were killed by decapitation exactly 60 seconds after the start of a 5 second lesion, the head being dropped immediately into liquid N_2 . A sharp chisel was used to remove the brain from the cranial vault and to separate as much of the lesioned area of the cortex as possible from the adjacent tissue. Because of small amount of the lesion area tissue samples from up to 4 animals were pooled. Samples were also chiselled out from cortex adjacent to the lesion and from the corresponding areas of the non-lesioned right hemisphere, as well as from unlesioned brains from rats decapitated into liquid nitrogen. The samples were kept frozen until homogenization.

Presentation of LCGU Results

    Since the changes in LCGU after lesion were most profound and uniform throughout the cortex of the traumatized hemisphere, in studies on effects of various drugs the results for the six cortical areas in each animal (visual, auditory, parietal, sensory-motor, olfactory and frontal) were expressed as percent of normal and averaged to give the "Average cortical LCGU". Mean $\pm$ SE of these "averages" are given in Tables.

Determination of Serotonin (5-HT) and 5-Hydroxyindoleacetic Acid (5-HIAA)

    5-HT and 5-HIAA were extracted with perchloric acid from frozen brain tissue samples and the supernatant was analyzed by HPLC with electrochemical detection (Anderson et al, 1981).

Behavioral Studies

    Sham-operated and lesioned rats were studied repeatedly for a week before and daily after craniotomy as follows:

    Bilateral Adhesive Removal (tactile extraction). This test has been shown to detect sensory-motor asymmetry independent of motor bias. Pieces of masking tape 1/4 cm x 1/4 cm were bilaterally placed on the radial aspect of rat forepaw. The latency to remove the tape for each paw was recorded. The animals were tested 3 times per day.

    Negative Geotaxis. Animals were placed head upwards on the grid of a runway that was positioned at a 45° angle. The runway was then rotated 180° such that the end of the runway, that was originally facing upwards, was now facing downward. The direction in which the animals rotated to right themselves was noted. Animals were tested 5 times per day.

    Tail Suspension. Animals were suspended by the base of the tail. The presence and direction of lateral or rotary movements of the body were noted. Animals were given five such tests daily.

    Limb Placement and Coordination. Animals were placed on a runway with a grid floor, the grids spaced 1.25 cm apart. The runway was placed 1 meter above the floor of the room. The animals were allowed to roam the runway for 3 minutes. The experimenters watched the limbs of the animals as they traversed the runway and noted the number of times as well as which limb fell through the grid. While the animals were at rest, the manner in which they grasped the grid was recorded.

DEVELOPMENT OF A WORKING HYPOTHESIS INCLUDING SOME RECENT RESULTS AND DISCUSSION

Trauma Model for Study of Functional Disturbances in Brain

It was shown in earlier studies (Pappius, 1981) that depression of LCGU developed with time after a focal freezing lesion, which was widespread but not uniformly distributed throughout the brain. The effects were not related to the location of the lesion which was standardly placed in the parietal cortex but could be made more frontally or caudally with the same results. Four groups of structures with distinct responses to trauma could be identified in awake animals. The most severely affected were all the cortical areas in the lesioned hemisphere, in which the metabolic depression reached its peak, approximately 50% of normal, 3 days after the lesion was made (see Tables 1,2,3). These changes were not associated with parallel alterations in blood flow which was slightly elevated throughout the whole brain at the time of the greatest decrease in LCGU. They also did not appear to be mediated by cerebral edema since their spatial distribution and time course was quite different from that of edema. The posttraumatic depression of cortical glucose utilization was interpreted as representing a manifestation of cerebral dysfunction, in keeping with the hypothesis that the functional state of cerebral tissue is closely coupled to its metabolism (Sokoloff, 1981).

These results suggested that the focal cortical freezing lesion in the rat could be a very useful model for the study of mechanisms underlying traumatically-induced functional disturbances, especially if the assumption that the demonstrated metabolic depression reflects a functional one could be validated by other means. For this purpose, with experimental psychologists L. Holmes and L. Colle of Concordia University in Montreal, a study was undertaken to determine the behavioral status of the lesioned rats. The animals were assessed before and after craniotomy with a battery of four standard tests for sensory-motor behavior (see Methods) by investigators unaware which rats were lesioned and which only sham-operated. The preliminary results summarized in Table 1 indicate that the lesioned rats were significantly but only mildly affected since the relatively low average score of 8 out of possible 24 (normal is 0) suggests that the sensory-motor deficits were minor. Whether other types of behavior are also affected, possibly more profoundly, is now under investigation.

In the meantime, the finding of definite behavioral dysfunction in the rats with a standard freezing lesion correlates with the previously demonstrated metabolic depression, and lends support to the interpretation that diminished LCGU reflects a functional disturbance in this model of brain injury.

TABLE 1

Metabolic Mapping and Behavioural Testing in Rats with Freezing Lesions

	Average Cortical LCGU[a] Percent of Normal		Behavioural testing score[b]
	Lesioned side	Contralateral side	
Sham operated (8)	93 ± 3	93 ± 3	2 ± 1
Lesioned (5)	55 ± 4^{c}	88 ± 4	8 ± 2^{d}

Mean $\pm$ SE. Number of animals in brackets. [a] LCGU determined 72 hrs after lesion. [b] Behavioural testing carried out 24, 48, and 72 hrs after lesion. Normal is 0. Highest possible score 24. Statistically significantly different from sham-operated c: $p < 0.01$; d: $p < 0.02$. Data from L. Colle, L. Holmes and H.M. Pappius, unpublished.

TABLE 3

Effect of PCPA and AMPT on Local Cerebral Glucose Utilization in Rat Brain Three Days after Focal Freezing Lesion

	Average Cortical LCGU Percent of Normal	
	No lesion	Lesioned hemisphere
Untreated	100 (11)	46 ± 4 (9)
72 hrs after PCPA (300 mg/kg)	99 ± 1 (5)	80 ± 4^{a} (5)
72 hrs of AMPT (250 mg/kg day)	95 ± 2 (4)	89 ± 5^{a} (5)

Mean $\pm$ SE. Number of animals in brackets. Drugs started before lesion was made. PCPA: p-chlorophenylalanine. AMPT: alpha-methyl-p-tyrosine. [a] Statistically significantly different from untreated $p < 0.01$. Data from Pappius and Wolfe (1983b).

Involvement of the Prostaglandin System in Traumatically Induced Cerebral
Depression and Effects of Steroids

Using the freezing lesion model, it was then shown that non-steroid-
al anti-inflammatory drugs (NSAIDs), known inhibitors of the prostaglandin synthetase complex in brain (Wolfe et al., 1976; Ramwell, 1980;
Wolfe, 1982), significantly diminished injury-induced changes in LCGU,
when given before or up to 24 hrs after the lesion was made (Pappius and
Wolfe, 1983c), but not when administered acutely before the deoxyglucose
study (Table 2). These findings suggested that some components of the
prostaglandin system must be involved in the development of the functional disturbances associated with injury. This was also another indication
that the metabolic and hence the functional depression was not directly
the result of cerebral edema, since it was shown earlier that one of the
NSAIDs, indomethacin, had no effect on the evolution of the edematous
process (Pappius and Wolfe, 1976).

To what extent cerebral edema is responsible for any neurological
abnormalities in traumatized brain is a constantly recurring question.
There are longstanding tacit assumptions that edema is the main cause
of clinical complications resulting from brain injury. Conversely, it is
considered that clinically beneficial effects of any empirically developed treatment, for example of steroid therapy, are due to edema
controlling properties (Katzman and Pappius, 1973; Pappius and Feindel,
1976; Cervos-Navarro and Ferszt, 1980). On the other hand, experimental
evidence as to the effects of steroids on cerebral edema has been
contradictory (Katzman and Pappius, 1973; Tornheim and McLaurin, 1978;
Mitamura et al., 1981). In contrast, dexamethasone in low doses (0.25
mg/kg day) was shown to significantly and unequivocally diminish the
widespread cortical depression of LCGU which developed after a freezing
lesion and which did not appear to be mediated by edema (Pappius, 1982).
As in the case of NSAIDs the effect of dexamethasone was greatest in regions where the depression was most profound in untreated animals. Thus
both types of drugs were active specifically in the areas most affected
by injury, while they had no effect in normal brain. Dexamethasone, simi-
larily to indomethacin and ibuprofen, was effective when given before or
up to 24 hrs after the lesion but not when given acutely just before the
measurement of LCGU, 3 days after the lesion (see Table 2).

Since corticoids have been reported to inhibit arachidonic acid release from some but not all non-neuronal tissues (Hong and Levine, 1976;
Flower, 1978; Hirata et al., 1980; Nelson, 1980; Majerus, 1983) the possibility that this was the mechanism of dexamethasone action in brain
subjected to a freezing lesion was investigated. Both arachidonic acid
release and prostaglandin ($PGF_{2\alpha}$ and E_2) synthesis were measured in the
area of the freezing injury, within one minute of making the lesion, and
found to be sharply elevated in untreated animals (Pappius and Wolfe,
1983a,c). Arachidonic acid release was found to be unaffected by dexamethasone or indomethacin pre-treatment, while indomethacin but not dexamethasone completely inhibited the formation of $PGF_{2\alpha}$, the only pro-

TABLE 2

Effect of Dexamethasone, Indomethacin and Ibuprofen on Local Cerebral
Glucose Utilization in Rat Brain Three Days after Focal Freezing Lesion

	Average Cortical LCGU Percent of Normal	
	No lesion	Lesioned hemisphere
Untreated	100 (11)	49 $\pm$ 8 (7)
Dexamethasone (0.25 mg/kg day)		
Started before lesion	98 $\pm$ 3 (4)	77 $\pm$ 2[a] (6)
Started 24 hrs after lesion	–	71 $\pm$ 2[a] (4)
Given 1 hr before DG study	–	45 $\pm$ 4 (2)
Indomethacin (7.5 mg/kg)		
Given before lesion	110 $\pm$ 2 (8)	87 $\pm$ 3[a] (7)
Given 1 hr after lesion	–	88 $\pm$ 3[a] (5)
Given 24 hrs after lesion	–	84 $\pm$ 3[a] (6)
Ibuprofen (36 mg/kg day)		
Started before lesion	95 (1)	100 $\pm$ 2[a] (3)
Started 24 hrs after lesion	–	80 $\pm$ 3[a] (6)
Started 3 hrs before DG study	–	42 $\pm$ 3 (3)

Mean $\pm$ SE. Number of animals in brackets.   [a] Statistically significantly
different from untreated $p < 0.01$. Data from Pappius (1982), Pappius and
Wolfe (1983c), and unpublished.

staglandin measured in the treated animals (Pappius and Wolfe, 1983a,c). This provided further support for a role of some components of the pro-staglandin system in functional disturbances resulting from trauma while indicating that steroid action in injured brain is mediated independently of the arachidonic acid cascade, at least in the freezing lesion model. The mechanism of action of dexamethasone remains to be elucidated.

Involvement of Biogenic Amines in Traumatically Induced Cerebral Depression

Both prostaglandins and dexamethasone are known to directly affect neuronal function by modulating neurotransmitter action and/or metabolism (Hedqvist, 1977; McEwen et al., 1979; Wolfe and Coceani, 1979). Thus the possibility was considered that their effects on the functional state of injured brain may be mediated through neurotransmitter systems. Biogenic amines appeared to be good candidates for such a role since in the rat both serotonergic and noradrenergic innervations are widely distributed to all parts of cerebral cortex (see General Considerations), the areas most affected in traumatized brain. Furthermore, changes in serotonin and catecholamine metabolism with cold injury and head trauma have been described (Bareggi et al., 1975; Vecht et al., 1975; Fenske et al., 1976).

Results summarized in Table 3 show that inhibition of serotonin synthesis with p-chlorophenylalanine (PCPA, 300 mg/kg) and of catecho-lamine synthesis with alpha-methyl-p-tyrosine (AMPT, 250 mg/kg day) significantly ameliorated the depresssion of LCGU normally seen as a re-sult of injury (Table 3), with nearly normal rates of glucose utilization being found even in immediate proximity of the lesion. Serotonin (5-HT) and its metabolite 5-hydroxy-indoleacetic acid (5-HIAA) were elevated in untreated animals in cerebral cortex surrounding the lesion area (Table 4) by about 50% within 1 minute of freezing. It is clear from the above studies that activation of both serotonergic and catecholaminergic sys-tems is involved in some way in functional disturbances associated with brain injury. It remains to be determined which of the catecholamines are involved.

Mechanisms Underlying Functional Disturbances in Traumatized Brain: A Working Hypothesis

Our working hypothesis on mechanisms of alterations of synaptic function resulting in disturbances in traumatized brain is summarized in Figure 2. It is clear that a release of several biologically active sub-stances is involved. The experiments discussed above have demonstrated that a prostaglandin type metabolite or metabolites of arachidonic acid must be involved. Serotonin is implicated in the chain of events since inhibition of its synthesis with p-chlorophenyl- alanine blocked the metabolic depression in injured brain, as did inhibition of formation of catecholamines in animals pre-treated with alpha-methyl-p-tyrosine. Which catecholamine is released and whether this release is mediated by the prostaglandin system remains to be determined. Further, the possibility must be considered that catecholamines in turn may be involved through mediation of serotonin. There is electrophysiological evidence that serotonin regulates catecholamine release in the rat neocortex (Ferron et al., 1982).

We postulate that **alteration** of calcium **distribution** is an integral part of the functional disturbances induced by injury. Calcium is generally thought to have a crucial role in coupling between excitation and functional response of the end organ (Rubin, 1970; Bolton, 1979) and under normal physiological conditions Ca^{++} entry is considered to be a trigger for release of neurotransmitters (Kelly et al., 1979; Akerman and Nicholls, 1983). More specifically influx of Ca^{++} into the cell is considered to be the main stimulus responsible for the mobilization of catecholamines (Cooper et al., 1978; Yarbrough et al., 1974; Lane and Aprison, 1977; Coyle and Snyder, 1981; Drapeau and Blaustein, 1983), while recently it has been shown as necessary for arachidonate release and prostacyclin synthesis (Wolfe et al., 1983; Whorton et al., 1984). At the same time certain external chemical messengers, norepinephrine and serotonin among them, are now known to modulate Ca^{++} channels (Hagiwara and Byerly, 1981). Massive Ca^{++} influx has been demonstrated across plasma membranes damaged by widely differing mechanisms and is envisaged as a starting point for a series of reactions that lead to cell death under a variety of pathological conditions (Schanne et al., 1979; Farber et al., 1981; Siesjö, 1981). This probably does not apply in our model since the demonstrated metabolic depression is reversible with time and thus is unlikely to reflect irreversible damage.

The development of so-called Ca^{++} modulators, drugs also referred to as Ca^{++} antagonists or Ca^{++} channel blockers, provided pharmacological tools for elucidation and characterization of Ca^{++} requiring processes. This led to considerable advances in the understanding of the role of Ca^{++} in vascular smooth muscle contractility and, as a result, to new therapy for heart disease (Fleckenstein, 1977; Godfraind et al., 1982). Cerebral vessel contractility can also be modified by Ca^{++} antagonists and their possible usefulness in cerebrovascular disease is under investigation (Hogestatt et al., 1982; Harris et al., 1982). Recently, physiologic activity in neuronal cell systems has been demonstrated for dihydropyridine type of Ca^{++} modulators (Ehlert et al., 1982; Toll, 1982; Takahashi et al., 1983) and the Ca^{++} blocker nitrendipine was shown by radioautography to bind to regions enriched in synaptic connections (Murphy et al., 1982). To determine if Ca^{++} entry into cells in the brain parenchyma plays a role in the sequence of events leading from a focal cortical lesion to the widespread disturbances of cerebral function, effects of Ca^{++} modulators of the dihydropyridine type are now being studied in the freezing lesion model of brain injury.

GENERAL CONSIDERATIONS

The release of arachidonic acid and formation of prostaglandins and eicosanoids after various types of cerebral injury has been well documented (Bazan, 1976; Kuwashima et al., 1976; Yoshida et al., 1980, 1982; Gardiner et al., 1981; Wolfe, 1982; Chan et al., 1983). The results summarized here represent, however, the first clear cut evidence of the direct involvement of this system in functional disturbances associated with brain damage. The particular metabolite of arachidonic acid involved remains to be identified. PGE_2 and thromboxane A_2 are possible candidates since they are known to alter calcium entry into cells (Hedquist, 1977; Gorman, 1979). Furthermore, lipoxygenase products should be examined in injured brain since 5-HETE was shown to be formed, albeit to a small extent, by cerebral cortex pieces stimulated by calcium ionophore A23187 (Wolfe et al., 1983). Much has been said about the possibility that products of cyclo-oxygenase and lipoxygenase pathways which are toxic per se may accumulate in injured brain and be the ultimate cause of damage (Del Maestro, 1980; Demopoulos et al., 1980; Siesjö, 1981; Yoshida et al., 1982). No evidence could be obtained that free radicals mediate the functional disturbances in brain injured by a freezing lesion, but the methods used did not unequivocally rule out such a mechanism (Pappius and Wolfe, 1983a). It must be pointed out, however, that the metabolic depression seen after a freezing lesion is to a large extent reversible with time. It may thus represent a milder type of injury than those in which accumulation of free radicals has been invoked to be a contributing factor to irreversible damage.

The noradrenergic and serotonergic neurotransmitter systems have the most divergent projections in the central nervous system and the wide distribution of their innervations to cerebral cortex in the rat has been well documented (Beaudet and Descarries, 1976; Descarries et al., 1977; Levitt and Moore, 1978; Moore and Bloom, 1979; Lidov et al., 1980; Morrison et al., 1981; Parent et al., 1981; Foote et al., 1983; Morrison and Magistretti, 1983). There appears to be some laminar complimentarity between the two systems in at least some cortical areas, while the major noradrenergic fibers terminate in all layers throughout the cortex and thus can exert their effects in many regions in a manner not related directly to local activity (Foote et al., 1983; Morrison and Magistretti, 1983). There is evidence that noradrenalin modulates actions of other neurotransmitters (Moises et al., 1979) and it has also been proposed that noradrenalin and serotonin act in the cortex as modulators of neuronal activity in a non-synaptic fashion (Beaudet and Descarries, 1978). The noradrenergic system originating in the locus coeruleus has been singled out as probably altering simultaneously and in a functionally relevant way the activity of its widely distributed target neurons (Foote et al., 1983). On the other hand, serotonin-containing cells are envisaged as exerting profound and global influences on cortical function in general (Lidov et al., 1980). Such interpretations are compatible with participation of serotonin in the widespread cortical effects of focal injury and make noradrenalin a candidate as the prime catecholamine involved in functional disturbances following freezing lesion.

The relationships outlined in Figure 2 are, in fact, quite complex as there are both spatial and time course differences in the perturbed systems. The scheme is, as usually in such cases, an over-simplification. Thus the release of arachidonic acid and the formation of PGs appear to be restricted to the area of the lesion and maximum levels were found within a minute of injury. The functional changes, as reflected by a metabolic depression, evolved slowly reaching its peak 3 days after the lesion. Thus important links in the actual chain of events are obviously missing in the postulated sequence, among them undoubtedly participation of some brain stem nuclei.

The results presented here do not preclude involvement of other mechanisms in the overall process of functional disturbances in injured brain. Several neurotransmitter systems, for example the cholinergic, GABAergic and peptidergic have equally widespread distribution in cortical areas as the noradrenergic and serotonergic systems (Emson and Lindvall, 1979; Morrison and Magistretti, 1983). It has recently been demonstrated that low levels of cerebral concussion led to activation of pontine cholinergic sites as inferred from increases in LCGU (Hayes et al., 1984). This was associated with reversible behavioral suppression and inhibition of LCGU in other parts of the brain. Finally, it is likely that the demonstrated neurochemical sequelae of the freezing lesion are not specifically restricted to this model, or even to the traumatic type of cerebral insult. Similar mechanisms can be envisaged as being activated under a variety of conditions in which brain damage leads to neurological malfunction. Poststroke changes in catecholamine neurotransmitters in areas of brain not injured by focal infarction have been implicated in behavioral changes which could not be adequately explained by local structure-function relationships (Robinson and Bloom, 1977).

CONCLUSION

These studies represent a novel approach to problems related to traumatic brain injury. Until now the emphasis has been on understanding and controlling the edematous process. Despite great strides made over the years in elucidation of the mechanisms of formation and resolution of vasogenic cerebral edema no striking advances in therapy have occurred. Osmotic agents and steroids remain the backbone of measures at the disposal of the clinician. If it is confirmed that cerebral injury leads to functional disturbances independently from cerebral edema and its consequences (e.g. herniation) and if the mechanisms underlying these disturbances are further elucidated the findings may provide a solid basis for the development of new rational modes of therapy for conditions in which cerebral edema has to-date been considered as the major cause of neurological complications.

ACKNOWLDGEMENTS

Supported in part by Grant MT-3021 (HMP) from the Medical Research Council of Canada and by The Canadian Donner Foundation. We are indepted to Hanna Szylinger, Klara Rostworowski, Michael McHugh and Ralph Dadoun for technical and to Linda Michel for clerical assistance.

REFERENCES

Akerman KEO, and Nicholls DG, Ca^{++} transport and the regulation of transmitter release in isolated nerve endings, Trends Biol Sci 8:63 (1983).

Anderson GM, Young JG, and Batter DK, Determination of indoles and catechols in rat brain and pineal using liquid chromatography with fluorometric and amperometric detection, J Chromatogr 223:315 (1981).

Bareggi SR, Porta M, Selenati A, Assael BM, Calderini G, Collice M, Rossandra M, and Morselli PL, Homovanillic acid and 5-hydroxy-indoleacetic acid in the CSF of patients after severe head injury, Eur Neurol 13:528 (1975).

Bazan NG, Free arachidonic acid and other lipids in the nervous system during early ischemia and after electroshock, Adv Exptl Med Biol 72:317 (1976).

Beaudet A, and Descarries L, Quantitative data on serotonin nerve terminals in adult rat neocortex, Brain Res 111:301 (1976).

Beaudet A, and Descarries L, The monoamine innervation of rat cerebral cortex: Synaptic and nonsynaptic axon terminals, Neurosci 3:851 (1978).

Bolton TB, Mechanisms of action of transmitters and other substances on smooth muscle, Physiol Rev 59:606 (1979).

Cervos-Navarro J, and Ferszt R, "Brain Edema: Pathology, Diagnosis and Therapy", Raven Press, New York (1980).

Chan PH, Longar S, and Fishman RA, Phospholipid degradation and edema development in cold-injured rat brain, Brain Res 277:329 (1983).

Cooper JR, Bloom FE, and Roth RH, "The Biochemical Basis of Neuropharmacology", Oxford University Press, New York (1978).

Coyle JT, and Snyder SH, Catecholamines, in: "Basic neurochemistry", Siegel GJ, Albers RW, Agranoff BW, and Katzman R, eds., Little Brown Co., Boston (1981).

Del Maestro RF, An approach to free radicals in medicine and biology, Acta Physiol Scand 492:153 (1980).

Demopoulos HB, Flamm ES, Pietronigro DD, and Seligman ML, The free radical pathology and the microcirculation in the major central nervous system disorders, Acta Physiol Scand 492:91 (1980).

Descarries L, Watkins KC, and Lapierre Y, Noradrenergic axon terminals in the cerebral cortex of rat. III. Topometric ultrastructural analysis, Brain Res 133:197 (1977).

Drapeau P, and Blaustein MP, Initial release of (^{3}H) dopamine from rat striatal synaptosomes: Correlation with calcium entry,

J Neurosci 3:703 (1983).

Ehlert FJ, Itoga E, Roeske WR, and Yamamura HI, The interaction of (^{3}H) nitrendipine with receptors for calcium antagonists in the cerebral cortex and heart of rats, Biochem Biophys Res Commun 104:937 (1982).

Emson PC, and Lindvall O, Distribution of putative neurotransmitters in the neocortex, Neurosci 4:1 (1979).

Farber JL, Chien KR, and Mittnacht S, The pathogenesis of irreversible cell injury in ischemia, Am J Pathol 102:271 (1981).

Fenske A, Sinterhauf K, and Reulen HJ, The role of monoamines in the development of cold-induced edema, in: "Dynamics of brain edema", Pappius HM, and Feindel W, eds., Springer-Verlag, Heidelberg (1976).

Ferron A, Descarries L, and Reader TA, Altered neuronal responsiveness to biogenic amines in rat cerebral cortex after serotonin denervation or depletion, Brain Res 231:93 (1982).

Flaim SF, and Zelis R, Effects of diltiazem on total cardiac output distribution in conscious rats, J Pharmacol Exptl Therap 222:359 (1982).

Fleckenstein A, Specific pharmacology of calcium in myocardium, cardiac pacemakers, and vascular smooth muscle, Ann Rev Pharmacol Toxicol 17:149 (1977).

Flower R, Steroidal antiinflammatory drugs as inhibitors of phospholipase A_2, Adv Prost Thromb Res 3:105 (1978).

Foote S, Bloom FE, and Aston-Jones G, Nucleus locus coeruleus: New evidence of anatomical and physiological specificity, Physiol Rev 63:844 (1983).

Gardiner M, Nilsson B, Rehncrona S, and Siesjö BK, Free fatty acids in the rat brain in moderate and severe hypoxia, J Neurochem 36:1500 (1981).

Godfraind T, Albertini A, and Paoletti R, "Calcium modulators", Elsevier Biomedical Press, Amsterdam-New York-Oxford (1982).

Gorman RR, Modulation of human platelet function by prostacyclin and thromboxane A_2, Fed Proc 38:83 (1979).

Hagiwara S, and Byerly L, Calcium channel, Ann Rev Neurosci 4:69 (1981).

Harris RJ, Branston NM, Symon L, Bayhan M, and Watson A, The effects of a calcium antagonist, nimodipine, upon physiological response of the cerebral vasculature and its possible influence upon focal cerebral ischaemia, Stroke 13:759 (1982).

Hayes RL, Pechura CM, Katayama Y, Povlishock JT, Giebel ML, and Becker DP, Activation of pontine cholinergic sites implicated in unconsciousness following cerebral concussion in the cat, Science 223:301 (1984).

Hedqvist P, Basic mechanisms of prostaglandin action on autonomic neurotransmission, Ann Rev Pharmacol Toxicol 17:259 (1977).

Hirata F, Schiffmann E, Venkatasubramanian K, Salomon D, and Axelrod J, A phospholipase A_2 inhibitory protein in rabbit neutrophils induced by glucocorticoids, Proc Natl Acad Sci USA 77:2533 (1980).

Hogestatt ED, Andersson KE, and Edvinsson L, Effects of nifedipine on potassium-induced contraction and noradrenaline release in cerebral and extracranial arteries from rabbit, Acta Physiol Scand 114:283 (1982).

Hong SL, and Levine L, Inhibition of arachidonic acid release from cells as the biochemical action of anti-inflammatory corticosteroids, Proc Natl Acad Sci USA, 73:1730 (1976).

Katzman R, and Pappius HM, "Brain electrolytes and fluid metabolism", Williams & Wilkins Co., Baltimore (1973).

Kelly RB, Deutsch JW, Carlson SS, and Wagner JA, Biochemistry of neurotransmitter release, Ann Rev Neurosci 2:399 (1979).

Kuwashima J, Fujitani B, Nakamura K, Kadokawa T, Yoshida K, and Shimizu M, Biochemical changes in unilateral brain injury in the rat: a possible role of free fatty acid accumulation, Brain Res 110:547 (1976).

Lane JD, and Aprison MH, Calcium-dependent release of endogenous serotonin, dopamine and norepinephrine from nerve endings, Life Sci 20:665 (1977).

Levitt P, and Moore RY, Noradrenaline neuron innervation of the neo-cortex in the rat, Brain Res 139:219 (1978).

Lidov HGW, Grzanna R, and Molliver ME, The serotonin innervation of the cerebral cortex in the rat - an immunohistochemical analysis, Neurosci 5:207 (1980).

Majerus PW, Arachidonate metabolism in vascular disorders, J Clin Invest 72:1521 (1983).

McEwen BS, Davis PG, Parsons B, and Pfaff DW, The brain as a target for steroid hormone action, Ann Rev Neurosci 2:65 (1979).

Mitamura JA, Seligman ML, Solomon JJ, Flamm ES, Demopoulos HB, and Ransohoff J, Loss of essential membrane lipids and ascorbic acid from rat brain following cryogenic injury and protection by methylprednisolone, Neurol Res 3:329 (1981).

Moises HC, Woodward DJ, Hoffer BJ, and Freedman R, Interactions of norepinephrine with Purkinje cell responses to putative amino acid neurotransmitters applied by microiontophoresis, Exptl Neurol 64:493 (1979).

Moore RY, and Bloom FE, Central catecholamine neuron systems: Anatomy and physiology of the norepinephrine and epinephrine systems, Ann Rev Neurosci 2:113 (1979).

Morrison JH, and Magistretti PJ, Monoamines and peptides in cerebral cortex. Contrasting principles of cortical organization, Trends Neurosci 6:146 (1983).

Morrison JH, Molliver ME, Grzanna R, and Coyle JT, The intracortical trajectory of the coeruleo-cortical projection in the rat: A tangentially organized cortical afferent, Neurosci 6:139 (1981).

Murphy KMM, Gould RJ, and Snyder SH, Autoradiographic visualization of (^{3} H) nitrendipine binding sites in rat brain: localization to synaptic zones, Eur J Pharmacol 81:517 (1982).

Nelson DH, Corticosteroid-induced changes in phospholipid membranes as mediators of their action, Endocrin Rev 1:180 (1980).

Pappius HM, Local cerebral glucose utilization in thermally traumatized rat brain, Ann Neurol 9:484 (1981).

Pappius HM, Dexamethasone and local cerebral glucose utilization in freeze-traumatized rat brain, Ann Neurol 12:157 (1982).

Pappius HM, and Feindel W, "Dynamics of brain edema", Springer-Verlag, Heidelberg (1976).

Pappius HM, and Wolfe LS, Some further studies on vasogenic edema, in: "Dynamics of brain edema", Pappius HM and Feindel W, eds., Springer-Verlag, Heidelberg (1976).

Pappius HM, and Wolfe LS, Functional disturbances in brain following injury: search for underlying mechanisms, Neurochem Res 8:61 (1983a).

Pappius HM, and Wolfe LS, Involvement of serotonin and catecholamines in functional depression of traumatized brain, J Cereb Blood Flow Metabol 3 (Suppl. 1) :S226 (1983b).

Pappius HM, and Wolfe LS, Effects of indomethacin and ibuprofen on cerebral metabolism and blood flow in traumatized brain, J Cereb Blood Flow Metabol 3:448 (1983c).

Parent A, Descarries L, and Beaudet A, Organization of ascending serotonin systems in the adult rat brain. A radioautographic study after intraventricular administration of (^{3}H)5-hydroxytryptamine, Neurosci 6:115 (1981).

Ramwell P, Prostaglandin synthetase inhibitors: New clinical applications, in: "Prostaglandins and related lipids", Ramwell P, ed., AR Liss, New York (1980).

Robinson RG, and Bloom FE, Pharmacological treatment following experimental cerebral infarction: Implications for understanding psychological symptoms of human stroke, Biol Psychiat 12:669 (1977).

Rubin RP, The role of calcium in the release of neurotransmitter substances and hormones, Pharmacol Rev 22:389 (1970).

Sakurada O, Kennedy C, Jehle J, Brown JD, Carbin GL, and Sokoloff L, Measurement of local cerebral blood flow with iodo (^{14}C)antipyrine, Am J Physiol 234:H59 (1978).

Schanne FAX, Kane AG, Young EE, and Farber JL, Calcium dependence of toxic cell death: A final common pathway, Science 206:700 (1979).

Siesjö BK, Cell damage in the brain: A speculative synthesis, J Cereb Blood Flow Metabol 1:155 (1981).

Sokoloff L, Reivich M, Kennedy C, DesRosiers MH, Patlak CS, Pettigrew KD, Sakurada O, and Shinohara M, The (^{14}C)-deoxyglucose method for measurement of local cerebral glucose utilization: Theory, procedure, and normal values in the conscious and anesthetized albino rat, J Neurochem 28:897 (1977).

Sokoloff L, Localization of functional activity in the central nervous system by measurement of glucose utilization with radioactive deoxyglucose, J Cereb Blood Flow Metabol 1:7 (1981).

Takahashi M, and Ogura A, Dihydropyridines as potent calcium channel blockers in neuronal cells, FEBS Letters 152:191 (1983).

Toll L, Calcium antagonists. High affinity binding and inhibition of calcium transport in a clonal cell line, J Biol Chem 257:13189 (1982).

Tornheim PA, and McLaurin RL, Effect of dexamethasone on cerebral edema from cranial impact in the cat, J Neurosurg 48:220 (1978).

Vecht CJ, Van Woerkom TCAM, Teelken AW, and Minderhoud JM, Homo-vanillic acid and 5-hydroxyindoleacetic acid cerebrospinal fluid levels, Arch Neurol 32:792 (1975).

Whorton AR, Willis CE, Kent RS, and Young SL, The role of calcium in the regulation of prostacyclin synthesis by porcine aortic endo-thelial cells, Lipids 19:17 (1984).

Wolfe LS, Eicosanoids: Prostaglandins, thromboxanes, leukotrienes, and other derivatives of carbon-20 unsaturated fatty acids, J Neurochem 38:1 (1982).

Wolfe LS, and Coceani F, The role of prostaglandins in the central ner-vous system, Ann Rev Physiol 41:669 (1979).

Wolfe LS, Ng Ying Kin NMK, and Spatz M, Metabolites of arachidonic acid after calcium ionophore stimulation of cultured cerebral capillary endothelial cells and brain tissue: Identification of lipoxygenase products, J Neurochem 41:S40C (1983).

Wolfe LS, Rostworowski K, and Pappius HM, The endogenous biosynthesis of prostaglandins by brain tissue in vitro, Can J Biochem 54:629 (1976).

Yarbrough GG, Lake N, and Phillis JW, Calcium antagonism and its effect on the inhibitory actions of biogenic amines on cerebral corti-cal neurons, Brain Res 67:77 (1974).

Yoshida S, Inoh S, Asano T, Sano K, Kubota M, Schimazaki H, and Ueta N, Effect of transient ischemia on free fatty acids and phospholipids in the gerbil brain. Lipid peroxidation as a pos-sible cause of postischemic injury, J Neurosurg 53:323 (1980).

Yoshida S, Abe K, Busto R, Watson BD, Kogure K, and Ginsberg MD, Influence of transient ischemia on lipid-soluble antioxidants, free fatty acids and energy metabolites in rat brain, Brain Res 245:307 (1982).

CHANGES IN BRAIN EDEMA MOVEMENT FOLLOWING EXPERIMENTAL

BRAIN INSULTS AND ACUTE THERAPY

Hector E. James[*)**)], R. Werner[**)], and M. del Bigio[*)]

University of California Medical Center, Div. Neurosurg. *
Dept. Pediatrics ** Univ. California, San Diego, USA

INTRODUCTION

Elevation of intracranial pressure in the clinical and experimental setting can be induced by a variety of brain insults. Cerebrovascular engorgement (swelling), may occur as a loss of autoregulation which would elevate intracranial pressure because of an increased blood volume. An increase of ICP may also result from disturbances at the capillary level by extravasation of serum fluid into the interstitium. Formation of vasogenic brain edema may continue as long as the propelling forces of the arterial pressure maintains an effective transmural pressure gradient (13,14). The area of increased permeability (13,14) and the duration of the opening of the blood-brain barrier (4,14) are additional variables. The increased interstitial fluid could lead to an increased brain volume and, hence, a further rise in intracranial pressure. Brain insults can affect glial and neuronal metabolism so that the cellular function is significantly altered. Examples comprise the processes that occur with ischemia and hypoxia. Blockade of cellular energy metabolism may lead to membrane permeability changes and alterations of the sodium pump. Cellular (cytotoxic) edema and swelling may follow (13). An increased water content in the intracellular compartment of the brain may also lead to an overall increase in brain volume and, hence, of the intracranial pressure.

Therapy of cerebral injuries should interrupt the chain of events that follow the primary insult (23). Multiple therapeutic modalities have been employed to this end (9,17,18). In an attempt to better understand the secondary processes and the effects of therapy, we studied the cerebral response to a vasogenic and cytotoxic insult including effects of acute treatment with C14-labeled dextran.

169

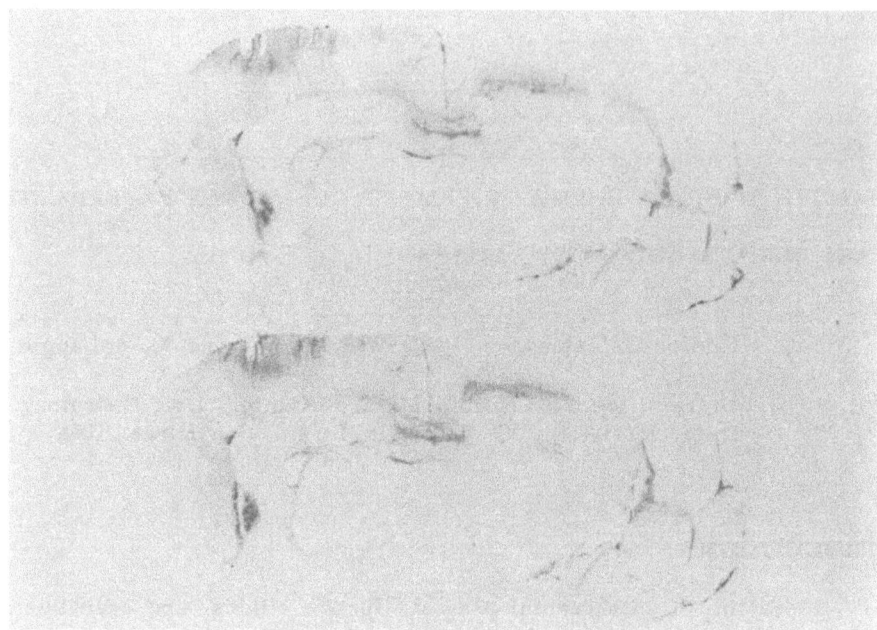

Fig. 1: Autoradiograph of a rabbit brain obtained by administration of
150 microcuries of C14-dextran. The animal was studied 24 hrs
after a left parieto-occipital cryogenic lesion. Duration of
freezing was 90 sec.

MATERIALS AND METHODS

Albino rabbits of 2.5 to 3.9 kg were anesthetized with sodium
thiopental (20 mg/kg) through the lateral ear vein and placed then on a
stereotactic head rest. The scalp was infiltrated with 1% lidocaine for
preparation of a 2.5 mm diameter circular trephine hole over the left
parieto-occipital cortex. A stainless steel probe previously cooled with
liquid nitrogen, was applied to the intact dura for 90 seconds. The skull
defect and wound were closed then. 1 ml of a 2% Evans Blue solution was
administered intravenously followed by 150 microcuries of C-14 labeled
dextran of 75.000 molecular weight. 50 mg/kg ampicillin were given by
intraperitoneal route. 24 h later, at the time of maximal brain edema
(10), the animals were re-anesthetized, sacrificed and bled by a 16-gauge
intracardiac cannula. The brains were promptly removed and frozen in 2-
methylbutane and freon. Prior to full hardening of the brain, a coronal
section was performed through the center of the surface lesion made visible by Evans Blue. Core samples were taken with a 19-gauge stainless
steel probe from the rostral and dorsal brain at three levels: subcortical (below the lesion), deep white matter, and periventricular.

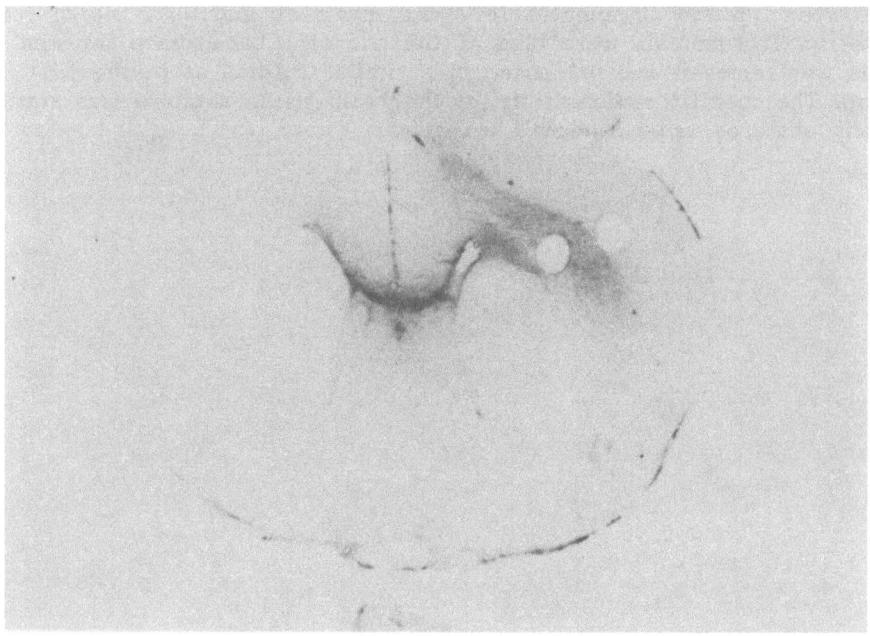

Fig.2: Occipital autoradiograph (150 microcuries of C14-dextran i.v.) to demonstrate the sites from where core samples were taken for determination of tissue radioactivity. Samples were obtained from a subcortical area, deep white and periventricular brain (less visible circle).

Figure 1 shows a coronal autoradiograph. Figure 2 shows an autoradiograph following sampling to demonstrate the sites of tissue removal. Each sample was weighed and dissolved by protosol, peroxide and cytoscint. The specimen were counted in the scintillation counter for two minutes and corrected for the background counts (blanks) to obtain the specific radioactivity. Autoradiographs were made then from the remaining ventral and dorsal portions of each hemisphere. The brains were sliced in a cryostat (American Optical) at 12° C. Slices of 40 microns were cut and mounted on x-ray film over 21 days for autoradiography. The above procedure was performed on eight animals. A second group (n=3) of animals submitted to the same procedure received intraperitoneally 120 mg/kg 6-aminonicotinamide as a metabolic inhibitor (2,11) immediately following the cryogenic injury. A third group underwent similar experimentation without receiving 6-aminonicotinamide. 24 hrs following the cryogenic injury the animals received acute treatment against brain edema and an increase of intracranial pressure. Treatment consisted of (a) a rapid in-

travenous bolus of mannitol (1 g/kg), (b) infusion of 20 mg/kg pentobarbital over 1 h and 20 minutes following mannitol, and (c) 1 mg/kg furosemide i.v. The animals were bled at the end of a barbiturate infusion. The
brain was removed and processed in a similar fashion as in the first
group. The specific radioactivity of the brain tissue samples was statistically analyzed using Student's t-Test.

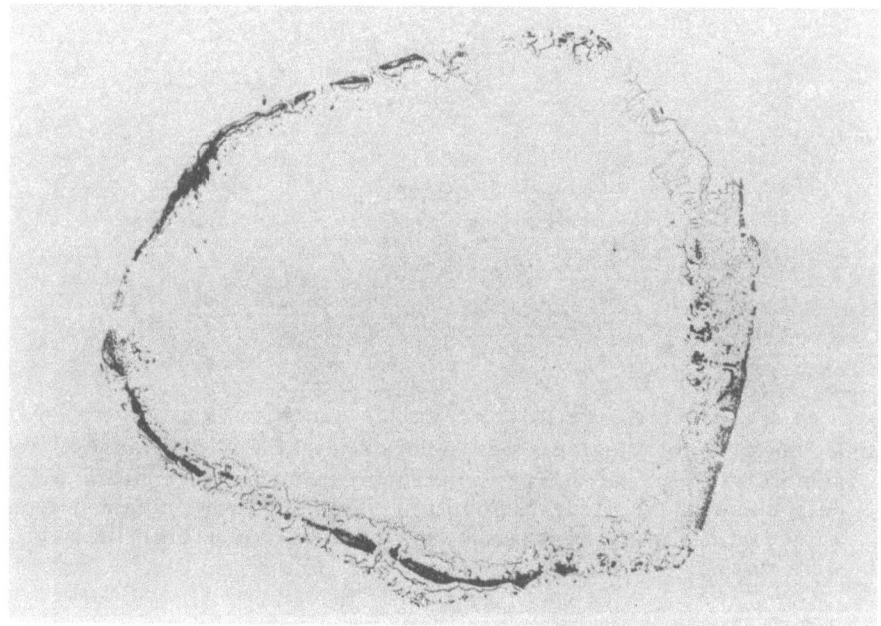

Fig.3: Cerebral autoradiograph obtained 24 hrs following a combined
cryogenic and cytotoxic (i.v. 6-aminonicotinamide) insult.

RESULTS

Gross Pathology

On autoradiography, as well as on gross sections of the brain the
C-14 activity matched the distribution of Evans Blue. Uniformly, dextran-
and Evans Blue exudation were maximal at the site of the lesion (cortical
and subcortical) with extension into ipsilateral deep white and periventricular areas. At times it extended into the central deep white
matter. Evans Blue staining could be seen also in the ventricular walls.
Figure 1 is an autoradiograph demonstrating the distribution of dextran.
The distribution of Evans Blue and C-14 dextran differed markedly between

the group with a combined cryogenic lesion and the group receiving 6-ANA. This is demonstrated in Figure 3. Dextran was seen in the area of the lesion, however, less than what was found in animals with a cryogenic lesion alone. Some dextran was also found in the pia-arachnoidea, but none in the deep white matter or in periventricular regions. In the subgroup receiving treatment, no gross pathological differences were observed as compared to the untreated group of animals with cold injury edema.

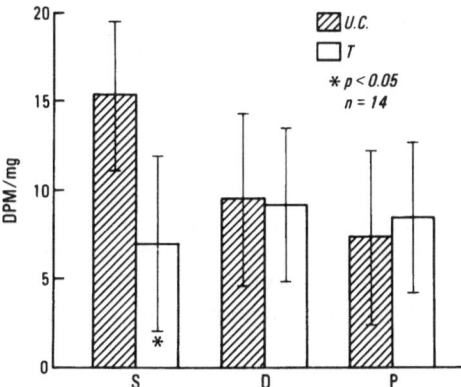

Fig.4: Distribution of radioactivity (DPM/mg) in subcortical (S), deep white (D), and periventricular (P) areas before (U.C.) and after therapy (T).

Distribution of C-14 Dextran

The distribution of C-14 dextran is analyzed in Figure 4. The amount of isotope in subcortical areas in animals with a cryogenic lesion alone (first group) was 15 + 8.8 DPM/mg, while in the deep white matter 9.3 + 7.5 DPM/mg. The difference is significantly different from the subcortical area ($p < 0.05$), but less different from the deep white matter (NS). In animals with therapy the subcortical radioactivity was 7.1 + 5.0, in the deep white matter 9.2 + 5.2, and in periventricular tissue 8.2 + 4.2 DPM/mg. There were no statistically significant differences.

DISCUSSION

Changes in Brain with Edema

The two forms of brain edema described by Klatzo (13), have been useful investigational tools in many studies and have added to the understanding of the disease process. The sequelae of vasogenic and cytotoxic brain edema can be described as a chain of events that spread to the surrounding tissue and bring about yet poorly understood secondary changes that lead to further disruption of function and - further edema (2,23,24). Reulen and Kreysch (25), have shown that the superimposition of a cytotoxic insult by administering hexachlorophene 3-5 days prior to a cryogenic lesion increases the water content in the white matter by about 4-5% over that what is found in a cryogenic lesion only. They postulated that the cytotoxic brain edema induced by the agent reduced the size of the extracellular space impeding spread of the vasogenic fluid through the white matter to the ventricular system. This was supported by a restriction of extracellular movement of an intravenously administered label in animals with a cryogenic lesion pretreated with hexachlorophene. The present study revealed a total lack of movement of an intravenously injected tracer through the cerebral parenchyma of animals simultaneously exposed to a cryogenic and cytotoxic insult (Fig. 3). This confirms previously mentioned observations of Reulen et al. (24) on a collapse of the extracellular space under these conditions.

Therapy and Secondary Changes In Brain Edema

The tissue changes that follow a brain insult are of importance as to the timing and type of therapy to halt and resolve the secondary processes (23). Therapy should be directed at the various stages of edema formation and resolution. First, reduction of the propelling force, i.e. the hydrostatic pressure gradient between the intra- and extravascular compartment may be considered by reducing blood pressure (14,17), second, enhancement of the clearance of edema fluid from the extra- and intracellular space. Osmotic and nephron-blocking diuretics may be appropriate (9,23). Increase of capillary flow in areas where edema led to collapse of the microvessels resulting in an increase of tissue pressure has also been considered as a method of treatment. An improved tissue perfusion would enhanced clearance of edema fluid into the capillary lumen (8).

Barbiturates may act favorably on brain edema due to the reduction of cellular metabolism, blood flow and blood volume. This in turn would lower intracranial pressure. Barbiturates may also directly reduce cerebral blood volume by constriction of cerebral arterioles. In a model of cryogenic edema, Go et al. (7) demonstrated that variations in the tone of cerebral arterioles affect the brain water content. Smith (26) observed in dogs with a cryogenic lesion a significant reduction of white matter edema by pentobarbital. Similar findings were reported by Matsumoto, et al. (16) who suggest that edema improved because of a lowering of the systemic blood pressure and, hence, the hydrostatic pressure differential across the blood-brain barrier.

Diuretic agents have been extensively employed for the reduction of intracranial pressure and brain edema. Osmotic diuretics extract water through the blood-brain barrier along an osmotic gradient. These substances function therefore only in areas where the blood-brain barrier is intact (9,20). In cryogenic injury in rabbits the water content of the edematous brain was not affected by acute administration of mannitol, although the intracranial pressure was found to decrease (9). On the other hand, in an experimental model of cytotoxic brain edema, peak reduction of intracranial pressure was associated with a significant decrease of the brain water content in edematous areas (11,17). The renal loop diuretic furosemide may work by a direct reduction of sodium transport into the brain, thereby reducing brain water content, and consequently tissue volume (27). Furosemide in high doses inhibits CSF production (21,22) and may thus reduce intracranial pressure. Reduction of the ventricular fluid pressure may enhance clearance of brain edema by increasing the pressure gradient between the site of edema entry and the ventricle, which would increase fluid drainage on extracellular routes (23). Experimental results on furosemide in cryogenic brain edema seem to indicate that treatment reduces the water content of edematous tissue areas (5,19). However, in previous work we failed to demonstrate a significant reduction of brain water at peak reduction of ICP in rabbits with cold injury edema (9), while furosemide led to a significant reduction of ICP in the cytotoxic modification (11). Likewise, in a model with combining a cytotoxic and cryogenic insult, furosemide alone did not reduce brain water content, while the combined administration of mannitol plus furosemide led to a significant decrease (17).

In the present studies a gradient of C14-dextran in a pattern similar to previous observations (3,24,25) was found in the brain of untreated animals. The edema front extended to the ventricle through the white matter. Obviously, edema was still propagating from the lesion at 24 hrs, where breakdown of the blood-brain barrier allowed for entry of serum constituents into the parenchyma. As seen in Figure 4, the combined treatment was mostly effective in the subcortex. Accumulation of C14-dextran was significantly lower there as compared to the untreated controls, while the deep white matter or periventricular areas were not influenced by treatment. It then follows that acutely administered barbiturates and diuretics affect the area where the propelling forces are generating tissue pressure gradients causing progression of edema fluid through the cerebral parenchyma. On the other hand, these agents seem only minimally, if at all to influence the secondary changes that have occurred in areas of spread of edema. This is in agreement with our previous work where acute treatment had only little influence on the brain water content in a cryogenic cerebral lesion (9,17,18). New forms of treatment as well as more effective combinations of existing modalities should be investigated for an inhibition of the secondary changes following a brain insult.

SUMMARY

Albino rabbits (n=8) were exposed to a cryogenic lesion of the left parieto-occipital cerebral cortex through the intact dura. A second group (n=3) was intraperitoneally administered with the metabolic inhibitor 6-aminonicotinamide in addition. 150 microcuries of C-14 labeled dextran were injected intravenously at the time of the lesion. The animals were sacrificed for removal of the brain 24 hrs later. Core samples of 1.5 mm width were obtained in the sagittal plane at three locations: subcortical, deep white matter, and periventricular for counting of radioactivity. The hemispheres were then sectioned and mounted for autoradiography. In the first group C-14 dextran extended from the lesion through the white matter of the hemisphere into periventricular regions. The radioactivity was higher in subcortical areas than in the deep white matter ($p < 0.01$), or periventricular samples ($p < 0.05$). In the second group, C-14 dextran was found in the lesion and in the pia-arachnoidea, but not in the white matter of the hemisphere. A third group (n=5) with a cryogenic lesion only received pentobarbital (20 mg/kg), mannitol (1 g/kg), and furosemide (1 mg/kg) acutely and 24 hours later. Although the distribution of C14-dextran was similar to that of untreated animals (group I), the radioactivity was significantly reduced in the subcortex when compared to the untreated controls ($p < 0.05$). It seems therefore that secondary tissue changes in cerebral edema cannot be modified on a qualitative basis by even the most aggressive treatment. However, therapy seems to act by reducing propagation of edema from the disrupted blood-brain barrier.

ACKNOWLEDGEMENTS

This work has been performed in part with the Research Funds of the Division of Neurosurgery. We are indebted to the excellent secretarial work of Sandra Bain for the preparation of this manuscript.

REFERENCES

1. Baethmann A, Oettinger W, Rothenfußer W, Geiger R, Pathophysiology of cerebral edema, in: "Pathophysiology of cerebral energy metabolism" , Mrsulja BB, Rakic LM, Klatzo I, Spatz M, eds., Plenum Publ. Corp. (1979).
2. Baethmann A, Van Harreveld A, Water and electrolyte distribution in gray matter rendered edematous with a metabolic inhibitor, J Neuropathol Exp Neurol 32:408 (1973).

3. Bruce DA, Ter Weeme C, Kaiser G, Langfitt TW, The dynamics of small and large molecules in the extracellular space and ce-rebro- spinal fluid following cold injury of the cortex, in: "Dynamics of Brain Edema", eds., Pappius H, Feindel W, Sprin-ger-Verlag, Heidelberg (1976).

4. Clasen RA, Cooke PM, Pandolfi S, Boyd D, Raimondi AJ, Experimental cerebral edema produced by focal freezing: I. An anatomic study utilizing vital dye techniques, J Neuropathol Exp Neurol 21:579 (1962).

5. Clasen RA, Pandolfi S, Casey D,Jr, Furosemide and pentobarbital in cryogenic cerebral injury and edema, Neurology (Minneap.), 24:642 (1974).

6. Frei HJ, Wallenfang T, Pöll W, Schubert M, Brock M, Regional cerebral blood flow and regional metabolism in cold injury induced brain edema, Acta Neurochir 29:15 (1973).

7. Go KG, Zijlstra WG, Flanderjin A, Zuiderveen F, Circulatory factors influencing exudation in cold-induced cerebral edema, Exp Neurol 42:332 (1974).

8. Harbaugh RD, James HE, Marshall LF, Shapiro HM, Laurin R, Acute therapeutic modalities for experimental vasogenic edema, Neurosurgery 5:656 (1979).

9. Herrmann HD, Neuenfeldt D, Development and regression of disturb-ances of the blood-brain barrier and edema in tissue surrounding a circumscribed cold lesion, Exp Neurol 34:115 (1972).

10. Hadjidimos A, Fischer F, Reulen HJ, Restitution of vasomotor autore-gulation by hypocapnia in brain tumors, in: "Advances in Neuro-surgery, Vol III", Penzholz H, Brock M, Hamer J, Klinger M, Spoerri O, eds., Berlin, Heidelberg, New York (1975).

11. James HE, Bruce DA, Welsh F, Cytotoxic edema produced by 6-amino-nicotinamide and its response to therapy, Neurosurgery, 3:196 (1978).

12. James HE, Laurin R, Intracranial hypertension and brain edema in albino rabbits: Part I: Experimental models, Acta Neurochir (Wien) 55:213 (1981).

13. Klatzo I, Neuropathological aspects of brain edema J Neuropathol Exp Neurol 26:1 (1967).

14. Klatzo I, Wisniewski H, Steinwall O, Streicher E, Dynamics of cold injury edema, in: "Brain edema", eds., Klatzo I, Seitelberger F, Springer-Verlag, New York (1967).

15. Lassen NA, The luxury perfusion syndrome and its possible relation to acute metabolic acidosis localized within the brain, Lancet 2:113 (1966).

16. Matsumoto A, Kogure K, Utsunomiya Y, Busto R, Scheinberg P, Rein-muth OM, Energy metabolism and CBF in cold-induced brain edema: Comparison of the effect of dexamethasone under nitrous oxide and pentobarbital anesthesia, in: "Blood flow and metabo-lism in the brain", eds., Harper AM, Jennett WB, Miller JD, Rowan JO, Edinburgh, Churchill-Livingstone (1975).

17. Millson CH, James HE, Shapiro HM, Laurin R, Intracranial hypertension and brain edema in albino rabbits. Part 2: Effect of acute therapy with diuretics, Acta Neurochir 56:167 (1981).
18. Millson CH, James HE, Shapiro HM, Laurin R, Intracranial hypertension and brain edema in albino rabbits. Part 3: Effect of acute simultaneous diuretic and barbiturate therapy, Acta Neurochir 61:271 (1982).
19. Pappius HM, Effects of steroids on cold injury edema, in: "Steroids and brain edema", eds., Reulen HJ, Schürmann K, Springer-Verlag, New York (1972).
20. Pappius HM, Dayes LA, Hypertonic urea: Its effect on the distribution of water and electrolytes in normal and edematous brain tissue, Arch Neurol 13:395 (1965).
21. Pollay M, Formation of cerebrospinal fluid: Relations or studies of isolated choroid plexus to the standing gradient hypothesis, J Neurosurg 42:665 (1975).
22. Reed DJ, The effects of furosemide on cerebrospinal fluid flow in rabbits, Arch Int Pharmacodyn Ther 178:324 (1969).
23. Reulen HJ, Vasogenic brain edema: New aspects in its formation, resolution and therapy. Br J Anesth 48:741 (1976).
24. Reulen HJ, Graham R, Spatz M, Klatzo I, Role of pressure gradients and bulk flow in dynamics of vasogenic brain edema, J Neurosurg 46:24 (1977).
25. Reulen HJ, Kreysch HG, Measurements of brain tissue pressure in cold induced cerebral edema, Acta Neurochir (Wien) 29:29 (1973).
26. Smith AL, Marque JJ, Anesthestics and cerebral edema, Anesthesiology 45:64 (1976).
27. Tutt HP, Pappius HM, Studies on the mechanisms of action of steroids in traumatized brain, in: "Steroids and brain edema", eds., Reulen HJ, Schürmann K, Springer-Verlag, New York (1972).

DISTURBANCES OF CELLULAR MEMBRANES AND MEMBRANE-BOUND ENZYMES IN CRYOGENIC BRAIN EDEMA

Francois Cohadon, N. Averet, and M. Rigoulet

Laboratoire de Neurochirurgie Experimentale et Neurobiologie, Universite de Bordeaux II, Service de Neurochirurgie A, Hopital Pellegrin-Tripode, Bordeaux, France

The hypothesis of cellular membrane dysfunction playing a key role in the pathophysiology of acute brain damage is not new. As early as 1966 Ishii (1) using an epidural balloon injury model has shown a loss of essential membrane phospholipids, namely of lecithin and gangliosides, and an impairment of ATPase function in brain tissue. Both phenomena were roughly proportional to the deterioration of the animals condition. In an attempt to explain the beneficial effect of corticosteroids, Demopoulos et al. (2) proposed an attack of free radicals on phospholipids as a mechanism of membrane damage. The fact that antioxidants were capable to diminish cold injury edema significantly (3) was a supporting argument. The "membrane theory" of brain damage has been extensively discussed since then for a number of acute brain conditions (cf.: 4).

MECHANISMS OF MEMBRANE ATTACK

From a biochemical point of view two basic mechanisms damage phospholipid layers of cellular membranes. One is enhancement of free radical reactions, the other the enzymatic action of phospholipases. Both mechanisms are likely to operate in acute insults of the brain parenchyma. During the course of ischemic injury, or more generally when ischemia is the predominant pathophysiological event with blood flow less than 6-7 ml/100g min, free fatty acids accumulate in the affected tissue (5,6) indicating break-down of phospholipids. It is widely accepted now that the intracellular rise of free Ca^{++} activates phospholipases, particularly phospholipase A_2 (7,8). This in turn causes an accumulation of free fatty acids which under these conditions can not be utilized again for lipid synthesis because the latter requires a functional energy metabolism. The detergent action of free fatty acids exerts deleterious effects on membrane structures with uncoupling of oxidative phosphorylation in mito-

179

chondria being the most serious. Hydrolysis by phospholipases seems to be
the major mechanism of membrane disruption in cerebral ischemia, whereas
a direct involvement of free radical reactions as advocated by Flamm (9)
remains highly controversial. In the late phase of ischemia, particularly
during recirculation it may play an important role, however (10,11).

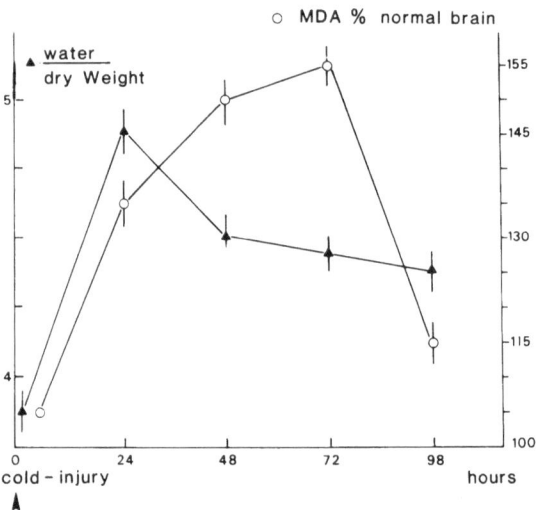

Fig.1: Cerebral wet-/dry-weight ratios and malonaldehyde (% normal) concentrations (± SEM) of rat brain after cold injury; n = 6-9 per
determination

In the course of various brain insults resulting in formation of
focal vasogenic edema, the mechanism of membrane disruption seems to
be different. Although regional blood flow is reduced in such lesions, in
the majority of cases it is not lowered below the threshold causing im-
balances of Ca^{++}, whereas free radical reactions seem to play a predominant role. This type of membrane damage is well documented in a
number of models. Since free radicals are very unstable, direct proof of
their action is difficult to obtain. We must rely on indirect evidence of
peroxidative damage. Malonaldehyde seems to be the best marker, which
is an end-product of phospholipid breakdown through peroxidative pathways. In the cryogenic brain injury model, malonaldehyde rises abruptly
during the first 24 hours following the lesion strictly parallel to the
increase of the tissue water content (Fig. 1). Similar results have been
reported elsewhere in the same model (12). More indirect evidence is also
available, as e.g. on protection by antioxidants (3), on reduced gluta-
thione concentrations (13), abnormal consumption of ascorbic acid, and on
a predominant loss of cholesterol and arachidonic acid (14). Cor-

responding observations have been made in compression edema, or spinal cord trauma (15,16). In traumatic insults, the presence of extravasated blood in the damaged parenchyma carrying metal complexes seems to be important for the initiation of free radical damage (2). Injection of iron compounds directly into brain tissue produces a typical focus of edema with abnormal production of malonaldehyde (17). Though further experimental studies and a more detailed analysis of phospholipid metabolism are necessary, it seems reasonable to conclude that membrane damage occurs in vasogenic injury, and that free radical reactions are the best explanation at present.

CONSEQUENCES OF MEMBRANE BREAK-DOWN FOR PHOSPHOLIPID DEPENDENT ENZYMES

In a severe lesion, e.g. within the core of an ischemic focus, abrupt and extensive damage of membranes rapidly leads to a complete loss of structure and function. In milder cases, especially around a focus surrounded by vasogenic edema, disorders of membrane function are conceivably more subtle. Using cryogenic injury of the brain in rats or rabbits as a model of a confined lesion, our group investigated the consequences of membrane damage on enzyme function. A wide number of enzymes are protein complexes undergoing configurational changes which strictly depend on the integrity of the lipid environment. We have been particularly interested in enzymes involved in energy production and cation transport. Since details of these studies have already been published, a brief overview is presented (18-22).

THE ENERGY PRODUCING SYSTEM

1. Krebs Cycle

The main form of available chemical energy of a cell is ATP mostly produced by oxidative phosphorylation in the mitochondria. ADP is phosphorylated via the Krebs cycle by two different reactions:

1. Substrate level phosphorylation. This concerns only one step of the Krebs cycle, where the chemical energy of succinyl-CoA is directly used for nucleotide phosphorylation without participation of ATP synthetase.
2. Oxidative phosphorylation where reduced coenzymes, e.g. NADH, FADH obtained at different steps of the Krebs cycle are reoxidized by the membrane ATP-synthetase complex for ADP-phosphorylation.

Both types of phosphorylation were analyzed 72 hrs after cold injury (Fig. 2).

(a) Alpha-ketoglutarate was used to measure substrate level phosphorylation. Oxidative decarboxylation of alpha-ketoglutarate catalyzed by alpha-ketoglutarate decarboxylase produces succinyl-CoA and NADH. Alpha-ketoglutarate decarboxylase is a matrix enzyme without membrane association. It was found slightly inhibited, i.e. to 80% of the control level.

(b) For analysis of the level of oxidative phosphorylation we measured the rate of ATP-synthesis using succinate as substrate. This reaction was drastically inhibited, i.e. to 35% of the control level. Inhibition could have resulted from an impairment of one or several enzymes involved in ATP-synthesis, such as the cytochromes, succinate dehydrogenase, and the ATP-synthetase complex. These enzymes are strictly dependent on the integrity of the lipid structure.

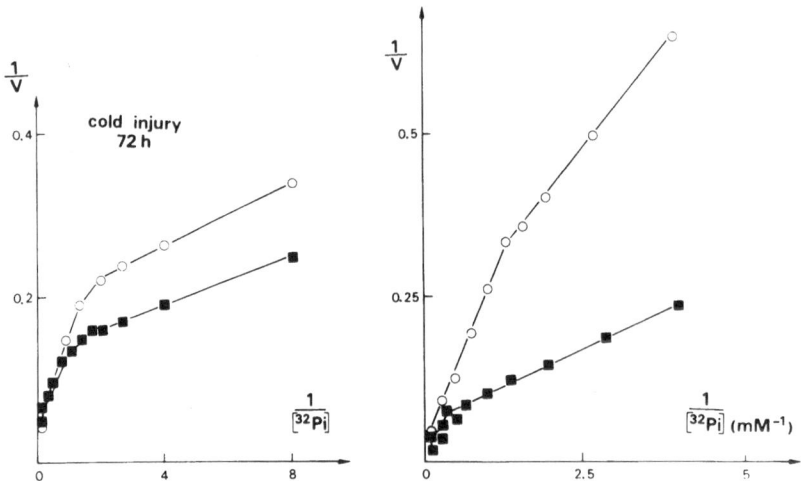

Fig.2: Phosphorylation rate as double-reciprocal plot of **V** and $^{32}P_i$. Left: substrate level phosphorylation with α-ketoglutarate as substrate plus oligomycin and carbonyl-cyanid-chlorophenylhydrazone (CCCP) as an uncoupler; right: oxidative phosphorylation with succinate as substrate plus rotenone. Filled squares: normal rabbit brain; open circles: injured brain 72 hrs after cold lesion.

Separate activity measurements of each of these enzymes revealed the succinate dehydrogenase reaction as limiting step. Succinate dehydrogenase was inhibited by 65% as compared to the controls. Under these circumstances the energy producing machinery was able to bypass the inhibiting step by using vicarious cycles, such as the aspartate-malate shuttle (Borst cycle; 22).

2. Mitochondrial ATPase (ATP-Synthetase)

We have measured total ATPase activity of mitochondria which were isolated at various intervals after cold injury. In injured hemispheres the activity was impaired to about 30% of normal suggesting losses of the ATPase-complex. Qualitative changes were observed in addition. The membrane-bound component of the ATPase-complex, which actually is involved in ATP-synthesis, can be determined by using oligomycin. The sensitivity to oligomycin was found markedly reduced. 48 hrs after cold injury, oligomycin-sensitive ATPase activity was reduced to 25% of the control level (21).

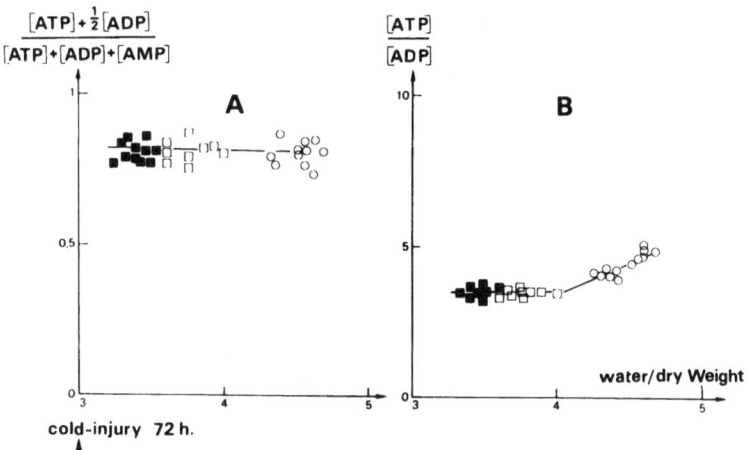

Fig.3: Relationship between energy-charge potential (left), or ATP/ADP ratio (right), respectively and the cerebral water content given as wet-/dry-weight ratio of rabbit brain with cold injury edema. Filled squares: normal brain; open circles: injured hemisphere; open squares: contralateral hemisphere

3. Available Energy

Although our in vitro studies on isolated mitochondria have demonstrated an impairment of several enzyme activities involved in energy production, the rate of ATP-production in vivo was not known. We measured therefore the energy charge potential as defined by: (ATP + 1/2 ADP)/(ATP+ADP+AMP). As seen in Fig. 3a, the energy charge potential was essentially normal. However, since this parameter is unlikely to provide an accurate information on the rate of oxidative phosphorylation, we studied the ATP/ADP-ratio in addition as the most straightforward index of the balance of regeneration and utilization of ATP in vivo. The

ATP/ADP ratio found in edematous tissue was 4.9 as compared to a control ratio of 3.5 (Fig. 3b). The control value is in good agreement with other findings.

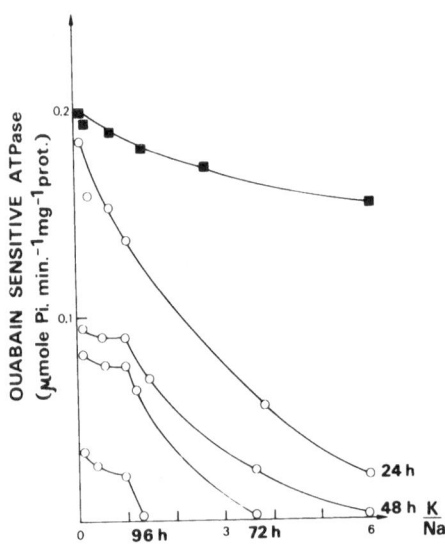

Fig.4: Effect of changes of the K^+/Na^+-ratio in the incubation medium on ouabain-sensitive Na^+-K^+-ATPase of normal brain (filled squares), or after cold injury (open circles) at 24, 48, 72 or 96 hrs after trauma.

4. Na^+-K^+-ATPase

The ouabain sensitive Na^+-K^+-ATPase activity decreased steadily with time after cold injury. The enzyme activity was measured as a function of the K^+/Na^+-ratio in the incubating buffer. The concentrations of K^+ and Na^+ were modified in such a manner that the sum remained constant. In samples of normal brain the activity fell slightly but significantly by raising the K^+/Na^+-ratio of the incubation medium. By increasing the K^+/Na^+-ratio the decrease of ATPase activity was far more pronounced in samples taken from edematous hemispheres indicative of a markedly changed affinity of the enzyme for these cations. Moreover, when studied at a physiological K^+/Na^+-ratio, the Na^+-K^+-ATPase activity was reduced to 20% of the control level (Fig. 4). Similar findings were obtained with a purified preparation of synaptosomes (18). There, Na^+-K^+-ATPase activity was also strongly inhibited by changing the K^+/Na^+-ratios (Fig. 5).

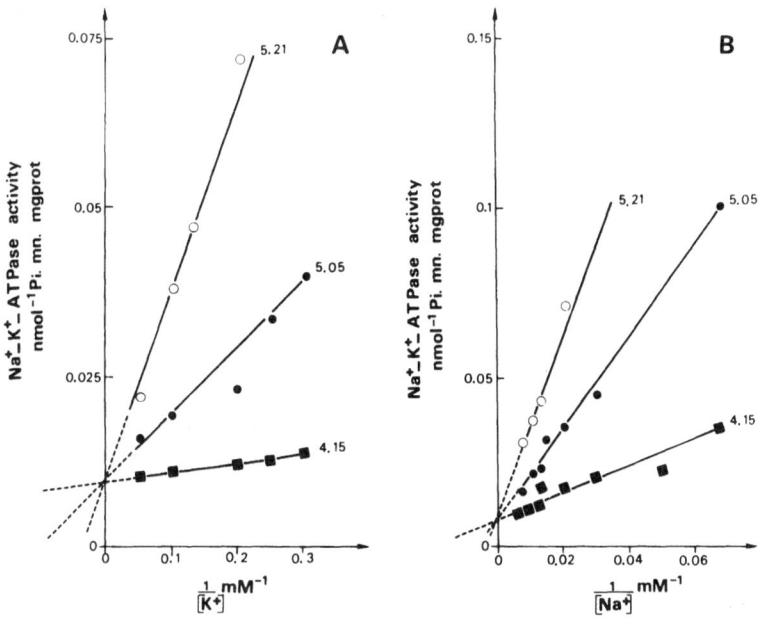

Fig.5: Double-reciprocal plot of Na^+ -K^+ -ATPase activity of synapto-somal preparations of cold-injured rabbit brain as a function of K^+ (left), or of Na^+ (right). Filled squares: normal brain; filled circles: moderate edema; open circles: severe edema. The cerebral water contents (wet-/dry-weight ratio) are indicated at the regression lines.

DISCUSSION

Two subjects emerge from these data. One is the normal level of cellular energy, the other the impairment of Na^+ -K^+ -ATPase activity. Results reported in the literature on the bioenergetics in vasogenic edema are somewhat controversial (23,24,25). Earlier studies have pointed out that regional cerebral blood flow is reduced in association with a decrease of the energy stores. Actually, a reduction of blood flow below 45%, which is the level considered as the threshold for disturbances of energy metabolism (26), has never been reported, even not in cases with a considerable increase of the brain water content. The tissue pO_2 was found essentially normal in edematous foci, and recent studies report normal energy levels (27,28). Despite the severe impairment of mito-chondrial enzyme activity (cf. above), no shortage of energy was detect-able in our studies. The apparently paradoxical result may be reconciled in part by assuming that the breakdown of some key enzymes was compensated for by metabolic regulation. The intervention of the aspartate-malate shuttle may serve as an example. Alternatively, the increased level of available ATP suggests energy utilization to be more affected than energy production. It is conceivable that despite a lowered energy production the energy stores remained filled, because hydrolysis of ATP was severely impaired as a result of inhibition of Na^+-K^+-ATPase.

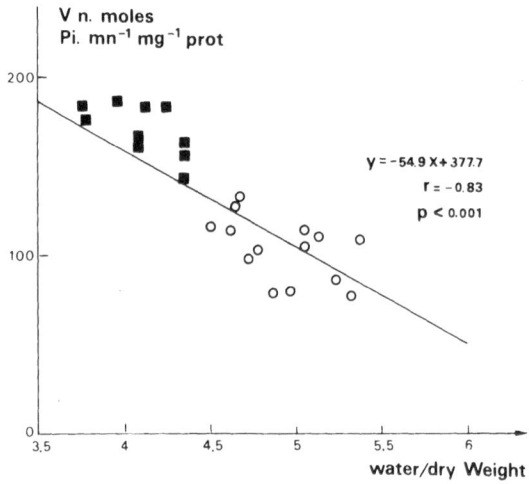

Fig.6: Regression of ouabain-sensitive Na^+-K^+-ATPase activity and cerebral water content (wet-/dry-weight ratio) of rabbit brain with cold injury edema. Filled squares: normal brain (wet-/dry-weight: 3.8 - 4.3); open circles: edematous brain (wet-/dry-weight: 4.5 - 5.4).

The impairment of Na^+-K^+-ATPase activity can be considered in addition to inhibit active cation transport, eventually leading to an intracellular retention of Na^+, Cl^- and, consequently, water. This mechanism probably accounts for the intracellular component of brain edema (1,29,30,31). The percentage of brain water localized in the cells has never been precisely assessed under these circumstances. Cryogenic edema is considered as an extracellular edema caused by the leakage of plasma constituents through the broken blood-brain barrier. The latter phenomenon would be independent from Na^+-K^+-ATPase. Still, a significant correlation between the decrease in activity of the enzyme and the increase of tissue water content was noticeable (Fig. 6). An impairment of Na^+-K^+-ATPase function has been described in similar models (32,33). It seems to be caused by peroxidative damage of membrane lipids. In vitro studies have shown inactivation of the enzyme by peroxidative reactions (34,35). In vivo, polyunsatured fatty acids and their degradation products inhibit Na^+-K^+-ATPase leading to a type of edema which is similar to the cryogenic type (36,37).

CONCLUSIONS

The cryogenic lesion and a number of comparable clinical conditions, such as focal injury, abscess etc. induce membrane damage which presumably is mediated predominantly by free radical reactions. The resulting breakdown of lipids disturbs the function of membrane-bound enzymes. Although these disturbances seem to be of little consequence for the mitochondria, e.g. the energy stores remain in excess of the need, they may have more serious consequences by inhibiting Na^+-K^+-ATPase resulting in formation of cytotoxic brain edema, and in disturbances of nerve function. On the other hand, this type of membrane damage per se does not seem to entail serious consequences. Severe disruption of membrane structure and function would only occur in areas where cerebral blood flow is reduced below the threshold causing disorders of the Ca^{++}-balance. This, however, would require an enormous degree of edema.

REFERENCES

1. Ishii S, Brain-swelling. Studies of structural, physiological and biochemical alterations, in: "Head injury", Caveness WE, Walker AE, eds., J.B. Lippincott Comp., Vol. 1 (1972).

2. Demopoulos HR, Milvy P, Dalari S, Ransohoff J, Molecular aspects of membrane structure in cerebral edema, in: "Steroids and brain edema", Reulen HJ, Schürmann K, eds., Springer, Berlin, Heidelberg, New York (1972).

3. Ortega BD, Demopoulos HB, Ransohoff J, Effect of antioxidants on experimental cold-induced cerebral edema, in: "Steroids and brain edema", Reulen HJ, Schürmann K, eds., Springer, Berlin, Heidelberg, New York (1972).

4. Cohadon F, Alternations des membranes cellulaires dans les situations d`agression aigue du parenchyme cerebral, Neurochirurgie 30:69 (1984).

5. Bazan NG, Bazan HEP, Kennedy, WG, Regional distribution and rate of production of free fatty acids in rat brain, J Neurochem 18:1387 (1971)

6. Bazan NG, Rodriguez de Turco EB, Membrane lipids in the pathogenesis of brain edema: Phospholipids and arachidonic acid, the earliest membrane components changed at the onset of ischemia, in: Adv Neurol 28: "Brain edema", Cervos-Navarro J, Ferszt R, eds., Raven Press (1980).

7. Hass WK, Beyond cerebral blood flow, metabolism and ischemic thresholds: an examination of the role of calcium in the initia-

tion of cerebral infarction in: "Cerebral vascular disease",
Meyer JM, Lechner H, Reivich M, Ott EO, Aranibar A, eds.,
Excerpta Medica (1981).

8. Goldman SS, The role of calcium on the cellular response following
   injury to the nervous system, in: "Head injury: Basic and clinical aspects", Grossman RG, Gildenberg PL, eds., Raven Press
   (1982).

9. Flamm ES, Demopoulos HB, Seligman ML, Poser RG, Ransohoff J,
   Free radicals in cerebral ischemia, Stroke 9, 5:445 (1978).

10. Rehncrona S, Westerberg E, Akesson B, Siesjö BK, Brain cortical
    fatty acids and phospholipids during and following complete and
    severe incomplete ischemia, J Neurochem 38:84 (1982).

11. Yoshida S, Inoh S, Asano T, Effect of transient ischemia on free
    fatty acids and phospholipids in the gerbil brain. Lipid peroxi-
    dation as possible cause of postischemic injury J Neurosurg
    53:323 (1980).

12. Suzuki O, Yagi K, Formation of lipoperoxide in brain edema induced
    by cold injury, Experienta 30:248 (1974).

13. Rap ZM, Wideman J, Changes in the sulfhydryl group level and in-
    fluence of exogenous glutathione on dynamics of vasogenic
    brain edema, in: "Dynamics of brain edema", Pappius HM,
    Feindel W, eds., Springer, Berlin, Heidelberg, New York (1976).

14. Mitamura JA, Seligman ML, Solomon JJ, Flamm ES, Demopoulos HB,
    Ransohoff J, Loss of essential membrane lipids and ascorbic
    acid from rat brain following cryogenic injury and protection by
    methylprednisolone, Neurol Research 3, 4:329 (1981).

15. Yoshida S, Busto R, Ginsberg MD, Kouichi ABE, Martinez E, Watson
    BD, Scheinberg P, Compression-induced brain edema: modifica-
    tion by prior depletion and supplementation of vitamin E,
    Neurol 33:166 (1983).

16. Demopoulos HB, Flamm ES, Pietronigro DD, The free radical pathology
    and the microcirculation in the major central nervous system
    disorders, Acta Physiol Scand, Suppl. 492:91 (1980).

17. Wilmore LJ, Rubin JJ, Formation of malonaldehyde and focal brain
    edema induced by subpial injection of $FeCl_2$ into rat isocortex,
    Brain Research 246:113 (1982).

18. Averet N, Rigoulet M, Cohadon F, Modifications of synaptosomal Na^+
    $-K^+$-ATPase activitiy during vasogenic brain edema in the rab-
    bit, J Neurochem 42:275 (1983).

19. Averet N, Rigoulet M, Cohadon F, Effet du naftidrofuryl sur l'oedeme
    cerebral vasogenique chez le lapin, Circulation et Metabolisme
    du Cerveau 2:145 (1983).

20. Cohadon F, Rigoulet M, Averet N, Alterations of membrane-bound
    enzymes in vasogenic edema, in: "Recent progress in the study
    and therapy of brain edema", Go KG, Baethmann A, eds.,
    Plenum Press, New York (1984).

21. Rigoulet M, Guerin B, Cohadon F, Vandendriessche M, Unilateral
    brain injury in the rabbit: reversible and irreversible damage
    of the membranal ATPases, J Neurochem 32:535 (1979).

22. Rigoulet M, Averet N, Cohadon F, Energy-producing machinery in

vasogenic brain edema, Neurochem Pathol 1:43 (1983).
23. Herrmann HD, Dittmann J, Examination of the metabolism of
    oedematous brain tissue, Acta Neurochir 22:167 (1970).
24. Frei HJ, Wallenfang Th, Poll W, Reulen HJ, Schubert R, Brock M,
    Regional cerebral blood flow and regional metabolism in cold
    induced oedema, Acta Neurochir 29:15 (1973).
25. Schmiedek P, Baethmann A, Sippel G, Oettinger W, Enzenbach R,
    Marguth F, Brendel W, Energy state and glycolysis in human
    cerebral edema, J Neurosurg 40:351 (1974).
26. Grubb RL, Raichle E, Phelps ME, Ratcheson RA, Effects of ICP on
    cerebral blood volume, blood flow and oxygen utilization in
    monkeys, J Neurosurg 43:385 (1975).
27. Grote J, Reulen HJ, Schubert R, Increased tissue water in the brain:
    influence on regional cerebral blood flow and oxygen supply,
    in: Adv Neurol 20: "Pathology of cerebrospinal microcircula-
    tion", Cervos-Navarro J, Betz E, Ebhardt G, Ferszt R, Wüllen-
    weber R, eds., Raven Press (1978).
28. Sutton LN, Welsh F, Bruce DA, Bioenergetics of acute vasogenic
    edema, J Neurosurg 53:470 (1980).
29. Torack R, Terry RD, Zimmermann HM, The fine structure of cere-
    bral fluid accumulation. 1 - Swelling secondary to cold injury,
    Am J Path 35:1135 (1959).
30. Blakemore WF, The ultrastructural appearance of astrocytes following
    thermal lesions of the rat cortex, J Neurol Sci 12:319 (1971).
31. Foroglou Ch, Dolivo M, Zander E, Foroglou G, Etude sur l'oedeme
    cerebral en microscopie electronique. Brain Research 38:267
    (1972).
32. Clendenon NR, Allen N, Gordon WA, Bingham WG, Inhibition of
    Na^+-K^+ activated ATPase activity following experimental spinal
    cord trauma, J Neurosurg 49:563 (1978).
33. Hall ED, Braughler JM, Effects of intravenous methylprednisolone on
    spinal cord lipid peroxidation and Na^+-K^+-ATPase activity.
    Dose-response analysis during 1st hour after contusion injury in
    the cat, J Neurosurg 57:247 (1982).
34. Sun AY, The effect of lipoxidation on synaptosomal Na^+-K^+
    -ATPase isolated from the cerebral cortex of squirrel monkey,
    Biochem Biophys Acta 266:350 (1972).
35. Kovachich GB, Mishra OP, Partial inactivation of Na^+-K^+-ATPase
    in cortical brain slices incubated in normal Krebs-Ringer
    phosphate medium at 1 and 10 ATM oxygen pressures,
    J Neurochem 36:333 (1981).
36. Chan PH, Fishman RA, Brain edema: induction in cortical slices by
    polyunsaturated fatty acids, Science 201:358 (1978).
37. Chan PH, Fishman RA, Caronna J, Schmidley JW, Prioleau G, Lee J,
    Induction of brain edema following intracerebral injection of
    arachidonic acid. Ann Neurol 13:625 (1983).

MODIFICATIONS OF cAMP AND CREATINE KINASE-ISOENZYMES IN CSF IN EXPERIMENTAL HEAD INJURY

Raffael Vara-Thorbeck and M. Ruiz-Morales

Catedra de Patologia Quirurgica, II. Facultad de Medicina
Universidad de Granada, Granada, Spain

INTRODUCTION

In previous papers we have demonstrated an increase in activity of various enzymes in CSF of patients suffering not only from traumatic brain injury but also from other cerebral diseases, such as tumors, meningoencephalitis, stroke etc. (21,27,29,30,31,37). Recent studies using computerized tomography have emphasized that brain swelling can occur immediately within 10 - 20 minutes after head injury (15,38). The etiology of acute brain swelling following head injury is the subject of many, also recently conducted studies (4,7,17). Vascular dilatation (engorgement), cerebral edema (Hirnödem), or a combination of both seem to offer a most likely explanation (17). Therefore it is our opinion that cold injury of the brain which is extensively used since the studies of Clasen (5) does not seem to be a valid model, because of the etiological dissimilarity with traumatic brain swelling of patients. Aim of the present study was to analyze immediate changes of cAMP and creatine kinase (CK) in CSF employing an experimental model of cerebral concussion and contusion.

MATERIAL AND METHODS

The experiments were carried out in 45 mongrel dogs of 16-20 kg bw. Three groups were formed, controls (n=15) and two experimental groups. In one group cerebral concussion was induced by a force of 56.95 J, in another a large contusion by application of 340.31 J (Table 1). Cerebral injury was produced by a gradual mechanical impact above the midline of the intraparietal suture of the animal's head, using a mechanical gun (Spitmatic, Decoletage and Extrusion, S.A. Ltd., Hospital de Llobregat, Barcelona) with a special projectile (29; Fig. 1). Under light anaesthesia by an intravenous pentobarbital drip plus 0.25 mg atropine, the animal was placed on the operating table in the horizontal position. The animal's head rested on foam rubber. This was to permit free movements of the skull upon impact and to avoid the head being crushed. Thereby we approached a model which is close to human head injury.

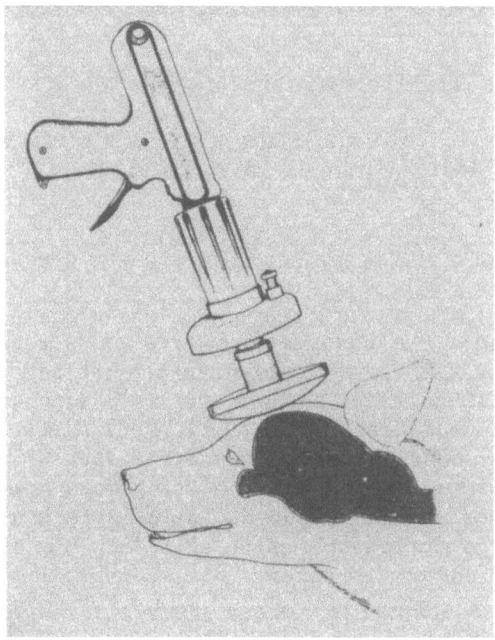

Fig.1: Method of induction of experimental concussion, or contusion to the unrestrained skull by mechanical impact.

The dynamic force of the impact was determined by considering the weight and maximal initial velocity of the projectile. The projectile weighed 2.5 g and reached a velocity of 225 m/sec in group 1, or even 550 m/sec in group 2. The energy which was absorbed by the skull was calculated as 63.3 J in group 1, and 378.1 J in group 2. We considered a transference coefficient of 0.9 to compensate for attenuation of the impact. The energy actually absorbed by the skull therefore was 59.9 J in group 1 with concussion, and 340.2 J in group 2 with contusion.

Table 1: Mechanical energy given in Joule (Julius) which was absorbed by the skull of experimental animals subjected to cerebral concussion (group 1), or cerebral contusion (group 2). The number of animals is given in the column at the right.

GROUPS	CRANIAL - ENCEPHALIC MECHANICAL IMPACT	ANIMALS
CONTROL	———	15
1	56,95 Julius	15
2	340,31 Julius	15

In order to determine the neurological sequelae produced by the impact the following criteria were employed:

1. Clinical:

a) Response to painful stimuli

b) Corneal and pupillary reflexes

c) Respiration was spontaneous throughout the experiment. All animals in group 2 had respiratory arrest immediately after injury requiring artificial ventilation for 5-10 minutes.

d) Intracranial pressure. All animals were monitored prior to and after impact using a Philips epidural pressure system.

2. Pathological Anatomy:

In group 2 the animals were subjected to perfusion fixation according to (24) when brain stem herniation had occurred. This was diagnosed by a failure to obtain CSF by cisternal puncture and respiratory arrest. Animals of group 1 were perfused at the end of the experimental observation period. Macro- and microscopical aspects of the lesion were studied in frontal sections cut through the maximal diameter of the brain. The results will be reported in a separate article.

3. Creatine kinase and cAMP:

Creatine kinase (CK) and cyclic adenosine 3'-5' monophosphate (cAMP) were measured in CSF samples of 1.4 - 1.6 ml obtained by percutaneous cisternal puncture. CSF was drawn 1, 24, 48, 72 and 168 hours after injury in both experimental groups, and at the same intervals in the controls. The CK-activity in CSF was spectrophotometrically measured at 25° C immediately after sampling. The CK isoenzymes were separated by mini-column chromatography on DEAD Sephadex A 50 according to (20). The values are given as international units per liter (U/1). cAMP was measured by a competitive protein binding assay described by Tovey et al. (35). Reagents were obtained freeze-dried from the Radiochemical Centre, Amersham, England. The values are given as nMol/1.

4. Statistical Analysis:

Results obtained on CK-activity and cAMP in CSF were subjected to factorial variance analysis. The t-test according to Newman and Keuls using Tukey's method, or the Scheffe test were employed.

Table 2: Epidural pressure (mmHg) of experimental animals with concussion (group 1), or contusion (group 2), respectively from 1 to 168 hrs after trauma. Only animals of group 2 with contusion developed an intracranial pressure rise. No animal of this group survived longer than 96 hrs.

SAMPLE HOUR / GROUP	1/1	2/24	3/48	4/72	5/168
CONTROL	7.80 ± 3.78	6.33 ± 3.04	7.33 ± 4.32	7.13 ± 3.74	5.87 ± 2.80
1	8.86 ± 2.92	7.20 ± 3.05	6.80 ± 2.51	6.40 ± 3.44	5.33 ± 2.79
2	○ ● 14.33 ± 4.63	○ ● 28.93 ± 12.16	○ ● 26.53 ± 8.25	○ ● 17.36 ± 5.36 (N = 14)	———— (N = 0)

RESULTS EXPRESSED AS MEAN ± STANDARD ERROR OF THE MEAN.

SIGNIFICANT DIFFERENCES (p<0,01) :

    ○   FROM GROUP 2 TO CONTROL VALUES

    ◉   "   "   1   "   "   "

    ●   "   "   2   "   GROUP 1

RESULTS

1. Clinical Observations:

(a) In group 1 with concussion the corneal and pupillary responses disappeared after trauma, and the animals did not react to painful stimulation. After recovery from anesthesia the animals remained stuporous for 10-15 minutes. Thereafter they behaved and fed normally. There was no evidence of focal or generalized neurological dysfunction. No animal died spontaneously in this group. Perfusion fixation was conducted for neuroanatomical investigations one week after the experiment. The intracranial pressure was never abnormal during the total experimental observation period.

Table 3: Creatine kinase activity in CSF (U/l) of control dogs and of experimental animals with concussion (group 1), or contusion (group 2) 1 to 168 hrs after trauma. Creatine kinase activity was significantly increased only in animals with contusion.

SAMPLE HOUR / GROUP	1/1	2/24	3/48	4/72	5/168
CONTROL	10.13 ± 3.23	8.53 ± 1.55	8.73 ± 3.17	7.86 ± 2.85	8.73 ± 2.25
1	16.66 ± 7.99	10.73 ± 4.01	10.46 ± 4.21	8.66 ± 5.04	9.73 ± 4.67
2	o ● 69.46 ± 15.08	o ● 484.73 ± 79.46	o ● 401.46 ± 82.57	o ● 339.07 ± 43.19 (N = 14)	———— (N = 0)

RESULTS EXPRESSED AS MEAN ± STANDARD ERROR OF THE MEAN.

SIGNIFICANT DIFFERENCES ($p < 0.01$):

    o   FROM GROUP 2 TO CONTROL VALUES

    ◒   "   "   1 "   "   "

    ●   "   "   2 " GROUP 1

(b) In group 2 with contusion, reflexes disappeared after trauma as in group 1, however, respiratory arrest occurred in addition for 30-60 seconds. Therefore artificial respiration was administered for 5-10 minutes. In animals with extended apnea subarachnoidal haemorrhage was found. These animals were considered as instantly dead and rejected from the study. All animals of group 2 died after 48-96 hours from brain stem herniation. This was concluded from (a) the increase of ICP, (b) the failure to obtain CSF by cisternal puncture, and (c) from respiratory arrest. In this group ICP rose significantly post injury (Table 2).

2. Creatine Kinase- and Isoenzyme-Activities in CSF:

Group 1: Neither total CK-activity nor the different isoenzymes (CK-BB, CK-MB and CK-MM) were significantly changed in CSF when compared with the control group.

Group 2: The total CK-activity in the sample drawn one hour after trauma was significantly increased. The increase of enzyme activity reached a plateau 24 hours later, which was maintained until death. The increase in total CK-activity largely resulted from an increase of the isoenzyme CK-BB, representing 86.8 % in the first sample, 94.6 % in the second, and 97.7 % in the third and fourth samples. The increase in total activity and of CK-BB was statistically significant when compared with group 1 and the controls (Table 3).

3. cAMP-Concentrations in CSF:

Group 1: cAMP was significantly increased already in the first sample drawn one hour after trauma, and remained at this level over more than 48 hours. The concentrations decreased then to normal between 72 and 168 hours after trauma.

Group 2: The cAMP levels in CSF increased dramatically upon injury attaining a statistically significant level, not only in comparison with the controls but also with group 1. The increase persisted during the first 24 hours, but fell later to a subnormal level before death (Table 4).

4. Pathological Anatomy:

Group 1: The brains of the animals of this group appeared normal upon macroscopical inspections. There was no evidence of swelling or haemorrhage. However, microscopical evidence of diffuse axonal injury in the white matter was obtained in accordance with the findings reported by Adams et al. (1).

Group 2: Macroscopical investigations revealed cortical contusions at the site of impact and at the temporal poles. Diffuse parenchymatose haemorrhages and large brain swelling were seen in addition which was confirmed by light-microscopy.

Table 4: cAMP (nM/l) in cerebrospinal fluid of control and experimental animals with concussion (group 1) or contusion (group 2), respectively. cAMP was significantly increased initially after trauma in both experimental groups, but fell to subnormal levels in animals with severe contusion (group 2).

SAMPLE HOUR / GROUP	1/1	2/24	3/48	4/72	5/168
CONTROL	18.26± 4.96	16.06±3.43	17.86±4.75	18.33± 2.16	18.66±3.24
1	33.33±8.67 ⊖	30.26±8.13 ⊖	24.26±5.38 ⊖	22.93±8.03	17.60±7.45
2	55.46±7.96 ○ ●	34.80±11.69 ○ ●	5.6±2.82 ○ ●	1.07±0.73 ○ ● (N=14)	——— (N=0)

RESULTS EXPRESSED AS MEAN ± STANDARD ERROR OF THE MEAN.

SIGNIFICANT DIFFERENCES (p<0.01):

○	FROM GROUP	2	TO	CONTROL	VALUES
⊖	" "	1	"	"	"
●	" "	2	"	GROUP 1	

DISCUSSION

The first CT-studies on head injury made clear that cold injury is not the most appropriate and comprehensive model to study traumatic brain damage experimentally. Therefore, we have designed an experimental dynamic model employing impact of the unrestrained skull similar to that of Tornheim and McLaurin (33). By controlling the impact force we are able to produce a concussion, or contusion, respectively. The clinical and neuropathological results are in good agreement with those of others (1,2,9,10,11,12,18,19,23,34). Our working hypothesis is that brain injury induces release of cAMP and of intracellular enzymes. Since studies of Reulen (26,36) have demonstrated that vasogenic edema resolves by clearance into CSF spaces, we expected changes of enzyme activity and of cAMP-levels in CSF obtained by cisternal puncture (21,27,37). In earlier studies we determined in CSF the activity of other enzymes, e.g. LDH, MDH, ALD, GOT, etc. Currently, we have considered measurements of creatine kinase of particular interest, since the isoenzyme CK-BB is not present in serum or CSF of normal subjects (14). Moreover, recent studies have demonstrated CSF-levels of the enzyme to correlate with various cerebral disorders and changes in intracranial pressure, etc. (6,13,15,29,30).

Further, the concentration of cyclic adenosine 3'-5' monophosphate (cAMP) is particularly high in central nervous system (16) indicative of an important role of this nucleotide in the brain. The cyclic nucleotide does not cross the blood-CSF barrier. The concentrations found may therefore reflect release of the compound from nervous tissue (3). Decreased levels of cAMP have been reported in the ventricular fluid of patients with prolonged coma after head injury (28). Data obtained in animal studies appear inconclusive, however. The changes of CK-activity in CSF, particularly of its isoenzyme CK-BB observed in the present study, may accurately reflect the amount of damaged brain tissue in agreement with former results obtained by others (6,13,25,29,30). As far as changes of the cAMP-concentration in CSF are concerned, we agree with Somlyo (32) that an increase of cAMP and the deterioration of the energy state in brain injury may affect the permeability characteristics of cell membranes. This could facilitate leakage of intracellular substances, such as cAMP from the brain parenchyma into CSF (8,22,28,29,31,39). This concept is supported by our findings and suggests the cAMP-concentration in CSF as an excellent prognosis factor in severe head injury.

Taken together, the results obtained allow us to draw the following conclusions:
1. Differences in concentrations of cAMP in cerebrospinal fluid may be utilized as an indicator to differentiate cerebral concussion from contusion.
2. Creatine kinase-activity, particularly of its isoenzyme CK-BB, together with cAMP-levels are considered as prognostic indicators of the evolution and progression of secondary brain damage in head injury.

SUMMARY

A dynamic head injury model using a traumatic impact to the unrestrained head was studied in 30 dogs. The energy absorbed by the skull was either 59.95 J causing cerebral concussion, or 340.31 J leading to severe contusion. The latter was associated with brain swelling and an increase of ICP resulting in death of the animal within 48-96 hours. In animals with concussion only, no significant changes of the ICP or of CK-activity in the CSF including of its isoenzyme CK-BB were found. However, concentrations of cAMP rose significantly in CSF which continued for 48 hours after trauma. In animals with severe contusion, CK-activity - largely representing the CK-BB isoenzyme - rose dramatically in CSF and remained elevated until death of the animals. On the other hand, cAMP concentrations in CSF increased significantly after trauma, but fell to subnormal levels 48 hours later. We conclude that CK-activity, particularly of the CK-BB isoenzyme and cAMP concentrations in CSF are useful indicators to assess the prognosis and extent of damaged brain tissue in severe head injury.

REFERENCES

1. Adams JH, Mitchell DE, Graham DI, and Doyle D, Diffuse brain damage of immediate impact type. Its relationship to "primary brain-stem damage" in head injury, Brain 100:489 (1977).
2. Bakay L, Lee JC, Lee GC, and Peng JR, Experimental cerebral concussion. Part 1: An electron microscopy study, J Neurosurg 47:525 (1977).
3. Brooks BR, Engel WK, and Sode J, Blood-to-cerebrospinal fluid barrier for cyclic adenosine monophosphate in man, Arch Neurol 34:468 (1977).
4. Bruce DA, Sutton LN, and Schut L, Acute brain swelling and cerebral edema in children, in: "Brain edema", de Vlieger M, de Lange SA, Beks JWF, eds., J. Wiley Sons, New York (1981).
5. Clasen RA, Cooke PM, Pandolfi S, Boyd D, and Raimondi AJ, Experimental cerebral edema produced by focal freezing, J Neuropathol Exp Neurol 21:579 (1962).
6. Cooper PR, Chalif DJ, Ramsey JF, and Moore RJ, Radioimmunoassay of the brain type isoenzyme of creatine phosphokinase (CK-BB): A new diagnostic tool in the evaluation of patients with head injury, Neurosurg 12:536 (1983).
7. Duckrow RB, LaManna JC, Rosenthal M, Levasseur JE, and Patterson JL, Oxidative metabolic activity of cerebral cortex after fluid-percussion head injury in the cat, J Neurosurg 54:607 (1981).
8. Fleischer AS, Rudman DR, Fresh CB, and Tindall GT, Concentration of 3'5' cyclic adenosine monophosphate in ventricular CSF of patients following severe head trauma, J Neurosurg 47:517 (1977).
9. Gennarelli TA, and Thibault LE, Biomechanics of acute subdural hematoma, J Trauma 22:680 (1982).
10. Gurdjian ES, Recent advances in the study of the mechanism of impact injury of the head - a summary, Clin Neurosurg 19:1 (1972).
11. Gurdjian ES, "Impact head injury. Mechanistic, clinical and preventive correlations", Thomas, Springfield, Ill. (1975).
12. Hamberger A, and Rinder L, Experimental brain concussion, J Neuropathol Exp Neurol 25:68 (1966).
13. Hans P, Born JD, Chapelle JP, and Milboum G, Creatine kinase isoenzymes in severe head injury, J Neurosurg 58:689 (1983).
14. Henry PD, Roberts R, and Sobel BE, Rapid separation of plasma creatine kinase isoenzymes by batch absorption on glass beads, Clin Chem 21:844 (1975).
15. Kobrine AI, Timmins E, Fajjoub RK, Rizzoli HV, and Davis DO, Demonstration of massive traumatic brain swelling within 20 minutes after injury. Case report, J Neurosurg 46:256 (1977).
16. Krishna G, Forn J, Voigt K, Pauls M, and Gessa GL, Dynamic aspects of neurohormonal control of cyclic 3',5'-AMP synthesis in brain, Adv Biochem Psychopharmacol 3:155 (1970).
17. Lanksch W, Baethmann A, and Kazner E, Computed tomography of

brain edema, in: "Brain edema", de Vlieger M, de Lange SA, Beks JWF, eds., J. Wiley Sons, New York, (1981).

18. Liu HC, Lee JC, and Bakay L, Experimental cerebral concussion. A histochemical study, Acta Neurochir 47:105 (1979).

19. McLaurin RL, and Tornheim PA, Changes in tissue density and vascular permeability in the cerebral cortex following experimental closed head trauma, International Conference on Recent Advances in Neurotraumatology, Edinburgh, Abstracts (1982).

20. Mercer DW, Separation of tissue and serum creatine kinase isoenzymes by ion exchange column chromatography, Clin Chem 20:36 (1974).

21. Morales-Valentin OI, Ruiz-Morales M, and Vara-Thorbeck R, Prognosis value of the modification in the activity of some enzymes in brain injury, Eur Surg Res 11:35 (1979).

22. Myllylä VV, Effect of cerebral injury on cerebrospinal fluid cyclic AMP concentration, Eur Neurol 14:413 (1976).

23. Ommaya AK, and Gennarelli TA, Cerebral concussion and traumatic unconsciousness. Correlation of experimental and clinical observations on blunt head injuries, Brain 97:633 (1974).

24. Palay SL, and Chan-Palay V, Cerebellar cortex. Cytology and organization, Springer-Verlag, Berlin-Heidelberg-New York (1974).

25. Rabow L, and Hedman G, CK-BB isoenzymes as a sign of cerebral injury,
    Acta Neurochir Suppl. 28:108 (1979).

26. Reulen HJ, Tsuyumu, M, Tack A, Fenske AR, and Prioleau GR, Clearance edema fluid into cerebrospinal fluid. A mechanism for resolution of vasogenic brain edema, J Neurosurg 48:754 (1978).

27. Ros-Die E, Morales-Valentin OI, Suarez-Paneda JR, and Vara-Thorbeck R, Modificaciones de la actividad enzimatica de algunas enzimas del liquido cefalorraquideo humano en los traumatismo craneo-encefalicos, Rev Esp 147:377 (1977).

28. Rudman D, Fleischer A, and Kutner HM, Concentration of 3',5'-cyclic adenosine monophosphate in ventricular cerebrospinal fluid of patients with prolonged coma after head trauma or intracranial hemorrhage, N Engl J Med 295:635 (1976).

29. Ruiz-Morales M, Oscilaciones en sangre y LCR del cAMP e isoenzimas CK y LDH en el TCE experimental gradual; MD thesis, Granada (1981).

30. Ruiz-Morales M, Herrero-Mateo LM, and Vara-Thorbeck R, Creatine kinase isoenzymes in serum and CSF after gradual experimental brain injury. Its relation to prognostic value or to extent of brain damage, in: Piotrowski W, Brock M and Klinger M, eds., Adv Neurosurg 12:307, Springer, Berlin (1984).

31. Ruiz-Morales M, and Vara-Thorbeck R, Changes in CSF cAMP after gradual experimental brain injury: Prognostic value, Eur Surg Res 14:111 (1982).

32. Somlyo Ap, Somlyo AV, and Smieesko V, Cyclic AMP and vascular smooth muscle, in: "Advances in cyclic nucleotide research", Greengard P, Robinson GA, eds., Raven Press, New York (1972).

33. Tornheim PA, and McLaurin RL, Acute changes in regional brain water content following experimental closed head injury,

J Neurosurg 55:407 (1981).

34. Tornheim PA, McLaurin RL, and Thorpe JF, The edema of cerebral contusion, Surg Neurol 5:171 (1976).

35. Tovey KC, Oldham KG, and Whelan JAM, A simple direct assay for cyclic AMP in plasma and other biological samples used an improved competitive protein binding technique, Clinica chim Acta 56:221 (1974).

36. Tsuyumu M, Reulen HJ, and Prioleau G, Dynamics of formation and resolution of vasogenic brain oedema. I. Measurement of oedema clearance into ventricular CSF, Acta Neurochir 57:1 (1981).

37. Vara-Thorbeck R, Modificaciones de la actividad enzimatica del liquido cefallorraquideo (LCR) humano en los procesos tumorales, inflamatorios, traumaticos y vasculares encefalicos, Rev Clin Esp 101:100 (1966).

38. Waga W, Tochio H, and Sakakura M, Traumatic cerebral swelling developing within 30 minutes after injury, Surg Neurol 11:191 (1979).

39. Watanabe H, and Passonneau JV, Cyclic adenosine monophosphate in cerebral cortex. Alterations following trauma, Arch Neurol 32:181 (1975).

CELL SWELLING MECHANISMS IN BRAIN

Oliver Kempski

Institute for Surgical Research, Klinikum Großhadern
Ludwig-Maximilians University, 8000 München 70, FRG

Brain function and survival depend critically on a sensitive regulation of brain volume. Three brain compartments - intra-, extracellular (e.c.), and intravascular - have to be accomodated in their individual size to ensure constancy of the total brain volume. Their volumes can be affected by various cerebral disorders. An enlargement of the intra-, or extracellular space resulting in an increase of cerebral tissue water is defined as brain edema (80). In case of an accumulation of intracellular water, edema is cytotoxic (60). The vasogenic edema type is characterized by a breakdown of the blood-brain barrier and accumulation of extracellular protein-rich fluid. This review is restricted to swelling mechanisms concerning the intracellular compartment. It should be noted that cell swelling may occur at the expense of a shrinking extracellular space (ECS) without increasing total water content of the brain. Such a process would therefore not fulfill the definition of brain edema (s. above).

Cell swelling is found in many pathological states of the brain. The cells mostly affected are glial elements, especially astrocytes (60). Cerebral ischemia and anoxia are the most common and clinically relevant causes. Cellular swelling due to an osmotic derangement (dialysis dysequilibrium syndrome, rapid rehydration of hyperosmotic patients (14)), intoxications (hexachlorophene, triethyltin (55,72)), or metabolic disorders (Reye syndrome, hepatic encephalopathy (77)) are severe complications in clinical patients. Cell swelling in brain has been induced experimentally by metabolic inhibitors, such as dinitrophenol, or 6-aminonicotinamide (5,6), or by ouabain which inhibits the Na^+-/K^+-pump (29,67). Pathophysiology as well as morphological aspects of cytotoxic edema have been reviewed by Baethmann (9).

A general understanding of the nature of cytotoxic brain edema has not been reached yet. Although detailed information is available on many isolated phenomena accompanying cell swelling in brain, the different mechanisms involved have apparently never been synthetized in a comprehensive model. This article is an attempt to provide a basis to understand cell swelling in brain. For the sake of clarity, large portions of the literature must remain undiscussed.

CELL VOLUME MAINTENANCE AND REGULATION

General Aspects

Prior to a thorough discussion of cell swelling mechanisms it is necessary to briefly consider the basic physiology of cell volume maintenance. Our knowledge is limited, particularly in the case of nerve- and glial cells, since most studies focus on kidney, white or red blood cells. A distinction should be made between **volume maintenance,** which describes control of the steady state and **volume regulation** reserved for volume readjustment, e.g. after anisotonic exposure (37). Current concepts on cell volume maintenance have been summarized by Macknight and Leaf (70,71). Briefly, intra- and extracellular solutes and fluids in the steady state exist in a double DONNAN equilibrium, where the osmotic pressure of intracellular impermeable anions (proteins) attracting positively charged cations is balanced and compensated for by active sodium extrusion and low Na^+-permeability of the cell membranes. Critical elements are (a) the amount of anionic charges on intracellular macromolecules, (b) selective membrane permeabilities for the different small ions, and (c), most important, an active sodium transport system, the Na^+/K^+-pump driven by the Na^+-, K^+-activated, Mg^{++}-dependent ATPase (82).

However, observations questioning the double DONNAN hypothesis are available. Inhibition of Na^+ - K^+ - ATPase by ouabain does not necessarily cause cell swelling (37,43). It has, moreover, been postulated that the energy available is not sufficient to fuel all pumping- and transport-activity implicit in the DONNAN equilibrium. Several proposals have been made to explain these discrepancies (cf: 37,70). A most revolutionary concept, the association-induction hypothesis has been introduced by Ling and coworker. It is based on the assumption that intracellular water exists in a more organized state than extracellular water, that it is absorbed to a lattice of intracellular macromolecules (65). Affinity of water for potassium would be higher in this state than for sodium, which would explain an establishment of intra-extracellular ion gradients. Ling's model was developed from physical calculations and experiments on interactions between macromolecules and water. It ignores, however, the critical function of the cell membrane completely. Obviously the double DONNAN- and association-induction hypotheses are mutually exclusive concepts.

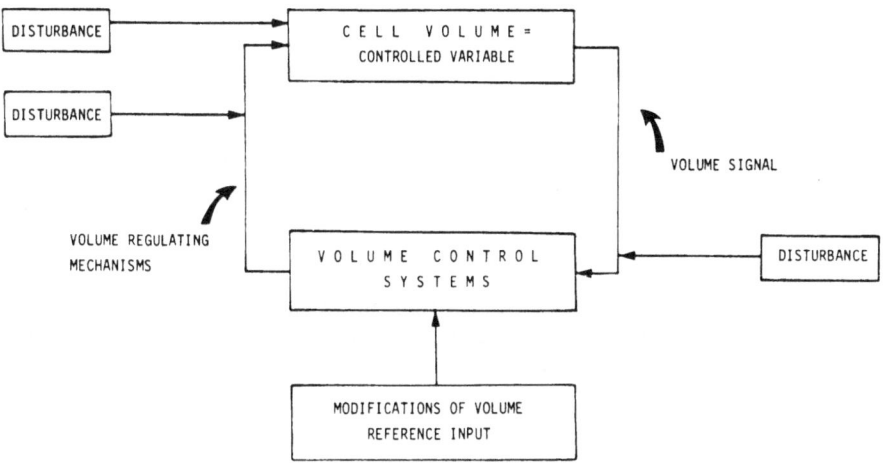

Fig. 1: Feedback loop of volume regulation, and potential sites where
disturbances may interact. Disturbances may interfere with the
volume control- and regulating mechanisms, distort the volume
signal, or induce modifications of the internal volume reference.
It may be noted that cell volume changes may have a function
in the maintenance of homeostasis, e.g. of low Ke^+ - and transmitter levels in the extracellular compartment.

Feedback Model of Volume Regulation

      The mechanisms which control cell volume efficiently can be viewed
as a feedback system continuously comparing the actual volume with an
internal reference. A respective model is proposed in Figure 1. A volume
signal of still undefined nature in a feedback loop communicates the
actual cell size to a volume control unit. Internal sodium concentration (82), cell pH (23), or intracellular Ca^{++} -levels (23) have been
proposed as volume signals. Ion pumps, or changes of membrane permeability are activated, if the actual volume differs from a given volume
standard. The nature of cell volume control is known only in fragments,
although its existence is evident from observations on osmotic swelling
and subsequent cell volume regulation (cf: below). In principle, four major mechanisms can be considered to be involved in cell swelling, or
shrinking according to the model shown in Figure 1:

1. DONNAN-swelling due to an increased membrane permeability for Na^+ -ions, or build-up of transcellular osmotic gradients. The feedback loop itself is not disturbed.
2. Inhibition of volume regulating mechanisms, e.g. blocking of the Na^+ / K^+ -pump, or of other transport systems.
3. Changes affecting the internal volume reference. As a consequence, the control system may actively initiate, or at least tolerate volume alterations. Respective examples are a volume increase of premitotic cells (34), or glial swelling during neuronal excitation.
4. Modifications, or distortions of the volume signal. Since the nature of the signal is not identified, it is premature to speculate about its disturbance. Nevertheless, modifications of the signal may be useful for therapeutic purposes.

Examples of disturbances of volume maintenance are discussed in the following sections to increase understanding of swelling mechanisms and volume regulation in the brain.

ANISOTONIC EXPOSURE

Hypotonic Swelling

A decrease of extracellular osmolality is always followed by inflow of water into cells. Due to a high conductivity of animal cell membranes to water, intra- and extracellular osmolalities equilibrate swiftly. Cells behaving like osmometers increase rapidly in volume. However, many cell types are capable to activate regulatory mechanisms within seconds after osmotic swelling causing reduction of cell volume back to the original size, although a low osmolality is maintained. This phenomenon is referred to as regulatory volume decrease (RVD, 22). It has been observed in many cell types (22,23,37,39,61,70) including glial and nerve cells (28,52,53). RVD in most cell types cannot be prevented by ouabain (23,52,61). Oxygen uptake of glial cells is stimulated suggesting RVD to require metabolic energy (52). RVD is accomplished in part by an increase of K^+ -permeability providing for cellular discharge of K^+ (22,53) and Cl^- (23). Measurements of the membrane potential in amphiuma red blood cells proved that ion fluxes involved in RVD are electroneutral (22, 23). Cala postulates (23) that cation fluxes occur in exchange for protons, while Cl^- is exchanged for HCO_3^- . Thus, changes in intra- and extracellular pH can be expected during RVD. Indeed, unbuffered medium becomes alkaline during RVD (22). Ca^{++} -ions probably play a role since phenothiazines which inhibit calmodulin block RVD in lymphocytes (29). This may indicate that Ca^{++} activates K^+ - channels (23). The existence of such channels has been demonstrated in neural tissue (33).

Hyperosmotic Shrinking

Some cells, but not glioma (52) are capable of regulating volume after hyperosmotic shrinking (regulatory volume increase, RVI; 22,23,61). Kregenow found RVI in duck red blood cells only, if e.c. K^+ exceeded 15 mM (cited from 23). During RVI, sodium is obviously exchanged for protons with the Cl^-/HCO_3^- -antiporter working in a reversed direction. Cl^- -ions enter cells while HCO_3^- is extruded. The mechanisms involved in RVI closely resemble phenomena observed during K^+ -induced swelling in the brain (see below).

SWELLING DURING NEURONAL EXCITATION

Shifts of water between the extra- and intracellular compartment have been reported during neuronal excitation. Light scattering indicative of cellular swelling was observed within 1 sec after administration of a train of excitatory stimuli to cerebral cortical slices (66). Cell swelling was reversible and accompanied by an enhancement of tissue respiration. It could be prevented by replacement of Cl^- with impermeant anions. Van Harreveld and Trubatch have shown in frog brain that stimulation by concentrated KCl results in a shift of water and electrolytes from extra- to intracellular space (92). In further studies the authors (88) used an atomizer to administer KCl vapor to the surface of blocks of frog brain, which were subjected then to instant freezing for electronmicroscopy. Postsynaptic structures increased from 4.8% to 14.2% of tissue volume. Swelling was explained by influx of Na^+ and Cl^- together with water into postsynaptic endings.

In 1980, Dietzel et al. reported repetitive stimulation of cat sensorimotor cortex to cause narrowing of the extracellular space of up to 50% (31). The authors used microelectrodes to measure concentrations of e.c. tetramethylammonium and choline for which cell membranes are largely impermeable. Shrinkage of the ECS was maximal where excitation-induced release of K^+ was highest. At superficial and very deep cortex layers shrinkage of the ECS was comparatively low. The authors explain this observation by a glial spatial buffering mechanism. This has already been proposed by Orkand et al.(reviewed in:76,78). According to this concept, extended or electrically coupled glial structures buffer inhomogenous elevations of Ke^+ (i.e. extracellular K^+) by complex transmembrane intra- and extracellular currents. Correspondingly, Dietzel found that stimulation attenuates the increase of Ke^+ by 35% during iontophoretic administration of potassium in standard amounts (31). Similar observations were made by Benninger (13). In another study, Dietzel measured Na^+ - and Cl^- -concentrations under identical conditions (32). They concluded that tissue osmolality increased for 30 mosm/l during stimulation, probably as a result of a break-down of large molecules secondary to an enhancement of tissue metabolism.

SWELLING INDUCED BY INCREASED EXTRACELLULAR POTASSIUM

Many investigators have reported pronounced swelling of glial elements during in-vivo, or in-vitro application of high K^+ -concentrations to brain tissue (cf: 43). Initially it was postulated that swelling can simply be attributed to uptake of K^+ by glial cells (16). Indeed, active glial uptake systems have been observed (44,86,95), and that K^+ -levels above 3 mM stimulate glial metabolism (79,87). ATP-concentrations of glial cells - but not of neurons - were reduced by 50% by increased Ke^+ -concentrations (83). Oxygen consumption reached maximal levels at 50 mM Ke^+ (43). Na^+ /K^+ -ATPase necessary for K^+ -pumping is present in glia (56). However, excessive K^+ -transport into cells may become limited by an increasing lack of intracellular Na^+ . Bourke, Nelson and Kimelberg studied K^+ -associated swelling of cerebral tissue for more than 15 years (16-21, 56-59). They provided experimental evidence which may require important modifications of the initially established concept. It was demonstrated that K^+ -induced swelling in the presence of HCO_3^- must be attributed to a net increase of Na^+ and Cl^- , rather than to an active accumulation of K^+ -ions. Furthermore, studies on Cl^- -movements during K^+ - stimulation (17-19, 57) revealed that uptake of 36-Cl^- is temperature dependent, a function of the extracellular K^+ -concentration, and that it can be inhibited by acetazolamide, an inhibitor of carbonic anhydrase. K^+ -dependent uptake of Cl^- is present in astrocytes but not in neuroblastoma cells (36). Recent findings demonstrate Cl^- uptake to be mediated by a Cl^- / HCO_3^- -antiporter in astroglia (1,57) similar to the anion exchanger observed in red blood cells (cf: 15,97). Kimelberg found a 1:1 exchange of Cl^- and HCO_3^- with saturation kinetics, which can be inhibited by the anion exchange blocker SITS (57), or acetazolamide (19). Evidence for the presence of a Na^+ / H^+ -antiporter was provided in the same study (57). The data suggest a complex interaction of ion fluxes and transport processes in glia upon exposure to K^+ -ions. The following model was derived from the currently available data.

Metabolically produced CO_2 readily enters glial cells where it is hydrated to HCO_3^- and H^+ by carbonic anhydrase. HCO_3^- is then exchanged for extracellular Cl^- ,and H^+ for Na^+ . This process should result in an electroneutral Na^+ -influx and, hence, cell swelling. However, provided energy supply is sufficient, intracellular Na^+ is extruded by the Na^+ /K^+ -pump. Similar systems are considered to control intracellular pH in other cell types (15,97). A cellular influx of Na^+ would enable the Na^+ /K^+ -pump to clear sudden elevations of Ke^+ - even at the expense of glial swelling. K^+ homeostasis of the extracellular space would range over volume homeostasis. In this case, swelling would be associated with a modification of the volume reference input (s. Figure 1).

It is conceivable that this mechanism plays a significant role in glial swelling under pathological conditions. However, Bourke and Kimelberg in recent publications raise a second possibility (21,58,59). They found that neurotransmitters, such as norepinephrine, adenosine and histamine mimick K^+ by inducing swelling via enhancement of Cl^- - and

Na^+ -influx (19-21, 58). The authors assume that K^+ -induced swelling is facilitated by a release of transmitters into the extracellular space followed by an interaction with their receptors on glial surface membranes. Actually, the activity of carbonic anhydrase of astrocytes can be regulated through such receptors (56). Carbonic anhydrase may have a control function over the antiporter systems and thereby influence cell swelling (21). Antiporters might also be involved in uptake of transmitter compounds by glial cells. This is concluded from experiments where accumulation of glutamate by glia, but not by synaptosomes was inhibited by SITS (96). The hypotheses notwithstanding, our understanding of K^+ -induced swelling is as yet unsatisfactory. Nevertheless, K^+ -uptake is ensured by a powerful system, since extracellular K^+ -homeostasis is very effective unter physiological as well as pathological conditions (43,76). It may be added that trialkyltin compounds causing severe cytotoxic brain edema in experimental animals (72) act as anion-exchangers in red blood cells (97). This may be instrumental for the cytotoxic activity.

OUABAIN-INDUCED SWELLING

Administration of ouabain to cerebral tissue is followed by drastic swelling (29). Ouabain inhibits Na^+ /K^+ -ATPase, which is responsible for Na^+ -K^+ -transport and, hence, volume maintenance. Swelling in affected areas may be explained by DONNAN-forces after failure of the Na^+ / K^+ -pump. However, Lowe (67) has recently shown in autoradiographic studies that swelling also occurs in areas which were never reached by 3-H-ouabain after superfusion of cerebral cortex with the compound. There, only glial cells were swollen. Lowe concluded that glial swelling resulted from "...some change in the chemical milieu surrounding the process". It appears likely that accumulation of Ke^+ and other substances was involved.

SPREADING DEPRESSION

During normal neuronal activity, repetitive stimulation, or even epileptic discharges, Ke^+ never rises above 12 mM (cf: 43,76). Higher extracellular K^+ -concentrations lead inevitably to spreading cortical depression (SD), a pathophysiological curiosity characterized by massive cell swelling, a dramatic increase of extracellular K^+ , electrical impedance, and a temporary flattening of the EEG. SD was first described by Leao (64). Mechanical irritation, topical application of KCl, or of glutamate induce a reversible wave of suppression of normal electrical activity spreading over the whole brain surface. Concomitantly, Ke^+ -concentrations rise to 60-80 mM (43,76). The rate of propagation is 2-6 mm/min. SD is accompanied by an increase of metabolism (87) and drastic cell swelling as shown by a decrease of the electrical conductance of the tissue (35,93) in proprotion to the shrinking of the extracellular compartment. Shrinking of the ECS during SD is similar to respective changes in cerebral ischemia (42). Recent studies on isolated retinas have shown that cell swelling during SD depends on presence of Cl^- (35). Although the mechanisms responsible for cell swelling in SD are not yet fully understood, an involvement of extracellular accumulation of K^+ (38), and glutamate (93,94) appears rather likely. Glutamate causes SD at considerably lower concentrations than K^+.

ANOXIA AND ISCHEMIA

Basic Pathophysiology

1 to 3 minutes after onset of complete cerebral ischemia, cell volume control is completely lost. The extracellular space shrinks within few minutes to at least 50% of its normal volume. This has been demonstrated by electronmicroscopy (90), impedance measurements (50,75,89, 93), and by studies of extracellular tracer substances (42). Extracellular K^+ - and Na^+ -concentrations change inversely (48, 85) reaching almost equilibrium concentrations between the intra- and extracellular compartment. Clipping of the middle cerebral artery of cats for 1, or 2 hours increases markedly tissue osmolality in grey matter of the affected territory (46,49) providing a basis of ischemic brain swelling.

Cytotoxic edema develops early during recirculation by influx of water from the vascular bed (47). However, if recirculation is effective edema may resolve, even after 1 hour of complete ischemia (47,48). Several hours later, the blood-brain barrier becomes defective and vasogenic edema enters the extracellular space of the parenchyma (cf: 9).

Extracellular Mediator Substances

Cell swelling has been attributed to a failure of active transport after breakdown of the cerebral energy supply. This does, however, not explain the short latency observed between onset of ischemia and collapse of ion-gradients and volume control. Respective in-vitro studies illustrate this point perfectly. E.g., isolated retinas incubated in a large extracellular compartment tolerated anoxia and substrate deprivation for 40 min without swelling, or irreversible damage (3). However, if the volume of the incubation medium was restricted to an extra-/intracellular volume ratio of 4:1 or 1:1, water content of retinas increased in anoxia. This was associated with a decrease of leucine incorporation, or of 2-deoxy-glucose uptake (4) indicative of inhibition of protein synthesis and metabolic depression. The response of retinas to anoxic exposure was significantly enhanced, if the tissue was incubated in medium conditioned by previous incubation with other retinas under anoxic conditions.

In experiments performed by ourselves, suspended glioma cells were incubated in a strictly controlled system to prevent changes of the extracellular fluid composition. Changes of the cell volume were measured

electrically by flowcytometry (28). The suspended glial cells did not
swell during 2 hours of anoxic exposure, or during anoxia plus inhibition
of glycolysis by iodoacetate, although intracellular K^+ was lost (54).
The observations illustrate that nervous tissue may tolerate long
periods of anoxia without swelling provided the extracellular space is
large. Hence, early onset of ischemic cell swelling in-vivo is likely to
result from an accumulation of swelling-inducing compounds in an extracellular compartment of limited dimensions. Such a mechanism has already
been discussed in the case of K^+ -induced swelling and spreading depression (s.above). Mediator compounds (cf: 8,9) must be released early in
cerebral ischemia into the extracellular space. As already mentioned,
ischemia causes a rapid increase of Ke^+ leading to levels close to the
intra- /extracellular concentration equilibrium. What, however, causes
the sudden release of K^+ ? Again, neurotransmitters are likely to be involved. An excessive release of transmitters and K^+ early during ischemia
is suggested by a sudden increase of cAMP (62), which is one of the first
metabolic changes (69). Release of excitatory transmitter compounds enhances efflux of K^+ which in turn stimulates further release of transmitters (43). A positive feedback cycle is entered.

A transmitter liable to be involved in this process is glutamate
(93). Glutamate is available in brain tissue in concentrations of up to
10-12 mM (40) with a strict intracellular localization under normal conditions. Cerebral administration of the amino acid leads to an accumulation of water and Na^+ ,while K^+ is lost (2,24,67). Iontophoretic application of glutamate to cerebral cortex, or ventriculo-cisternal perfusion
with this compound cause massive swelling (91), or cytotoxic brain edema,
respectively (7, 50). Evidence is available that glutamate is released
during cerebral ischemia (13a,73). Edema fluid collected after focal
freezing of cat brain contained markedly enhanced glutamate levels, particularly in those animals suffering in addition from cerebral hypoperfusion secondary to an increase of the intracranial pressure (73). Glutamate causes depolarization and swelling of postsynaptic endings together
with an efflux of K^+ -ions (93). The perisynaptic glial elements accumulate against a steep concentration gradient not only K^+ -ions but also
transmitter compounds released during excitation (45, 96). Uptake of 1 mM
glutamate is associated with uptake of 2 mM Na^+ . Clearance of K^+ and
glutamate requires metabolic energy rendering glial cells dependent on
continuous supply with O_2 and substrate and a functioning energy metabolism. A close dependency on energy metabolism raises the probability of
cell swelling under critical conditions, such as ischemia. Suspended
glial cells increase oxygen consumption and cell volume during exposure
to glutamate (51). This holds also for other cell types, e.g. hepatocytes, which swell during uptake of amino acids (10). Glutamate is certainly not the only transmitter causing disturbances of cell volume control. Norepinephrine, adenosine and histamine may also induce glial
swelling by enhancement of Cl^- -influx and activation of carbonic
anhydrase (19,20,21,58).

Another class of putative edema mediators are free fatty acids. They are formed early in the course of ischemia (12). Fatty acids enhance cell swelling and eventually cause cell death (25,26). Among others, these compounds uncouple oxidative phosphorylation, inhibit Na^+-K^+ -ATPase and interfere with the extracellular clearance of glutamate and GABA (27). Metabolization of arachidonic acid is associated with production of free radicals, which also occur early during ischemia (cf: 30). Formation of free radicals induces a further release of free fatty acids. The tissue pH is another pertinent factor in cerebral ischemia. Manipulation of cerebral lactate levels to influence tissue pH by administration of glucose in ischemia supports this point. Hyperglycemic animals had a more extensive fall of the extracellular pH in cerebral ischemia than normo-, or hypoglycemic animals which correlated with the neurological outcome (81,84).

OTHER CAUSES OF CELL SWELLING IN BRAIN

Swelling Secondary to Vasogenic Edema

Swelling of glial cells is regularly found in areas with primary vasogenic brain edema (9) probably resulting from a release of mediator substances. The extracellular space is flooded under these circumstances with highly active material, such as free fatty acids, transmitters, kinins and proteolytic enzymes released from the damaged cells. Development of cytotoxic swelling in the affected parenchyma is therefore not surprising (s. above, cf: 8). Vasogenic edema fluid obtained from perifocal brain tissue of cats with a freezing lesion had highly abnormal levels of glutamate and free fatty acids (73). The abnormalities were markedly enhanced, when cerebral blood flow was compromised on account of an increased intracranial pressure.

Hepatic Encephalopathy

Metabolic disorders, such as Reye's Syndrome, or other forms of hepatic encephalopathy are associated with glial swelling (63,77). The pathophysiology of this process is, however, unclear. Various substances are held responsible, particularly elevated ammonia concentrations (cf: 77). Inhibition of glutamine synthetase (GS) causes glial swelling in experimental animals (41). In centralnervous tissue this enzyme is only found in glial- but not in nerve cells (74). It catalyzes formation of glutamine under consumption of ammonia and glutamate (s. above). Glial swelling in hepatic encephalopathy may be attributed to an inhibition of glutamine synthetase (11) resulting in a failure of glial cells to clear the extracellular compartment from glutamate.

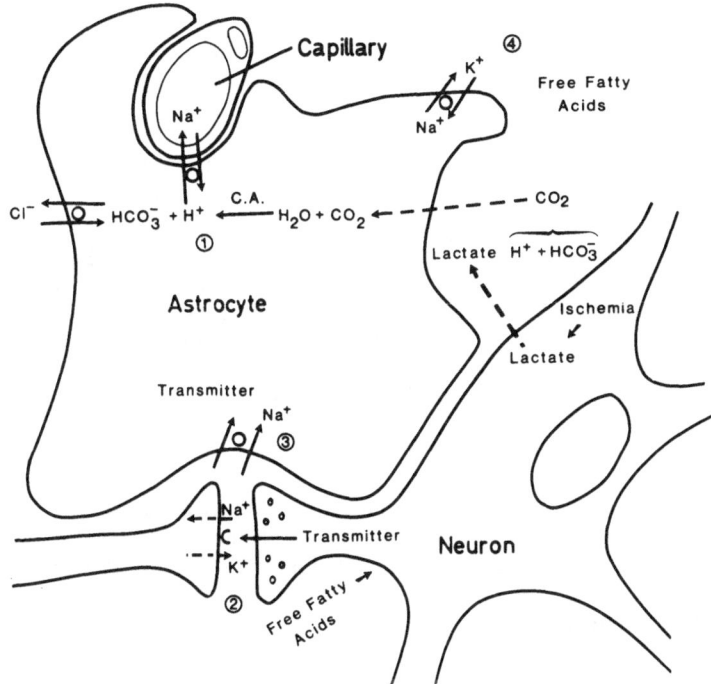

Fig. 2: Pertinent aspects of cell swelling occurring under physiological
and pathophysiological conditions:
1. Removal of high extracellular K^+ causing activation of Na^+/H^+
   and Cl^-/HCO_3^- antiporter systems leading to an intracellular
   accumulation of Na^+, Cl^- and water.
2. Swelling of postsynaptic endings by a transmitter dependent
   increase of membrane permeability to Na^+.
3. Swelling of glial cells upon neuronal excitation by intracellular uptake of released K^+ and transmitter compounds from the
   extracellular space raising cell osmolality.
4. Abnormal accumulation of active mediator compounds in the
   tissue, e.g. fatty acids, or others under pathophysiological
   conditions (ischemia) which directly interfere with functional
   and structural components of the cell membrane. In the case
   of arachidonic acid, inhibition of Na^+-K^+-ATPase, uncoupling
   of oxidative phosphorylation, and direct membrane lesions can
   be considered as pathological mechanisms.

SUMMARY AND CONCLUSIONS

Cell volume control under physiological conditions depends on a low membrane permeability to Na^+ - ions, and on active extrusion of Na^+ -ions to maintain the DONNAN equilibrium. Volume regulation secondary to osmotic swelling is accomplished by cellular discharge of K^+ - and Cl^- -ions involving various active, yet not clearly defined processes. Many pathological conditions of the brain are associated with cell swelling. Formation, or release of mediator substances seem to play a major role in the swelling process. The most pertinent concepts on swelling mechanisms in brain tissue are summarized in Figure 2:

1. Clearance of extracellular K^+ together with control of pH through exchange systems (antiporter) of Na^+ against H^+, and Cl^- against HCO_3^- via carbonic anhydrase. However, cell swelling resulting from intracellular accumulation of Na^+ and Cl^- under normal conditions is compensated for by the active Na^+/K^+ -pump.
2. Extracellular release of transmitter compounds. This causes swelling of postsynaptic endings by increase of membrane permeability to Na^+ -ions associated with an influx of Cl^- and water.
3. Uptake of transmitter compounds together with Na^+ -ions during neuronal excitation increasing the osmotic load of glial cells.
4. Release of active mediator compounds, such as free fatty acids in ischemia, or other pathological conditions which among others interfere with membrane Na^+ -K^+ -ATPase, and other mechanisms involved in cell volume control.

Obviously, homeostasis in brain tissue is rather complex. Competition of Ke^+ homeostasis and volume control of glial cells may serve as an example. Constancy of cell volume may transiently be sacrificed to maintain a low Ke^+ to ensure neuronal function. Under pathological conditions, e.g. ischemia release of transmitters causes more pronounced disturbances of the extracellular homeostasis which cannot be controlled then by the glial elements due to the shortage of metabolic energy. Cellular accumulation of osmotically active molecules after breakdown of macromolecules in ischemia and a release of free fatty acids may accelerate cellular dysfunction and explain the rapid development of cell swelling after onset of cerebral ischemia.

Important progress of the understanding of cell swelling in centralnervous tissue has been accomplished, although many details remain speculative. Further efforts are mandatory to provide a basis for the development of more specific and, hence, more effective forms of treatment of cytotoxic cell swelling in damaged brain tissue.

REFERENCES

1. Ahmad HR, and Loeschke HH, Evidence for a carrier mediated exchange diffusion of HCO_3^- against Cl^- at the interphases of the central nervous system, in: "Central neurone environment", Schlaefke ME, Koepchen HP, and See WR, eds., Springer, Berlin, Heidelberg (1983).
2. Ames A, Tsukuda Y, and Nesbett B, Intracellular Cl^-, Na^+, K^+, Ca^{++}, Mg^{++}, and P in nervous tissue. Response to glutamate and to changes in extracellular Ca^{++}, J Neurochem 18:259 (1967).
3. Ames A, and Nesbett FB, Pathophysiology of ischemic cell death. I. Time of onset of irreversible damage; importance of different components of the ischemic insult, Stroke 14:219 (1983).
4. Ames A, and Nesbett FB, Pathophysiology of ischemic cell death. III. Role of extracellular factors, Stroke 14:233 (1983).
5. Baethmann A, and Van Harreveld A, Water and electrolyte distribution in gray matter rendered edematous with a metabolic inhibitor, J Neuropath Exp Neurol 32:408 (1973).
6. Baethmann A, and Sohler K, Electrolyte- and fluid-spaces of rat brain in situ after infusion with dinitrophenol, J Neurobiol 6:73 (1975).
7. Baethmann A, Oettinger W, Rothenfusser W, Kempski O, and Unterberg A, CSF-pressure of rats during ventriculo-cisternal perfusion with potential brain edema factors, in: "ICP-IV", Shulman K, Marmarou A, Miller JD, Becker DP, Hochwald GM, Brock M, eds., Springer, Berlin, Heidelberg, New York (1980).
8. Baethmann A, Oettinger W, Rothenfusser W, Kempski O, Unterberg A, and Geiger R, Brain edema factors. Current state with particular reference to plasma constituents and glutamate, in: "Advances in neurology, Vol 28: Brain edema", Cervos-Navarro J, Ferszt R, eds., Raven Press, New York (1980).
9. Baethmann A, Pathophysiological and pathochemical aspects of cerebral edema, Neurosurg Rev 1:85 (1978).
10. Bakker-Grunwald T, Potassium permeability and volume control in isolated rat hepatocytes, Biochim Biophys Acta 731:239 (1983).
11. Baldessarini RJ, and Yorke C, Uptake and release of possible false transmitter amino acids by rat brain tissue, J Neurochem 23:839 (1974).
12. Bazan NG, Effects of ischemia and electroconvulsive shock of free fatty acid pool in the brain, Biochim Biophys Acta 218:1 (1970).
13. Benninger C, Kadis J, and Prince DA, Extracellular calcium and potassium changes in hippocampal slices, Brain Res 187:165 (1980).
13a. Benveniste H, Drejer J, Schousboe A, and Diemer NH, Elevation of the extracellular concentrations of glutamate and aspartate in rat hippocampus during transient cerebral ischemia monitored by intracerebral microdialysis, J Neurochem 43:1369 (1984).
14. Bivins BA, Hyde GL, Sachatello CR, and Griffen WO, Physiopathology and management of hyperosmolar hyperglycemic nonketotic dehydration, Surg Gynecol Obstet 154:534 (1982).
15. Boron F, Transport of H^+ and ionic weak acids and bases,

J Membr Biol 72:1 (1983).

16. Bourke RS, Studies of the development and subsequent reduction of swelling of mammalian cortex under isosmotic conditions in vitro, Exp Brain Res 8:232 (1969).

17. Bourke RS, Evidence for mediated transport of chloride in cat cerebral cortex in vitro, Exp Brain Res 8:219 (1969).

18. Bourke RS, and Nelson KM, Further studies on the K^+-dependent swelling of primate cerebral cortex in vivo: The enzymatic basis of the K^+-dependent transport of chloride, J Neurochem 19:663 (1972).

19. Bourke RS, Kimelberg HK_2 and Daze MA, Effects of inhibitors and adenosine on (HCO_3^- /CO_2)-stimulated swelling and Cl^--uptake in brain slices and cultured astrocytes, Brain Res 154:196 (1978).

20. Bourke RS, Waldman JB, Kimelberg HK, Barron KD, San Filippo BD, Popp AJ, and Nelson LR, Adenosine-stimulated astroglial swelling in cat cerebral cortex in vivo with total inhibition by a nondiuretic acylaryloxyacid derivative, J Neurosurg 55:364 (1981).

21. Bourke RS, Kimelberg HK, Daze M, and Church G, Swelling and ion uptake in cat cerebrocortical slices: Control by neurotransmitters and ion transport mechanisms, Neurochem Res 8:5 (1983).

22. Cala PM, Volume regulation of amphiuma red blood cells, J Gen Physiol 76:683 (1980).

23. Cala PM, Volume regulation by red blood cells: Mechanisms of ion transport, Molec Physiol 4:33 (1983).

24. Chan PH, Fishman RA, Lee JL, and Candelise L, Effects of excitatory neurotransmitter amino acids on swelling of rat brain cortical slices, J Neurochem 33:1309 (1979).

25. Chan PH, Fishman RA, Lee JL, and Quan SC, Arachidonic acid-induced swelling in incubated rat brain cortical slices, Neurochem Res 5:629 (1980).

26. Chan PH, and Fishman RA, Alterations of membrane integrity and cellular constituents by arachidonic acid in neuroblastoma and glioma cells, Brain Res 248:151 (1982).

27. Chan PH, Kerlan R, and Fishman RA, Reductions of aminobutyric acid and glutamate uptake and (Na^+ K^+)-ATPase activity in brain slices and synaptosomes by arachidonic acid, J Neurochem 40:309 (1983).

28. Chaussy L, Baethmann A, and Lubitz W, Electrical sizing of nerve and glia cells in the study of cell volume regulation, in: "Cerebral microcirculation and metabolism", Cervos-Navarro J, Fritschka E, eds., Raven Press, New York (1981).

29. Cornog JL, Gonatas NK, and Feierman JR, Effects of intracerebral injection of ouabain on the fine structure of rat cerebral cortex, Am J Path 51:573 (1967).

30. Demopoulos HB, Flamm ES, Seligman ML, Mitamura JA, and Ransohoff J, Membrane perturbations in central nervous system injury: Theoretical basis for free radical damage and a review of the experimental data, in: "Neural trauma", Popp J, ed., Raven Press, New York (1979).

31. Dietzel I, Heinemann U, Hofmeier G, and Lux HD, Transient changes in the size of the extracellular space in the sensorimotor cortex of cats in relation to stimulus-induced changes in potassium concentration, Exp Brain Res 40:432 (1980).

32. Dietzel I, Heinemann U, Hofmeier G, and Lux HD, Stimulus-induced changes in extracellular Na^+ and Cl^- concentration in relation to changes in the size of the extracellular space, Exp Brain Res 46:73 (1982).

33. Eckert R, and Tilotson D, Potassium activation associated with intraneuronal free calcium, Science 200:437 (1978).

34. Elvin P, and Evans CW, Cell adhesiveness and the cell cycle: Correlation in synchronized balb/c 3T3 cells, Biol Cell 48:1 (1983).

35. Ferreira-Filho CR, and Martins-Ferreira H, Electrical impedance of isolated retina and its changes during spreading depression, Neuroscience 7:3231 (1982).

36. Gill TH, Young OM, and Tower DB, The uptake of 36-Cl into astrocytes in tissue culture by a potassium-dependent, saturable process, J Neurochem 23:1011 (1974).

37. Gilles R, Volume maintenance and regulation in animal cells: Some features and trends, Molec Physiol 4:3 (1983).

38. Grafstein B, Mechanisms of spreading cortical depression, J Neurophysiol 19:154 (1956).

39. Grinstein S, Dupre A, and Rothstein A, Volume regulation of human lymphocytes. Role of calcium, J Gen Physiol 79:849 (1982).

40. Guroff G, Transport and metabolism of amino acids, in: "Basic neurochemistry", Albers RW, Katzman R, Siegel GJ, Agranoff BW, eds., Academic Press, Boston (1972).

41. Gutierrez JH, and Norenberg MD, Alzheimer II astrocytosis following methionine sulfoximine, Arch Neurol 32:132 (1975).

42. Hansen AJ, and Olsen CE, Brain extracellular space during spreading depression and ischemia, Acta Physiol Scand 108:355 (1980).

43. Hertz L, Drug-induced alterations of ion distribution at the cellular level of the central nervous system, Pharmacol Rev 29:35 (1977).

44. Hertz L, An intense potassium uptake into astrocytes, its further enhancement by high concentrations of potassium, and its possible involvement in potassium homeostasis at the cellular level, Brain Res 145:202 (1978).

45. Hoesli E, and Hoesli L, Uptake of l-glutamate and l-aspartate in neurones and glial cells of cultured human and rat spinal cord, Experientia 32:219 (1976).

46. Hossmann KA, and Takagi S, Osmolality of brain in cerebral ischemia, Exp Neurol 51:124 (1976).

47. Hossmann KA, Development and resolution of ischemic brain swelling, in: "Dynamics of brain edema", Pappius HM, and Feindel W, eds., Springer, Berlin, Heidelberg, New York (1976).

48. Hossmann KA, Sakaki S, and Zimmermann V, Cation activities in reversible ischemia of the cat brain, Stroke 8:77 (1977).

49. Hossmann KA, and Matsuoka Y, Influence of tissue osmolality on intracerebral fluid shifts and the development of ischemic brain

   edema, in: "Cerebrovascular diseases", Reivich M, and Hurtig HI,
   eds., Raven Press, New York (1983).

50. Kempski O, and Baethmann A, Localization of cerebral edema induced
   by glutamate by measurement of tissue impedance, Proc. of the
   Vth Int. Conf. Bio-Impedance, Tokyo (1981).

51. Kempski O, Gross U, and Baethmann A, An in vitro model of cytotoxic brain edema: Cell volume and metabolism of cultivated
   glial and nerve cells, in: Advances in neurosurgery, Vol. 10,
   Driesen W, Brock M, and Klinger M, eds., Springer, Berlin,
   Heidelberg (1982).

52. Kempski O, Chaussy L, Gross U, Zimmer M, and Baethmann A, Volume
   regulation and metabolism of suspended C6-glioma cells: An in
   vitro model to study cytotoxic brain edema, Brain Res 279:217
   (1983).

53. Kempski O, Zimmer M, Chaussy L, and Baethmann A, Volume and
   metabolism of C6-glioma cells suspended in hypotonic medium:
   An in-vitro model to study cytotoxic brain edema, in: "Recent
   progress in the study and therapy of brain edema", Go KG, and
   Baethmann A, eds., Plenum Press, New York (1984).

54. Kempski O, Zimmer M, and Baethmann A, in preparation

55. Kimbrough RD, Review of the toxicity of hexachlorophene including
   its neurotoxicity, J Clin Pharmacol 13:439 (1973).

56. Kimelberg HK, Biddlecome S, Narumi S, and Bourke RS, ATPase and
   carbonic anhydrase activities of bulk-isolated neuron, glia and
   synaptosome fractions from rat brain, Brain Res 141:305 (1978).

57. Kimelberg HK, Biddlecome S, and Bourke RS, SITS-inhibitable Cl$^-$
   transport and Na$^+$-dependent H$^+$ production in primary astroglial
   cultures, Brain Res 173:111 (1979).

58. Kimelberg HK, Glial enzymes and ion transport in brain swelling, in:
   "Neural trauma", Popp J, ed., Raven Press, New York (1979).

59. Kimelberg HK, Bourke RS, Stieg PE, Barron KD, Hirata H, Pelton EW,
   and Nelson LR, Swelling of astroglia after injury to the nervous
   system: Mechanisms and consequences, in: "Head injury, basic and
   clinical aspects", Grossman RG, and Gildenberg PL, eds., Raven
   Press, New York (1982).

60. Klatzo I, Neuropathological aspects of brain edema, J Neuropath Exp
   Pathol 20:1 (1967).

61. Kregenow FM, Osmoregulatory salt transporting mechanisms: Control of
   cell volume in anisotonic media, Ann Rev Physiol 43:493 (1981).

62. Krivanek J, A different mode of action of potassium ions and ver-
   atridine on the formation of cyclic adenosine monophosphate in
   the cerebral cortex, J Neurochem 3:333 (1978).

63. Laursen H, Cerebral vessels and glial cells in liver disease, Acta
   Neurol Scand 65:381 (1982).

64. Leao AAP, Spreading depression of activity in the cerebral cortex,
   J Neurophysiol 7:359 (1944).

65. Ling GN, Miller C, and Ochsenfeld MM, The physical state of solutes
   and water in living cells according to the association-induction
   hypothesis, Ann NY Acad Sci 204:6 (1973).

66. Lipton P, Effects of membrane depolarization on light scattering by

cerebral cortical slices, J Physiol 231:365 (1973).

67. Lowe DA, Morphological changes in the cat cerebral cortex produced by superfusion of ouabain, Brain Res 148:347 (1978).

68. Lund-Andersen H, and Hertz L, Effects of potassium content in brain cortex slices from adult rats, Exp Brain Res 11:199 (1970).

69. Lust WD, and Passonneau JW, Cyclic nucleotide levels in brain during ischemia and recirculation, in: "Neuropharmacology of cyclic nucleotides", Palmer GC, ed., Urban & Schwarzenberg, Baltimore, Munich (1979).

70. Macknight ADC, and Leaf A, Regulation of cellular volume, Physiol Rev 37:510 (1977).

71. Macknight ADC, Cellular response to injury, in: "Edema", Staub NC, Tayler AE, eds., Raven Press, New York (1984).

72. Magee PN, Stoner HB, and Barnes JN, The experimental production of edema in the central nervous system of the rat by triethyl tin compounds, J Path Bact 73:107 (1957).

73. Maier-Hauff K, Lange M, Schürer L, Guggenbichler C, Vogt W, Jacob K, and Baethmann A, Glutamate and free fatty acid concentrations in extracellular vasogenic edema fluid, in: "Recent progress in the study and therapy of brain edema, Go KG, and Baethmann A, eds., Plenum Press, New York (1984).

74. Martinez-Hernandez A, Bell K, and Norenberg MD, Glutamine synthetase: Glial localization in brain, Science 195:1356 (1977).

75. Matsuoka Y, and Hossmann KA, Cortical impedance and extracellular volume changes following middle cerebral artery occlusion in cats, J Cereb Blood Flow Metabol 2:466 (1982).

76. Nicholson C, Dynamics of the brain cell microenvironment, Neurosci Res Prog Bull 18:175 (1980).

77. Norenberg MD, The astrocyte in liver disease, Adv Cell Neurobiol 2:303 (1981).

78. Orkand RK, Nicholls JG, Kuffler SW, Effect of nerve impulses on the membrane potential of glial cells in the central nervous system of amphibia, J Neurophysiol 29:788 (1966).

79. Orkand PM, Bracho H, and Orkand RK, Glial metabolism: Alteration by potassium levels comparable to those during neural activity, Brain Res 55:467 (1973).

80. Pappius HM, Fundamental aspects of brain edema, in: Handb. Clin. Neurol. Vol.16, Part I, Vinken PJ, Bruyn GW, eds., North Holland Publ., Amsterdam American Elsevier (1974).

81. Pulsinelli WR, Waldman S, Rawlinson D, and Plum F, Moderate hyperglycemia augments ischemic brain damage: A neuropathologic study in the rat, Neurol 32:1239 (1982).

82. Robinson JD, Regulating ion pumps to control cell volume, J Theoret Biol 19:90 (1968).

83. Schousboe A, Booher J, and Hertz L, Content of ATP in cultivated neurons and astrocytes exposed to balanced and potassium-rich media, J Neurochem 17:1501 (1970).

84. Siemkowicz E, and Hansen AJ, Clinical restitution following cerebral ischemia in hypo-, normo-, and hyperglycemic rats, Acta Neurol

Scand 58:1 (1978).

85. Siemkowicz E, and Hansen AJ, Brain extracellular ion composition and EEG activity following 10 minutes ischemia in normo- and hyperglycemic rats, Stroke 12:236 (1981).

86. Somjen GG, Electrophysiology of neuroglia, Ann Rev Physiol 37:163 (1975).

87. Somjen GG, Rosenthal M, Cordingley G, LaManna J, and Lothman E, Potassium, neuroglia, and oxidative metabolism in central gray matter, Fed Proc 35:1266 (1976).

88. Trubatch J, Loud AV, and Van Harreveld A, Quantitative stereological evaluation of KCl-induced ultrastructural changes in frog brain, Neurosci 2:963 (1977).

89. Van Harreveld A, and Ochs S, Cerebral impedance changes during circulatory arrest, Amer J Physiol 187:180 (1956).

90. Van Harreveld A, Crowell J, and Malhotra SK, A study of extracellular space in central nervous tissue by freeze-substitution, J Cell Biol 25:117 (1965).

91. Van Harreveld A, and Fifkova E, Light- and electronmicroscopic changes in central nervous tissue after electrophoretic injection of glutamate, Exp Mol Pathol 15:61 (1971).

92. Van Harreveld A, and Trubatch J, Conditions affecting the extracellular space in the frog's forebrain, Anat Rec 178:587 (1974).

93. Van Harreveld A, Brain tissue electrolytes, in: "Molecular biology and medicine series", Bittar EE, ed., Butterworths, London (1966).

94. Van Harreveld A, Two mechanisms of spreading depression in the chicken retina, J Neurobiol 9:419 (1978).

95. Walz W, and Hertz L, Ouabain-sensitive and ouabain-resistant net uptake of potassium into astrocytes and neurons in primary cultures, J Neurochem 39:70 (1982).

96. Waniewski RA, Martin DL, Selective inhibition of glial versus neuronal uptake of l-glutamic acid by SITS, Brain Res 268:390 (1983).

97. Wieth JO, Brahm J, and Funder J, Transport and interaction of anions and protons in the red blood cell membrane, Ann NY Acad Sci 341:394 (1980).

PROGRESSION AND IRREVERSIBILITY IN BRAIN ISCHAEMIA

Lindsay Symon

Gough Cooper Department of Neurological Surgery
Institute of Neurology, Queen Square, London WC1N 3BG
United Kingdom

In recent years the twin concepts of thresholds of ischaemia and of the ischaemic penumbra have proved useful hypotheses in the investigation and management of cerebrovascular disease. This paper presents the current status of these concepts. The notion of an ischaemic threshold arose from clinical observation. It has long been clear in clinical neurosurgery, for example, that patients recovering from anaesthesia may show the progressive clearance of a neurological deficit, that some patients who develop a neurological deficit with lower blood pressure have that deficit promptly cleared when blood pressure is elevated and that patients with established cerebrovascular occlusion and dense neurological deficits may show quite evident improvement over months or years. While in this last example some of the potential for re-learned circuitry in the nervous system may play a part, it is clear that in more acute circumstances neurons which at one time are non-functioning, may under improved conditions of perfusion return to normal function.

The effects of ischaemia on the electrical function of the central nervous system have been assessed from a variety of standpoints. Studies have been made of the differential effects of ischaemia on the electrophysiological activity of presynaptic terminals and synaptic transmission (1,2), of the concomitants of ischaemia such as tissue impedance, surface potentials, transmembrane ionic- and water shifts, and histological damage (1,3,4,5), and of correlation between cerebral blood flow, intracranial pressure and the direct cortical response (6). Clinical studies have related regional cerebral blood flow and EEG (7,8). The well-established sequence of changes in the EEG caused by various forms of anoxia has recently been reviewed (9). There is also a developing interest in the survival of neuronal function following ischaemia (10,11). However, in most of the experimental studies the ischaemia produced has been total and widespread. This is so at variance with the clinical situ-

221

ation, where complete ischaemia is most unusual, for example in cerebrovascular disease. Therefore, it would be of great interest to analyze the
effects of partial ischaemia upon cerebral function experimentally and to
relate the quantitative changes in electrical activity to different
degrees of cerebral ischaemia.

THE ELECTRICAL THRESHOLD

Experimentally, occlusion of the middle cerebral artery simulates
the production of an acute clinical stroke (12,13,14,15) and induces an
ischaemic lesion which is restricted to known regions of the cerebral
hemisphere and, therefore, interferes only selectively with the neural
pathways. Moreover, it has recently been shown (16) that the density of
the ischaemia produced by occlusion of the middle cerebral artery is
graded over the surface of the hemisphere. Thus, different levels of residual flow result at different positions on the hemisphere soon after
the occlusion. In particular, such a gradient is present along the sensorimotor strip. The thalamic nuclei are not significantly affected by
occlusion of the middle cerebral artery (17,18). Any changes in the
peripherally stimulated somatosensory evoked potential observed in the
strip after occlusion can therefore be directly related to the changes in
tissue flow in the region of the recording electrode provided the physiological state of the preparation in all other relevant respects is held
constant. We have now used middle cerebral artery occlusion to produce
a reduction in blood flow. Flow was measured by the highly focal technique of hydrogen clearance (19) in the immediate area of the evoked
potential electrode. In this way we have been able to relate electrical
activity to flow at selected positions on the cortex.

We selected the evoked potential as a quantifiable index of neuronal function in preference to single unit potentials chiefly on technical grounds. There are major difficulties (principally brain movement)
associated with monitoring, even extracellularly, the activity of single
cells for periods of time sufficiently long to encompass the range of
events which accompany experimentally induced ischaemia. We produced
the evoked potential by peripheral rather than local cortical stimulation
(6,10) because of the possibility that a direct electrical stimulus of
the brain might affect cerebral blood flow itself. From previous studies
(16) of the distribution of ischaemia after middle cerebral artery occlusion, we did not expect that the visual evoked potential would change
nearly as much as the somatosensory evoked potential. Although the audi-
tory evoked potential was another possible response, we chose the somatosensory pathway for technical simplicity and its closer relevance to the
clinical problems of stroke in view of the known homology between sensory and motor strips.

Our objectives were to study the effects of experimentally induced
partial ischaemia produced by acute middle cerebral artery occlusion in
the primate. Our experimental observations indicated that the primary

positive/negative wave of the somatosensory evoked response was fully
maintained down to flow levels of around 20 ml/100g min from a control
value of 50 ml/100g min. At this point there was a fairly sharp decline
in evoked response with complete disappearance by 12 ml/100g min. The
half value for this threshold appeared to be 16 ml/100g min in our experimental primates. Using intracellular recording, Heiss et al. (20) suggested in the cat a threshold of 18 ml/100g min for spontaneous neuronal activity. These carefully controlled observations accorded well with
clinical findings (21) in carotid disobliteration, suggesting that the
EEG was critically impaired at flow levels below 20 ml/100g min. In all
our subsequent experimental work, this approximate level of 15-20 ml/100g
min for an electrical threshold has been sustained. This work has been
directly transferred to the clinic because of the ease of interpretation.

We have recently carried out a series of experiments to clarify the
question of the relative ischaemic sensitivity between regions, measuring
SEPs and CBF in thalamus and brain stem (medial lemniscus). With middle
cerebral artery occlusion alone, all regions showed some reduction in
flow from control. In the most affected area, the opercular region on the
side of occlusion, flow decreased to less than half, while in the region
of the somatosensory evoked response it decreased below the SEP threshold. Smaller and insignificant flow decreases were found in the contralateral hemisphere and in both thalami. Occlusion abolished the early
components of the response in cortex on the affected side, while the
ventrolateral thalamus and medial lemniscal responses were unchanged.

Progressive hypotension was thereafter used to reduce and finally
abolish the thalamic response. The medial lemniscus proved most resistant
to ischaemia.A plot of the SEP in the lemniscus in relation to flow made
it clear that the response was sustained down to at least 10ml/100g min,
while cerebral cortex had a flow threshold around 18ml/100g min. With
the corresponding ventrolateral thalamic plot the contrast with cortex
was not as great as with that of the medial lemniscus. There were several data points well above the zero level of about 10ml/100g min. The
situation was clearer in studies using hypotension. The lemniscal volley
was preserved down to a blood pressure of 20-30 torr. Comparing the data of CBF and mean systemic blood pressure in bands of equal width, we
found that between 15-20 ml/100g min the cortical SEPs were significantly reduced, while VPL and lemniscus remained in the control range. Between 10-15 ml/100g min VPL was significantly reduced, the cortical response almost abolished. Only below 10ml/100g min the lemniscal response
was similarly reduced.

The principal result of this study is the indication that as one
descends the neuraxis there is an increasing resistance of electrophysiological function to systemic hypotension, together with a decreasing
threshold for local ischaemia. It has been pointed out (22) that failure
of the cortical SEP could arise from a reduction in CBF in subcortical
segments of the afferent pathway as well as in cortex. Our present re-

sults show that, at least up to the thalamus, the possibility of prior
subcortical transmission failure is unlikely in progressive hypotension.
Since the VPL and lemniscus are less sensitive to ischaemia than cortex,
the cortically generated components of the SEP would begin to tail, while
the subcortical afferent volley was still intact to a large extent. This
study has also provided information on how central transmission times depend on local conditions of perfusion at various levels. Our results show
that with MCA occlusion followed by a general reduction in flow due to a
decreased blood pressure, or with hypotension alone, increases in conduction time (CT) recorded at the cortex arise primarily in structures above
the thalamus. At present, the contribution is unknown of the internal
capsule and thalamocortical radiation to changes of conduction time, as
distinct from that due to cortical failure. The small but significant decrease in CT measured at flows above 20 ml/100g min on the verge of
cortical failure, may arise from a reduction in the temporal dispersion
of the subcortical afferent volley, or from interaction of the cortical
N10 and P15 components by superimposition within the cranial volume
conductor. Such a decrease has not been found in clinical data.

IONIC THRESHOLDS

It is clear that electrical function is closely related to rCBF. It
has been suggested that complete electrical failure below the ischaemic
threshold is due to synaptic depolarization (23) subsequent to a release
of potassium into the extracellular space (24). Other conditions, however, such as a change in neurotransmitter metabolism or tissue lactacidosis, have recently been discussed in relation to complete electrical
failure (24). The aim of this study, therefore, was to test the association between electrical function and extracellular activities of K^+ and
H^+ (K_e and pH_e) at the ischaemic threshold of electrical failure. For
that purpose, rCBF was lowered by middle cerebral artery occlusion ap-
proximating the ischaemic threshold of the somatosensory evoked response (23). The regional blood flow was manipulated then about the
threshold by varying systemic blood pressure. K_e and pH_e were continuously recorded using extracellular microelectrodes.

In the control phase, mean extracellular potassium was maintained
for at least half an hour before occlusion at 5.5 mM (SD= 1.3). The normal tissue flows were in a range of 105 ml/100g min (SD=24; 19). The
middle cerebral artery was occluded from 37 to 290 min (mean=136 min;
SD=63). In the clip phase, extracellular potassium increased to 24-92 mM
at 26 electrode sites (12 animals) and to 11-17 mM at 3 sites (2 animals). Flow fell below 11ml/100g min at 22 sites (13 animals) and below
15ml/100g min in the remainder. When the flow threshold of electrical
failure was first described, it was not clear how the energy state and,
hence, ion homeostasis were affected. Since oxygen uptake at the electrical threshold supposedly was reduced, energy- and pump failure with
efflux of potassium and membrane depolarization were suspected as the
cause of electrical failure. Subsequent studies in the baboon, however,

clearly showed that the extracellular potassium concentration in the cortex was normal or only slightly elevated at the threshold where electrical function ceased (25).

Increase in extracellular potassium concentration indicative of "pump failure" did not occur unless local blood flow was further reduced. In these (25) and subsequent studies on baboons with middle cerebral artery occlusion, a critical ischaemic flow threshold of about 10 ml/100g min was determined, below which extracellular potassium massively increased. Further studies in the rat with bicuculline-induced seizures have confirmed this observation (26). In these studies, progressive brain ischaemia was induced by controlled hypotension allowing a correlation between the EEG, extracellular potassium, and cerebral energy metabolism. Electrical failure appeared as cessation of seizure discharges, while extracellular potassium concentration was normal or only slightly elevated. At the point of seizure interruption, the extracellular potassium concentration decreased indicating that sufficient energy remained for ion pumping. This was verified by direct tissue analyses. Although lactic acid was elevated and phosphocreatine decreased, ATP was close to normal.

Inspite of the maintenance of cerebral energy stores at the threshold of electrical failure, the rate of ion pumping was affected even earlier. This was concluded from the reduction in potassium clearance during the interictal periods (26), which may reflect a failing oxygen supply and a decline in ATP production. Reduction in blood pressure below the threshold for electrical failure caused massive increase in extracellular potassium, indicating ion pump failure. Metabolic studies showed depletion of ATP at this point. These studies did not allow measurements of blood flow. Yet, the significant difference in blood pressure levels separating the thresholds of electrical ion pump failure (i.e. 46 and 32 mmHg, respectively), suggests presence of progressive ischaemia and not merely a natural course of severe ischaemia. The close correspondence with the experiments in baboons concerning electrical failure and potassium homeostasis, further emphasizes this view.

The vital role of calcium in normal and pathological cellular function has become apparent. The "calcium paradox" as reported by Zimmermann and Hulsmann (27) showed that heart perfused with a calcium-free solution, although losing its contractility, was able to maintain its electrical activity. However, when the normal perfusate was restored after more than 2 min of calcium-free perfusion, the electrocardiogram disappeared in association with a rapid, irreversible loss of contractility. Shen and Jennings (28,29) have shown that depending upon the severity of the ischaemic insult, reperfused myocardium can accumulate up to 10 times its normal concentration of calcium. The myocardial changes occurring during reperfusion have recently been reviewed by Hearse (30). It has been suggested that the mechanisms of damage by calcium uptake are, first, a reduction of the ability of the intracellular membranes to withstand or control certain ion fluxes, and second, a failure of the mitochondria by the physical effects of calcium uptake and deposition. In the light of such work, a fuller understanding of calcium mechanisms in the cerebral cortex in ischaemia and on reperfusion is of critical value in determining survival and function of cerebral tissue.

Changes of calcium in the extracellular space of the brain are known to occur. Nicholson has shown relationships between extracellular potassium (K_e) and calcium activity (Ca_e) in spreading depression and terminal anoxia in cerebral cortex and cerebellum (31,32). The aim of this study was to measure Ca_e and to relate the changes evolving during partial ischaemia in the primate cerebral cortex to K_e and local cerebral blood flow (lCBF). To aid analysis of the changes of any ion, we used triple-barrelled, double-ion sensitive microelectrodes to measure both K_e and Ca_e at the same point simultaneously. K_e and Ca_e were measured at 12 sites in 6 animals. The respective base-line values measured during the control period were for K_e :3.95 $\pm$ 0.80 mM and for Ca_e :1.31 $\pm$ 0.01 mM . All values are quoted as means $\pm$ SD. The values of $\bar{K}_e$ and $\bar{Ca}_e$ were taken from each electrode at each lCBF determination to obtain activity/flow relationships. A flow threshold of 10 ml/100g min was found below which major changes occurred in extracellular ion activities. Although flow thresholds for changes of potassium and calcium could not be statistically separated, it could be shown in other ways that these were not identical.

Spontaneous K_e and Ca_e transients were also seen which shed further light upon the relationship between both ions in the extracellular space. As in ischaemia, K_e began to rise before Ca_e fell, with an average interval of 13.7 $\pm$ 3.7 s. K_e rose to 10.5 $\pm$ 3.0 mM, when Ca_e started to move. This is not significantly different from the K_e level associated with Ca_e movement in ischaemia. The spontaneous and transient changes of potassium and calcium described here appear to be similar in the characteristics of the onset to spreading depression studied by Kraig and Nicholson (31). The authors found also that Ca_e started to change at a potassium level of about 10 mM. At this point, the extracellular slow potential moved in a negative direction. They explained this as an increase in membrane permeability by depolarization allowing ions to flow down their concentration gradients.

In the present study, the levels of K_e at which Ca_e began to fall were not significantly different between the spontaneous transients and ischaemia. Therefore, it may concluded that depolarization-induced changes in membrane permeability are the cause of calcium movement in both situations. An intracellular influx of calcium upon depolarization is well known, for example, in smooth muscle contraction (33), electrical transmission (34) and the release of neurotransmitters (35). A K_e of 13 mM appears to be critically associated with a fall in Ca_e . This relationship has been further substantiated by the demonstration that the delay between the changes of potassium and calcium depends on blood flow although the value of K_e associated with the fall in Ca_e remains constant. The increase of extracellular potassium in ischaemia is undoubtedly due to a progressive overload of the potassium clearance mechanisms with increased leakage from the cells (25,36,37). It might be expected, therefore, that the time elapsing until the threshold of 13 mM K_e is reached depends upon blood flow. The increased leakage of intracellular potassium before any change occurs in Ca_e indicates a differential sensitivity of the ion homeostatic mechanisms to ischaemia.

The rates of potassium efflux expressed as minutes per pK can be described by two functions K_1 and K_2. Ca_1 describes the rate of fall of extracellular Ca_e as minutes per pCa (37). The relationships between K_1, K_2 and Ca_1 suggest that the availability of cellular energy may provide an alternative to depolarization as the explanation for calcium movement. K_2 is greater than K_1 in every case. It may be that efflux of potassium increases at a time when the reduced cellular energy levels can no longer maintain ion homeostasis. This is also the point when Ca_e begins to fall exponentially. The relationships between K_1 and both K_2 and Ca_1 strengthen this suggestion. The relationship between K_2 and Ca_1 suggests that the cell membrane becomes more permeable to both ions in the same proportion. The rise in extracellular potassium and fall in subarachnoid sodium activity in ischaemia as described by Hossmann and co-workers (38) have been explained as a redistribution down their concentration gradients. In view of the dramatic fall in extracellular calcium it is reasonable to offer the same explanation. Movement of calcium into the cells is corroborated to some extent by a calculation of the Ca_e that would occur if all calcium were evenly redistributed between the intra- and extracellular compartments in ischaemia. If we allow for the changes in compartment size as reported by Hossmann (38) and presume that there is no net change in total calcium, the value attained by Ca_e should be 0.25 mM. The final Ca_e found in densely ischaemic tissue (lCBF: 6 ml/100g min) in this preparation was 0.28 mM indicating that the calcium activity in the intra- and extracellular compartments were in equilibrium. Intracellular movement of calcium is further substantiated by Adey (39) who showed that, although calcium is complexed in the membrane, this effect is probably modest. As K_e rises to the threshold of 13 mM, Ca_e increases only slightly. The explanation for this may be a change in the binding coefficient for calcium in the extracellular space. A decrease in the extracellular space (ECS) without any net-loss in calcium is more likely, however. This would result in a concentration of Ca_e. Rapid changes of the ECS in ischaemia have been demonstrated by impedance measurements (40), or the use of choline and tris buffer as markers (41).

Hass (42) has recently suggested a hypothesis for the molecular mechanism of cellular damage by an increased intracellular calcium level. Shanne et al. (43) showed that certain agents caused toxic cell death in cultured rat hepatocytes only, if Ca_e was normal. If Ca_e was reduced to the level of its intracellular activity, these agents had no effect. Hass suggests that raised intracellular calcium stimulates breakdown of the phosphatidylinositol pool with serious consequences for neurotransmitter metabolism during reperfusion. Raised intracellular calcium activity would also stimulate phospholipase A_2 activity with release of free-fatty acids. A major free-fatty acid is arachidonic acid, the precursor of the prostaglandins and other related substances, which may have further deleterious effects upon ischaemic cerebral tissue. The results of this study show that the changes of calcium in ischaemia, although related to reduced blood flow, are closely associated with and follow the rise in K_e . We do not feel that it is possible to distinguish between depolarisation and a critical reduction in energy levels as the cause for the fall in Ca_e when K_e reaches 13 mM. The decrease of extracellular calcium is probably due to a progressive entry into the intracellular compartment until both compartments are in equilibrium.

THE RELATIONSHIP OF FLOW TO TISSUE VIABILITY

In many animals the somatosensory evoked response can be abolished by middle cerebral artery occlusion without added hypotension. It is particularly interesting to note that the area of functionally suppressed cortex is larger than the area which eventually infarcts in chronic experiments. A group of comparable animals was maintained over three years to study the characteristics of their clinical stroke. Detailed pathological investigations after perfusion fixation have been performed by Brierley and Symon (44). They indicate that the ultimate area of infarction is confined to tissue where blood flow in the acute stage of stroke is certainly reduced below 10 ml/100g min. Thus, in the acute stage of infarction, loss of function will assume a much wider distribution than the ultimate tissue destruction. Accordingly, Morawetz et al. (45) found histopathological signs of structural infarction following a 2 to 3 hour period of focal ischaemia in the monkey only at sites where local blood flow was below 10-12 ml/100g min.

We developed the concept from these observations that the area of ischemic tissue destruction in stroke is surrounded for some time in the acute phase by an area of functional neuronal suppression. The structural integrity of the neurons can immediately, and even permanently, be preserved in this area. This we termed "ischaemic penumbra". We have now some evidence that in the penumbra certain physiological mechanisms of the neuron remain intact, particularly those concerned with ionic homeostasis. Potassium-sensitive microelectrodes developed by Walker (46) and modified by Astrup and ourselves (25) enabled us to determine movements of extracellular potassium during ischaemia. Thus, changes of regional cerebral blood flow could be related to the failure of the ionic pumps and the flux of potassium into the extracellular space. Potassium has long been known to accumulate in the extracellular space during hypoxia. But in our experiments, control levels of extracellular potassium ranging from 3-9 mM (mean:5.7 $\pm$ 1.5 mM) were maintained with only minor changes at about the level of the threshold for electrical failure. Significant movements of potassium into the extracellular space occurred only when blood flow fell to 7-11 ml/100g min, which is significantly lower than the flow level leading to failure of electrical function. Therefore, in the penumbra a differential failure of neuronal metabolism must be assumed. There, synaptic transmission is impaired but the energy state and ionic balance are maintained. The metabolic rate of oxygen is presumably reduced to a minimal level by the arrest of function. It has been found that during progressive ischemia and termination of seizure activity, ATP is almost normal until a point at which potassium is released from the cells.

NEUROTRANSMITTER FUNCTION

We attempted with Davison and Bowen (47) to link the degree of ischaemia to the function of subcellular components, particularly to the uptake of neurotransmitters by synaptosomes. The evidence so far suggests that the synaptosomal uptake of GABA declines in relation to flow, which probably commences at a flow level higher than that required to interfere with electrical function. Cholinergic neurons, however, showed no such sensitivity. Their uptake remained unimpaired at much lower flow levels.

WATER MOVEMENT

Work from our own laboratory (48), and from Hossmann (49), has indicated a definite relationship between the development of cerebral swelling and the intensity of ischaemia. In our experimental model, we assessed brain water content in-vivo by measurements of brain impedance and intracranial pressure. Following sacrifice of the animal, density gradient columns of kerosene-bromobenzene were employed to study specific gravity of the edematous tissue. We find that significant ischaemia of 1 1/2 hours is associated with an increase in water content in the most densely ischaemic zones and in the penumbra. Thus, a significant relationship between water content and blood flow is evident. Uptake of water occurs when flow falls below 20 ml/100g min. The initial depth of ischaemia appears to act as a trigger for the influx of water into the ischemic brain, which thereafter advances through the hemisphere as described for cold oedema by Klatzo and Reulen.

Experiments indicate that reperfusion after 1 1/2 hours of ischaemia is associated with an increase rather than a decrease in oedema. Restoration of blood flow after a significant period of ischaemia may thus compound brain swelling. The evident thresholds of flow necessary to evoke brain swelling are significantly higher than those concerned with impairment of cell membrane permeability. Increase of vascular permeability to large molecules is unlikely to be the cause. There is clear evidence that the ischemic vasogenic oedema generally develops only after 4-6 hours and reaches its maximum after a few days. Leakage of large molecules such as pertechnetate or R.I.H.S.A. is maximal in an ischaemic stroke only after ten days to three weeks. O'Brien (50) has reported an increase in water content of cerebral cortex to 90% a few days after an experimental stroke. It seems probable that such oedema results from tissue necrosis, since it does not seem to be evoked by vascular occlusion of less than 6 hours. According to Morawetz, this would be expected to produce significant tissue loss. The early phase of brain oedema is cytotoxic rather than vasogenic. Our own and Hossmann's observations indicate that the degree of oedema is determined by the initial reduction of flow after vascular occlusion. Hossmann has also shown that one hour of complete ischaemia causes a significant increase in tissue osmolality creating a gradient of about 50 mOsm between brain and blood (51).

REPERFUSION AND RECOVERY

This study has shown that under various conditions of focal ischaemia reducing blood flow to a level sufficient to release potassium into the extracellular space, recovery of potassium toward the control level occurs upon reperfusion in the majority of cases. Complete failure of potassium to decrease was only noted at 1 electrode site of a total of 29. Recovery to the control level occurred in about half of the cases. Complete recovery was found in areas with significantly greater density and duration of ischemia as compared to incomplete recovery. Normalization of K_e was also associated with significantly higher flows during reperfusion, probably a reflection of the condition of the ischaemic phase.

The rate of recovery was not significantly correlated with the duration of occlusion, the minimum flow, or the maximal extracellular potassium level recorded during the clip. However, it was inversely correlated ($r = 0.5$, $p < 0.04$) with the total time of flow levels below 10 ml/100g min during occlusion. This flow threshold is close to the level, at which significant increases in K_e are first seen during progressive ischaemia (36). We have termed the time variable, thus defined, the "flow deficit" associated with that site (the units are in minutes). The slow component of the biexponential clearance of K_e did not appear to depend on the flow deficit. The slope of the regression curve was zero. When the MCA was occluded according to our experimental conditions, the degree of ischaemia in a given region determined whether the EP became depressed. Severe electrophysiological depression occurs when the local blood flow falls below 16 ml/100g min. This was previously established (23) and has been confirmed in the present study. It is in agreement with data on EEG monitoring and flow over wider brain areas with less localization (52,53). Additional corroborative data on cortical neurons has been presented by Heiss et al. (54). Our data indicate further, that maintenance of this degree of ischaemia for longer than 15 minutes results in incomplete recovery of the EP during reperfusion, at least for an hour following removal of the clip. During this period the local blood flow has been restored, but tissue pO_2 on average remains reduced well below control.

On the other hand, if the critical level of ischaemia is not attained, the EP is much less depressed. In the subsequent post-clip phase,
the EP recovers together with flow with a strong suggestion of tissue hyperoxia. These data may demonstrate a condition similar to the
"luxury-perfusion syndrome" according to Lassen (55). During EP recovery,
tissue pO_2 tends to be higher than normal, probably due to vasodilatation
accompanied or induced by local metabolic acidosis (56,57). The tissue
does not appear to use the abundant O_2 immediately, since functional recovery is gradual. However, an abundant O_2 -supply may be ultimately
necessary for complete recovery of function. The fact that the tissue
flow on average is not above normal throughout recovery does not
contradict this interpretation. Waltz (58) has pointed out that luxury
perfusion may only be relative. Flows associated with hyperoxia may well
be higher than elsewhere, yet lower than normal.

CLINICAL FEATURES OF AN EXPERIMENTAL INFARCT

Baboons subjected to middle cerebral artery occlusion had a fairly
typical clinical course. Over the first 24 to 36 hours the animals were
generally quiet. They could be handled without anaesthesia. A limited
neurological examination was possible, provided the approach was made
from the side of the hemiparesis. In all instances, a dense facial weakness was evident which, as a rule, persisted up to three years, although
tone appeared to recover after four months. We observed, however, that
food was retained much longer in the cheek pouch on the paralyzed side,
suggesting that complete muscle tone was never regained. The animals in-
variably showed some degree of hemianopia which persisted over the entire period of observation. It is impossible to state with confidence,
how far this represents an inattention defect or complete loss of the visual half field, since the testing was restricted to the reaction to me-
nace. Evidence of this could be obtained in all the animals as late as
three years, although the clinical testing of a substantially recovered
20-kg baboon without anesthesia leaves much to be desired. In all the
animals who recovered, leg weakness was minimal and evident only in the
first day or two. Thereafter, the animals were observed free-walking,
which was possible in the majority within three to four months. It was
almost impossible to detect any weakness of the leg. The animals would
leap to a height of 3 to 4 feet without difficulty and remain quite
straight.

In the first few weeks after operation, when unrestrained, the animals tended to circle. Within the first three to four days, circling was
in some instances almost continuous, resembling forced circling (59) in
hemiplegic dogs and monkeys. Later on, however, circling became a more
casual affair and appeared to arise only from visual inattention. The
animal walked straight if its attention was fixed. But when walking along
a corridor, it would tend to deviate as its attention became drawn to
something in its retained visual field. Therefore, forced circling was
not a characteristic of any of the animals over a prolonged period of

time, although a tendency to circle could be seen up to four months in many animals. From four months onward it was impossible to allow these animals unrestrained movement. The judgement of the occurrence of circling thereafter was very difficult. The activity of the animals could be well determined by the presence or absence of leaping when unrestrained. Characteristically, the animals, when allowed free from their cage would rush from one end to the other within the confines of a run leaping toward the heavily barred windows. This enabled the judgement of the use of the paralyzed hand. Only the most severely affected animals showed an absence of leaping lasting more than a few days.

The hemiparesis produced by MCA occlusion with involvement of the perforator bearing segment usually was fairly dense in the arm. Yet, complete abolition of arm movements, even in the first few days following surgery, was rare. It occurred in the two animals who died and in two of the more severely affected survivors. From ten days onward, however, all animals showed reasonable movement in the proximal muscles of the arm and evident recovery in the muscles of the elbow and wrist. A differentiation of stroke density could be performed best by considering the time scale of recovery of finger movements and of the reaching and placing reactions. Although these could not be tested in detail, the recovery of use of the forelimb could be well assessed, even by studying finger movements in reaching and placing. None of the animals recovered a completely normal forelimb. The least affected would by four months hold onto bars as they leaped in free-range. Movement of the fingers was evident on cage observation. The limb, however, was in no instance used as freely as its fellow. In one animal it was possible to watch for more than a few minutes and yet remaining uncertain from its upper limb movements which was the hemiparetic side. From four months on, the neurological result obtained did not change further.

HUMAN CLINICAL CORRELATES

Ischaemic thresholds in the clinic were first observed by correlating neurological function with hemispheric perfusion. In a number of aneurysm cases, postoperative hemiparesis associated with reduced cerebral blood flow could be abolished by elevation of blood flow using hypervolemia and hypertension. Induction of hypertension could be readily established. Clinical estimates of cerebral blood flow have, however, a large standard error. An accurate determination of focal blood flow in relation to function is difficult. We thus elaborated a technique of somatosensory evoked response recording in the clinic, using median nerve stimulation. We measured primarily the transit time from the dorsal column nuclei at the base of skull to the cortex. Thus, an acceptable relationship could be established between the outcome in aneurysm cases and the central conduction time (60). Translation to the circumstances of the operating theatre enabled us to monitor somatosensory evoked responses closely while the cerebral circulation was temporarily compromised. The monitoring was employed, e.g. during temporary clipping of major arteries in cerebrovascular reparative work, such as the obliteration and evacuation of giant intracranial aneurysms.

As in the primate, we found (61) that acute focal ischaemia in the cortex of man was associated with a failure of the evoked response. Central conduction time is more easily measured under these circumstances than the evoked response amplitude, which tends to be much more variable. Prolongation of conduction time from a normal level of 5.4 $\pm$ 0.4 msec to 10 msec was quite tolerable and would be readily reversed on release of the ischaemia. Complete disappearance of the evoked response could apparently be tolerated for periods of up to ten minutes with restoration and only minor and reversible impairment of neurological function. Protracted disappearance of the evoked potential with perhaps the reappearance of a poor and ill-sustained response would be associated with hemiparesis or hemiplegia. The attainment of clinical certainty must await more clinical data which, of course, is difficult to obtain, since nobody would wish to evoke permanent disappearance of an evoked response. It seems clear from observations already reported, however, that the threshold concept is valid within the terms of acute ischaemia in man and that the evoked response is a useful monitor of reversible ischaemia during aneurysm surgery.

It has become apparent also from detailed CT scan studies, that flow thresholds for the development of ischemic brain oedema in man as in animals, are significantly higher than the flow thresholds of irreversible tissue destruction or even functional loss. Following permanent ligation of an anterior cerebral artery during the excision of a giant aneurysm, the CT scan at twenty-four hours showed an area of diminished attenuation in the peripheral distribution of the cerebral vessel. A diminished attenuation is common in the immediate region of craniotomy, where its significance has scarcely been appreciated. However, the interpretation is plain in cases where the hypodensity extends in the distribution of a peripheral artery clearly beyond the area of operative intervention. The area of the anterior cerebral artery, functioning perfectly with no leg weakness nor neurological abnormality, showed a decreased density. This can be interpreted as an increased brain water content evolving at flow levels well above those of functional failure.

In a further case, changes of blood-brain barrier permeability occurred following the occlusion of a middle cerebral artery for 45 minutes during excision and clipping of a giant aneurysm. Reperfusion was followed by complete restoration of the normal neurological function. Some days later, the enhanced CT-scan showed clear evidence of permeability changes of the blood-brain barrier. Efflux of contrast medium was found into the region of the aneurysm, despite the maintenance of a perfectly normal function. Thus, changes of the permeability of the blood-brain barrier apparently occur at higher flow levels than those responsible for the failure of electrical function or ionic homeostasis.

REFERENCES

1. Collewijn H, and Van Harreveld A, Membrane potential of cerebral cortical cells during spreading depression and asphyxia, Exp Neurol 15:425 (1966).
2. Collewijn H, and Van Harreveld A, Intracellular recording of spinal motoneurones during acute asphyxia, J Physiol 185:1 (1966).
3. Van Harreveld A, and Tachibana S, Recovery of cerebral cortex from asphyxiation, Amer J Physiol 202:59 (1962).
4. Hossmann KA, Cortical steady potential, impedance and excitability changes during and after total ischemia of cat brain, Exp Neurol 32:163 (1971).
5. Brierley JB, Salford LG, Siesjö BK, and Plum F, Moderate hypoxic ischemia irreversibly damages rat brain in 30 minutes, Stroke 4:339 (1973).
6. Grossman RG, Turner JW, Miller JD, and Rowan JO, The relationship between cortical electrical activity, cerebral perfusion pressure and cerebral blood flow during increased intracranial pressure, Stroke 4:346 (1973).
7. Ingvar DH, Normal and postanoxic regulation of the regional cerebral blood flow, in: "Recent advances in the study of cerebral circulation", Taveras et al., eds., Thomas, Springfield, Illinois (1970).
8. Boysen G, Engell HC, and Trojaborg W, Effect of mechanical rCBF reduction on EEG in man, Stroke 4:361 (1973).
9. Meyer JS, and Marx P, The pathogenesis of EEG changes during cerebral anoxia, in: "Handbook of EEG and clinical neurophysiology", Remond A, ed., Elsevier, Amsterdam (1972).
10. Hossmann KA, and Sato K, Effect of ischemia on the function of the sensorimotor cortex in cat, Electroenceph Clin Neurophysiol 30:535 (1971).
11. Przybylski A, Activity pattern of visceral cortex neurons during asphyxia, Exp Neurol 32:12 (1971).
12. Harvey J, and Rasmussen T, Occlusion of the middle cerebral artery, Arch Neurol 66:20 (1951).
13. Symon L, Studies of leptomeningeal collateral circulation in macacus rhesus, J Physiol 159:68 (1961).
14. Symon L, Dorsch NWC, and Ganz JC, Lactic acid efflux from ischaemic brain, J Neurol Sci 17:411 (1972).
15. Yamaguchi T, Waltz AG, and Okazaki H, Hyperemia and ischemia in experimental cerebral infarction: Correlation of histopathology and regional blood flow, Neurology 21:565 (1971).
16. Symon L, Pasztor E, and Branston NM, The distribution and density of reduced cerebral blood flow following acute middle cerebral artery occlusion: An experimental study by the technique of hydrogen clearance in baboons, Stroke 5:355 (1974).
17. Kaplan HA, and Ford DH, "The brain vascular system", Elsevier, London (1966).
18. Lazorthes G, and Campan L, "La cirulation cerebrale", Editions Sandoz, Paris (1964).

19. Pasztor E, Symon L, Dorsch NWC, and Branston NM, The hydrogen clearance method in assessment of blood flow in cortex, white matter and deep nuclei of baboons, Stroke 4:556 (1973).
20. Heiss WD, Hayakawa T, and Waltz AG, Cortical neuronal function during ischemia - Effects of occlusion of one middle cerebral artery on single-unit activity in cats, Arch Neurol 33:813 (1976).
21. Trojaborg W, and Boysen G, Relation between EEG, regional cerebral blood flow and internal carotid artery pressure during carotid endarterectomy, Electroenceph Clin Neurophysiol 34:61 (1973).
22. Gregory PC, McGeorge AP, Fitch W, Graham DI, Mackenzie ET, and Harper AM, Effects of hemorrhagic hypotension on cerebral circulation. II. Electrocortical function, Stroke 10:719 (1979).
23. Branston NM, Symon L, Crockard HA, Pasztor E, Relationship between the cortical evoked potential and local cortical blood flow following acute middle cerebral artery occlusion in the baboon, Exp Neurol 45:195 (1974).
24. Marshall LF, Welsh F, Durity F, Lounsbury R, Graham DI, and Langfitt TW, Experimental cerebral oligemia and ischemia produced by intracranial hypertension. Part 3: Brain energy metabolism, J Neurosurg 43: 323 (1975).
25. Astrup J, Symon L, Branston NM, and Lassen NA, Cortical evoked potential and extracellular K^+ and H^+ at critical levels of brain ischemia, Stroke 8:51 (1977).
26. Astrup J, Blennow G, and Nilsson B, Effects of reduced cerebral blood flow on EEG pattern, cerebral extracellular potassium, and energy metabolism in the rat cortex during bicuculline-induced seizure, Brain Res 177:115 (1979).
27. Zimmerman ANE, and Hulsman WC, Paradoxical influence of calcium ions on the permeability of the cell membranes of the isolated rat heart, Nature 211:646 (1966).
28. Shen AC, and Jennings RB, Myocardial calcium and magnesium in acute ischemic injury, Am J Pathol 67:417 (1972).
29. Shen AC, and Jennings RB, Kinetics of calcium accumulation in acute myocardial ischemic injury, Am J Pathol 67:441 (1972).
30. Hearse DJ, Reperfusion of the ischemic myocardium, J Mol Cell Cardiol 9:605 (1977).
31. Kraig RP, and Nicholson C, Extracellular ionic variations during spreading depression, Neurosciences 3:1045 (1978).
32. Nicholson C, Dynamics of the brain cell microenvironment, Neurosci Res Prog Bull 18:177 (1980).
33. Van Breemen C, Calcium requirement for activation of intact aortic smooth muscle, J Physiol 272:317 (1977).
34. Baker PF, Hodgkin AL, and Ridgway EB, Depolarization and calcium entry in squid giant axons, J Physiol 218:709 (1971).
35. Katz B, and Miledi R, Further study of the role of calcium in synaptic transmission, J Physiol 207:789 (1970).
36. Branston NM, Strong AJ, and Symon L, Extracellular potassium activity, evoked potential and tissue blood flow, relationship during progressive ischaemia in baboon cerebral cortex,

J Neurol Sci 32:305 (1977).

37. Harris RJ, Symon L, Branston NM, and Bayhan M, Changes in extra-
    cellular calcium activity in cerebral ischaemia, J Cereb Blood
    Flow Metabol 1:203 (1981).

38. Hossmann KA, Sakaki S, and Zimmermann V, Cation activities in
    reversible ischaemia of the cat brain, Stroke 8:77 (1977).

39. Adey WR, Evidence for cerebral membrane effects of calcium derived
    from current gradient impedance and intracellular records,
    Exp Neurol 30:78 (1971).

40. Branston NM, Strong AJ, and Symon L, Impedance related to local
    blood flow in cerebral cortex, J Physiol 275:81P (1978).

41. Hansen AJ, and Olsen CE, Brain extracellular space during spreading
    depression and ischaemia, Acta Physiol Scand 108:355 (1981).

42. Hass WK, Beyond cerebral blood flow, metabolism and ischaemic
    thresholds: an examination of the role of calcium in the initia-
    tion of cerebral infarction, in: "Cerebral vascular disease 3",
    Meyer JS et al., eds., Excerpta Medica, Amsterdam (1981).

43. Schanne FAX, Kane AB, Young EE, and Farber JL, Calcium depen-
    dence of toxic cell death: A final common pathway,
    Science 206:700 (1979).

44. Brierley JB, and Symon L, The extent of infarcts in baboon brain
    three years after division of the middle cerebral artery,
    J Neuropath Appl Neurobiol 3:217 (1977).

45. Morawetz RB, De Girolami U, Ojemann RG, Marcoux FW, and Cro-
    well RM, Cerebral blood flow determined by hydrogen clearance
    during middle cerebral artery occlusion in unanaesthetized mon-
    keys, Stroke 9:143 (1978).

46. Walker JL, Specific liquid ion exchanger micro electrodes,
    Analytical Chemistry 43:83a (1971).

47. Bowen MD, Goodhard NJ, Strong AJ, Smith SM, White B, Branston NM,
    Symon L, and Davison AN, Biochemical indices of brain struc-
    ture, function, and hypoxia in cortex from baboons with middle
    cerebral artery occlusion, Brain Research 177:503 (1976).

48. Symon L, Branston NM, and Chikovani O, Ischaemic brain oedema fol-
    lowing middle cerebral artery occlusion in baboons. Relationship
    between regional cerebral water content and blood flow at 1-2
    hours, Stroke 10:184 (1979).

49. Hossmann KA, and Schuier FJ, The metabolic (cytotoxic) type of
    brain oedema following middle cerebral artery occlusion in cats,
    in: "Cerebrovascular diseases", Price T, Nelson E, eds., Raven,
    New York (1979).

50. O'Brien MD, Waltz AG, and Jordan NM, Ischaemic cerebral oedema.
    Distribution of water in brains of cats after occlusion of the
    middle cerebral artery, Arch Neurol 30:456 (1974).

51. Hossmann KA, and Takagi S, Osmolality of brain in cerebral ischaemia,
    Exp Neurol 51:124 (1976).

52. Sharbrough FW, Messick JM, and Sundt TM, Correlation of continuous
    electroencephalograms with cerebral blood flow measurements
    during carotid endarterectomy, Stroke 4:674 (1973).

53. Trojaborg W, and Boysen G, Relation between EEG, regional cerebral

blood flow and internal carotid artery pressure during carotid endarterectomy, Electroenceph Clin Neurophysiol 34:61 (1973).

54. Heiss WD, Waltz AG, and Hayakawa T, Neuronal function and local blood flow during experimental cerebral ischaemia, in: "Blood flow and metabolism in the brain", Harper AM, Jennett WB, Miller JD, Rowan JO, eds., Churchill Livingstone, London (1975).

55. Lassen NA, The luxury-perfusion syndrome and its possible relation to acute metabolic acidosis localized within the brain, Lancet 2:113 (1966).

56. Zwetnow NN, Effects of increased cerebrospinal fluid pressure on the blood flow and on the energy metabolism of the brain, Acta Physiol Scand Suppl 339:1 (1970).

57. Symon L, Ganz JC, and Dorsch NWC, Experimental studies of hyperaemic phenomena in the cerebral circulation of primates, Brain 95:265 (1972).

58. Waltz AG, Red venous blood: Occurrence and significance in ischemic and non-ischemic cerebral cortex, J Neurosurg 31:141 (1969).

59. Harvey J, and Rasmussen T, Occlusion of the middle cerebral artery: An experimental study, Arch Neurol Psychiat 66:20 (1951).

60. Symon L, Hargadine J, Zawarski M, and Branston NM, Central conduction time as an index of ischaemia in subarachnoid haemorrhage, J Neurol Sci 44:95 (1979).

61. Wang AD, Cone J, Symon L, and Costa de Silva IE, Somatosensory evoked potential monitoring during the management of aneurysmal subarachnoid haemorrhage, J Neurosurg 60:264 (1984)

THE ROLE OF RECIRCULATION FOR FUNCTIONAL AND METABOLIC RECOVERY AFTER CEREBRAL ISCHEMIA

Konstantin-A. Hossmann

Max-Planck-Institut für neurologische Forschung, Abteilung für experimentelle Neurologie, Cologne, FRG

Disturbances of brain function and metabolism induced by ischemia are primarily caused by a reduction of the supply of oxygen and nutrients from the blood to the brain, and a reduction of the removal of metabolic waste products from the brain into the blood. There are indications, however, that this is not the only reason for the development of ischemic brain lesions. When blood flow is restored to the brain, damage may continue in a self-propagating way, and secondary disturbances may appear at a later time, even after the brain had already started to recover from the primary ischemic impact.

There is considerable controversy about the pathogenesis of these post-ischemic effects. Some authors stress the importance of early biochemical changes which are evoked during ischemia but lead to irreversible cell damage after a latent "maturation phase" which may last for several days (Klatzo, 1975). The best known example of such processes is delayed neuronal death in the CA1 sector of hippocampus after ischemia as short as 5 min (Kirino, 1982). A basically different process causing permanent or delayed lesions is that associated with post-ischemic disturbances of blood flow (Ames et al., 1968; Hossmann et al., 1973). Such lesions are not the direct consequence of the primary ischemic impact but develop at a later phase when blood flow within the previously ischemic territory is secondarily impaired. At present, little is known about the relationship between early biochemical and delayed circulatory disturbances, and even less about the actual significance of such disturbances for the final outcome of an ischemic insult. It is obvious that post-ischemic circulation disturbances must prevent eo ipso post-ischemic recovery but it remains unclear, if such disturbances are merely an epiphenomenon of irreversible brain damage rather than its cause.

As long as this question has not been solved, recirculation disturbances must be considered to be potentially limiting for the resistance of the central nervous system to ischemia, and therefore have received considerable attention. In the following, a brief survey of current knowledge about recirculation disturbances is given, and discussed in respect to the functional and metabolic recovery process.

DEFINITIONS

Ischemia is the reduction of cerebral blood flow below the requirements of brain tissue for maintenance of normal functional and biochemical integrity. Depending on the degree and extent of ischemia, complete or incomplete, and global or local ischemia are distinguished. The effects of reduced blood flow can be enhanced by additional reduction of oxygen- or substrate content of the arterial blood (hypoxic-anoxic ischemia, hypoglycemic ischemia, etc.). Pre-ischemic events are those occurring before the onset of ischemia, and post-ischemic events those which occur after the cause of ischemia - vascular occlusion, hypotension, etc. - has been reversed. During the post-ischemic period, circulation disturbances and, hence, "secondary ischemia" may appear which is only indirectly linked to the primary ischemic impact. Such disturbances may be present at the beginning of the post-ischemic period (no-reflow phenomenon) or may develop after a period of unimpaired or even increased reperfusion (post-ischemic hypoperfusion). These two forms of post-ischemic recirculation disturbances are basically different and will be described separately.

EARLY RECIRCULATION DISTURBANCES (NO–REFLOW PHENOMENON)

The term "no-reflow phenomenon" was introduced by Ames et al. (1968) who observed reperfusion deficits of the rat brain after strangulation ischemia of more than 7-10 min. In earlier investigations of Hirsch and Müller (1962) a vascular pattern of histological damage was also noted but not considered to be of pathogenetic relevance because the revival time of the brain was thought to be shorter than 10 min. In the following, however, numerous authors have stressed the importance of homogeneous reperfusion of the tissue for reactivation of metabolic activity, not only in the brain (Ginsberg and Myers, 1972; Gurvitch et al., 1976; Deb et al., 1978) but also in peripheral organs such as heart (Kloner et al., 1974; Tranum-Jensen et al., 1981) and kidney (Flores et al., 1972; Brodman et al., 1974). The pathogenesis of the no-reflow phenomenon is multifactorial. Most important factors are post-ischemic hypotension, changes in blood viscosity, intravascular disseminated coagulopathy, and brain edema.

Post-ischemic hypotension depends on the type and duration of ischemia. If blood flow of the brain is interrupted without affecting cardiac function, blood pressure initially tends to increase because reduction of oxygen supply to the brain stem evokes a Cushing response. This response may last up to 10 min and is followed by a return of blood-pressure to or below the pre-ischemic level when ischemia persists. Beginning of recirculation in such instances causes a further decrease of blood pressure because the release of acid equivalents from the ischemic tissue into the systemic circulation causes vasodilation of the peripheral bed. After ischemia of 30 to 60 min, blood pressure without pharmacological support may decrease to values below 50 mmHg, precluding adequate recirculation of the ischemic brain (Hossmann and Kleihues, 1973). If cerebral blood flow ceases as a consequence of cardiac arrest, even shorter periods of ischemia may be complicated by post-ischemic hypotension because the ischemic heart does not resume immediately its function. This explains why the revival time of the brain after cardiac arrest is shorter than that after selective ischemia of the brain (Hirsch et al., 1957).

Disseminated intravascular coagulopathy occurs shortly after the onset of ischemia and further increases in severity during the early recirculation period (Stullken and Sokoll, 1976; Hossmann and Hossmann, 1977). The main reason for intravascular coagulopathy is a massive ischemia-induced sympathotonic discharge, which is further aggravated by administration of sympathomimetics necessary for stabilization of blood pressure during the early recirculation period. The effect of intravascular coagulation on post-ischemic recirculation is at least threefold. Aggregation of platelets in the ischemic vascular bed causes mechanical obstruction (Hekmatpanah, 1973; Dougherty et al., 1977), the release of serotonin from aggregated platelets induces vasoconstriction (Welch et al., 1972), and the additional involvement of peripheral organs causes general disturbances with adverse effects on the brain (Hossmann et al., 1980). The most important is pulmonary distress causing impairment of blood oxygenation and, consequently, an increase of post-ischemic brain edema (see below) if not adequately treated (Hossmann and Hossmann, 1977). Intravascular coagulation is further aggravated by changes of blood viscosity which increases as blood flow slows down (Schmid-Schönbein, 1977). This increase is rapidly reversed as soon as blood starts flowing, but for initiation of recirculation a higher perfusion pressure is necessary than for maintaining flow.

The effect of ischemic brain edema on post-ischemic recirculation is complex (Hossmann, 1976). The main reasons for the development of ischemic brain edema are an increase of intracellular osmolality (Hossmann and Takagi, 1976; Bandaranayake et al., 1978) and the breakdown of membrane potentials (Astrup et al., 1977) which results in an equilibration of intra/extracellular ion gradients. As a consequence osmotic and ionic gradients build up between blood and extracellular fluid, causing uptake of water and sodium from blood into brain. If ischemia is complete, development of brain edema is negligible because cessation of blood flow also interrupts supply of water and sodium (Hossmann, 1976). However, in such instances formation of brain edema starts at the moment of blood recirculation and continues until either osmotic and

ionic gradients are equilibrated or until restitution of metabolic activity restores ionic and osmotic homeostasis of brain tissue (Hossmann,
1976). If ischemia is incomplete, and sodium and water supply to the
brain persists, edema develops already during ischemia and may impede cerebral blood flow from the beginning of recirculation. Such disturbances
are most severe in hyperglycemic animals because the high concentration
of lactate increases tissue osmolality (Ginsberg et al., 1978). The situation is further complicated by the fact that ischemia causes an increase
of extracellular potassium which, in turn, produces vasoconstriction (Wade et al., 1975; Hart et al., 1978; Hansen et al., 1980). During ischemia
this effect is counteracted by the simultaneous decrease of pH, but
during the early recirculation phase normalization of extracellular potassium and pH may proceed with different time courses. All these factors
make it almost impossible to foresee, in a given experiment, to what extent edema will interfere with blood recirculation. It also explains that
blood recirculation after ischemia may be initially restored but ceases a
few minutes later, when ischemic brain edema becomes critical. The no-
reflow phenomenon, in a wider sense, may therefore be defined as the
phenomenon of early recirculation disturbances.

Treatment of no-reflow is possible by interfering with these pathogenetic factors. Increasing post-ischemic blood perfusion pressure reduces the area of no-reflow and improves homogeneity of microcirculation
(Cantu et al., 1969; Appelgren, 1972; Tweed et al., 1977; Ito et al.,
1980). A beneficial effect is also obtained by preventing aggregation of
platelets with indomethacin and prostacyclin (Hallenbeck and Furlow,
1979), by diminishing intravascular coagulation with heparin (Stullken
and Sokoll, 1976) or streptokinase (Lin, 1978), by reducing blood viscosity with hemodilution (Fischer and Ames, 1972; Siemkowicz, 1980) and by
treating ischemic brain edema with osmotic dehydration (Hossmann and
Takagi, 1976). In our laboratory a combination therapy consisting of an-
ticoagulation with heparin, osmotherapy with 20 % sorbit, controlled
equilibration of blood acidosis with Tris-buffer, and induced hypertension with norepinephrine or dopamine is able to prevent no-reflow in most
of the experiments even after 1 hour complete ischemia (Hossmann and
Zimmermann, 1974). In such instances reactive hyperemia develops and
metabolic and electrophysiological functions begin to recover as early as
8 to 10 min after beginning of recirculation (Hossmann et al., 1973). In
animals with manifest no-reflow, on the other hand, recovery is absent.
This is why after prolonged ischemia a close correlation is found between
the degree of post-ischemic hyperemia and the rate and quality of post-
ischemic recovery (Hossmann et al., 1973). It should be noted, however,
that this opinion is not shared by all authors because, under certain experimental conditions, neuronal damage has been observed in the absence
of no-reflow (Levy et al., 1975; Harrison et al., 1975).

POST-ISCHEMIC HYPOPERFUSION

Development of post-ischemic hypoperfusion is independent of the state of early post-ischemic recirculation. In fact, it develops regularly in animals with pronounced reactive hyperemia, although a direct relationship between the degree of hyperemia and post-ischemic hypoperfusion does not seem to exist (Hossmann et al., 1973; Snyder et al., 1975; Levy et al., 1979; Miller et al., 1980; Pulsinelli et al., 1982; White et al., 1983). It is interesting to note, however, that also in other instances of transiently increased blood flow, such as hypoglycemia (Siesjö and Abdul-Rahman, 1979; Abdul-Rahman et al., 1980), epilepsy (Meldrum and Nilsson, 1976; Ingvar et al., 1981), spreading depression (Lauritzen et al., 1982), or subarachnoid hemorrhage (Mendelow et al., 1981), a secondary reduction of blood flow below control has been observed. Angiographic findings as well as vital microscopy of the pial vasculature have revealed that post-ischemic hypoperfusion, other than the no-reflow phenomenon, is due to an increase in vascular tone (Hossmann et al., 1973; Takagi et al., 1977). Post-ischemic brain edema, intravascular coagulation or hypotension are not involved, as evidenced by normalization of water and ion homeostasis, disappearance of intravascular platelet aggregates, and normalization of blood pressure before post-ischemic hypoperfusion begins to develop (Hossmann and Kleihues, 1973).

The leading pathophysiological symptom of post-ischemic hypoperfusion is the disappearance of CO_2 -reactivity in the presence of a normal autoregulatory response to increasing blood pressure (Hossmann et al., 1973; Nemoto et al., 1975; Miller et al., 1980). This dissociated disturbance of flow regulation results in stabilization of blood flow at a subnormal level which cannot be influenced by either changes in cerebral perfusion pressure or activation of metabolism. The latter is of particular importance because failure of metabolic regulation may result in tissue hypoxia when arterial oxygen supply does not or only partially cover the oxygen demands of the recovering brain. In this situation anaerobic glucose utilization is stimulated, as evidenced by an increase of the glucose/oxygen uptake ratio of the brain, leading to secondary brain edema (Hossmann, 1979). Interestingly, anaerobic glycolysis begins to appear already at a time, when only 50 % of available blood oxygen is extracted (Hossmann, 1979), indicating that during post-ischemic hypoperfusion a substantial degree of non-nutritional flow must exist. There is evidence that post-ischemic hypoperfusion is not an epiphenomenon but a limiting factor for progression of recovery of the brain after ischemia. In our laboratory, we regularly observe that following ischemia of up to 1 hour duration evoked potentials and EEG activity steadily improve during the initial 2 to 3 hours of recirculation but then frequently deteriorate and even secondarily disappear after 4 to 6 hours. Autopsy of such brains reveals severe edema with secondary circulatory arrest. In other instances in which post-ischemic hypoperfusion is less severe, EEG continues to improve and may even normalize, if the animals survive for more than 12 hours (Hossmann and Zimmermann, 1974).

Improvement of blood flow during the phase of post-ischemic hypoperfusion succeeded up to now only by lowering viscosity with hemodilution (Hossmann et al., 1973; Hossmann et al., 1981; Bleyaert et al.,
1980). However, the increase of flow is paralleled by a decrease of the
oxygen-binding capacity of arterial blood. Therefore, the calculated oxygen availability to the brain does not improve. Vasoactive substances
such as papaverine, phenoxybenzamine, and prostacyclin did not improve
blood flow or the disturbed regulation of blood circulation (Hossmann et
al., 1973; van den Kerckhoff et al., 1983). Attempts have also been made
to decrease metabolic activity of the brain during post-ischemic hypoperfusion in order to ameliorate the disparity between oxygen supply and requirements of the tissue. For this purpose, pentobarbital, thiopental or
hypothermia were used, but none of these approaches had a reproducible
effect on post-ischemic recovery, mainly because of a further reduction
of blood flow (Snyder et al., 1978; Snyder et al., 1979; van den Kerckhoff et al., 1980; Gisvold et al., 1982). Recently, it has been hypothesized, that blood flow and the efficiency of mitochondrial respiration
can be improved by reducing ischemic and post-ischemic intracellular calcium uptake (Hass, 1981; Siesjö, 1981). In some experiments the calcium
antagonists lidoflazine and nimodipine had, in fact, a beneficial effect
on both post-ischemic blood flow and functional recovery (Kazda et al.,
1982; White et al., 1982; Steen et al., 1983). However, another calcium
entry blocker, flunarizine, failed to reduce post-ischemic calcium accumulation in the brain after 60 min global ischemia and had no beneficial effect on biochemical or functional recovery (Hossmann et al.,
1983). The therapeutic possibilities for improvement of post-ischemic hypoperfusion, in consequence, are very poor, and further progress in brain
resuscitation can only be expected when this pertinent problem has been
solved. Without adequate treatment of post-ischemic hypoperfusion it will
also be impossible to determine whether delayed biochemical alterations
will interfere with progression of recovery after prolonged ischemia in a
similar way as in CA1 neurons of hippocampus after 5 min circulatory
arrest. It, therefore, appears to be of great importance to develop experimental procedures which allow the maintenance of unimpaired blood
supply to the brain during this critical period in order to solve this
pertinent question.

REFERENCES

Abdul-Rahman A, Agardh CD, and Siesjö BK, Local cerebral blood flow in
        the rat during severe hypoglycemia, and in the recovery period
        following glucose injection, Acta Physiol Scand 109:307 (1980).
Ames A III, Wright RL, Kowada M, Thurston JM, and Majno G, Cerebral
        ischemia. II. The no-reflow phenomenon, Am J Pathol 52:437
        (1968).
Appelgren KL, Effect of perfusion pressure and hematocrit on capillary
        flow and transport in hyperemic skeletal muscle of the dog,
        Microvasc Res 4:231 (1972).
Astrup J, Symon L, Branston NM, and Lassen NA, Cortical evoked poten-

tial and extracellular K^+ and H^+ at critical levels of brain ischemia, Stroke 8:51 (1977).

Bandaranayake NM, Nemoto EM, and Stezoski SW, Rat brain osmolality during barbiturate anesthesia and global brain ischemia, Stroke 9:249 (1978).

Bleyaert A, Safar P, Nemoto EM, and Stezoski SW, No amelioration of brain damage after global brain ischemia (GBI) in monkeys by hemodilution or heparinization, Crit Care Med 8:251 (1980).

Brodman RF, Hackett RL, Finlayson B, and Pfaff WW, Microangiography of the renal vasculature following total renal artery occlusion, Surgery 75:734 (1974).

Cantu RC, Ames A III, DiGiacinto G, and Dixon J, Hypotension: A major factor limiting recovery from cerebral ischemia, J Surg Res 9:525 (1969).

Deb S, Sensharma GC, and Singh S, Neuronal-vascular relationship in experimental ischaemic anoxia, Acta Antom 102:254 (1978).

Dougherty JH, Levy DE, and Weksler BB, Experimental cerebral ischemia produces platelet aggregates, Neurology 27:382 (1977).

Fischer EG, and Ames A III, Studies on mechanisms of impairment of cerebral circulation following ischemia: effect of hemodilution and perfusion pressure, Stroke 3:538 (1972).

Flores J, DiBona DR, Beck CH, and Leaf A, The role of cell swelling in ischemic renal damage and the protective effect of hypertonic solute, J Clin Invest 51:118 (1972).

Ginsberg MD, and Myers RE, The topography of impaired microvascular perfusion in the primate brain following total circulatory arrest, Neurology 22:998 (1972).

Ginsberg MD, Welsh FA, and Budd WW, Effect of glucose infusion on the brain's response to diffuse ischemia, Stroke 9:98 (1978).

Gisvold SE, Safar P, Hendrickx HH, and Alexander H, Failure of thiopental (Th) to ameliorate brain damage after global brain ischemia (GBI) in pigtail monkey, Crit Care Med 10:206 (1982).

Gurvitch AM, Blinkov SM, Valanchute AL, and Nikolayenko EM, Types of no-reflow phenomenon observed during arrest of cerebral circulation and postischemic period, Crit Care Med 4:132 (1976).

Hallenbeck JM, and Furlow TW Jr, Prostaglandin I_2 and indomethacin prevent impairment of post-ischemic brain reperfusion in the dog, Stroke 10:629 (1979).

Hansen AJ, Gjedde A, and Siemkowicz E, Extracellular potassium and blood flow in the post-ischemic rat brain, Pflügers Arch 389:1 (1980).

Harrison MJ, Sedal L, Arnold J, and Russel RWR, No-reflow phenomenon in the cerebral circulation of gerbil, J Neurol Neurosurg Psychiatry 38:1190 (1975).

Hart MN, Sokoll MD, Davies LR, and Henriquez E, Vascular spasm in cat cerebral cortex following ischemia, Stroke 9:52 (1978).

Hass WR, Beyond cerebral blood flow, metabolism and ischemic thresholds: An examination of the role of calcium in the initiation of cerebral infarction, in: "Cerebral vascular disease 3", Meyer JS, Lechner H, Reivich M, Ott EO, Aranibar A, eds., Excerpta Medica, Amsterdam, Oxford, Princeton (1981).

Hekmatpanah J, Cerebral blood flow dynamics in hypotension and cardiac arrest, Neurology 23:174 (1973).

Hirsch H, Euler KH, and Schneider M, Über die Erholung und Wiederbelebung des Gehirns nach Ischämie bei Normothermie, Pflügers Arch 265:281 (1957).

Hirsch H, und Müller HA, Funktionelle und histologische Veränderungen des Kaninchengehirns nach kompletter Gehirnischämie, Pflügers Arch 275:277 (1962).

Hossmann KA, Development and resolution of ischemic brain swelling, in: "Dynamics of brain edema", Pappius HM, Feindel W, eds., Springer, Berlin, Heidelberg (1976).

Hossmann KA, Cerebral dysfunction related to local and global ischemia of the brain, in: "Brain function in old age", Hoffmeister F, Müller C, eds., Springer, Berlin-Heidelberg-New York (1979).

Hossmann KA, and Hossmann V, Coagulopathy following experimental cerebral ischemia, Stroke 8:249 (1977).

Hossmann KA, van den Kerckhoff W, and Matsuoka Y, Treatment of cerebral ischemia by hemodilution, Bibliotheca Haemat 47:77 (1981).

Hossmann KA, and Kleihues P, Reversibility of ischemic brain damage, Arch Neurol 29:375 (1973).

Hossmann KA, Lechtape-Grüter H, and Hossmann V, The role of cerebral blood flow for the recovery of the brain after prolonged ischemia, Z Neurol 204:281 (1973).

Hossmann KA, Paschen W, and Csiba L, Relationship between calcium accumulation and recovery of cat brain after prolonged cerebral ischemia, J Cereb Blood Flow Metabol 3:346 (1983).

Hossmann KA, and Takagi S, Osmolality of brain in cerebral ischemia, Exp Neurol 51:124 (1976).

Hossmann KA, and Zimmermann V, Resuscitation of the monkey brain after 1 h complete ischemia. I. Physiological and morphological observations, Brain Res 81:59 (1974).

Hossmann V, Hossmann KA, and Takagi S, Effect of intravascular platelet aggregation on blood recirculation following prolonged ischemia of the cat brain, J Neurol 222:159 (1980).

Ingvar M, Nilsson B, and Siesjö BK, Local cerebral blood flow in the brain during bicuculline-induced seizures and the modulating influence of inhibition of prostaglandin synthesis, Acta Physiol Scand 111:205 (1981).

Ito U, Ohno K, Yamaguchi T, Tomita H, Inaba H, and Kashima M, Transient appearance of "no-reflow" phenomenon in Mongolian gerbils, Stroke 11:517 (1980).

Kazda S, Garthoff B, Krause HP, and Schloßmann K, Cerebro-vascular effects of the calcium antagonistic dihydropyridine derivative nimodipine in animal experiments, Drug Res 32:331 (1982).

Kerckhoff van den W, Hossmann KA, and Hossmann V, No effect of prostacyclin on blood flow and blood coagulation following global cerebral ischemia, Stroke 14:724 (1983).

Kerckhoff van den W, Matsuoka Y, Paschen W, and Hossmann KA, Influence of barbiturates, hypothermia and hemodilution on post-ischemic metabolism and functional recovery following cerebro-circulatory arrest in cats, in: "Circulatory and developmental aspects of brain metabolism", Spatz M, Mrsulja BB, Rakic LJ, eds., Plenum, New York (1980).

Kirino T, Delayed neuronal death in the gerbil hippocampus following ischemia, Brain Res 239:57 (1982).

Klatzo I, Pathophysiologic aspects of cerebral ischemia, in: "The nervous system, 1", Tower DB, ed., Raven Press, New York (1975).

Kloner RA, Ganote CE, and Jennings RB, The "no-reflow" phenomenon after temporary coronary occlusion in the dog, J Clin Invest 54:1496 (1974).

Lauritzen M, Jorgensen MB, Diemer NH, Gjedde A, and Hansen HJ, Persistent oligemia of rat cerebral cortex in the wake of spreading depression, Ann Neurol 12:469 (1982).

Levy DE, Brierley JF, and Plum F, Ischaemic brain damage in the gerbil in the absence of "no-reflow", J Neurol Neurosurg Psychiatry 38:1197 (1975).

Levy DE, van Uitert RL, and Pike CL, Delayed post-ischemic hypoperfusion: A potentially damaging consequence of stroke, Neurology 29:1245 (1979).

Lin SR, The effect of dextran and streptokinase on cerebral function and blood flow after cardiac arrest. An experimental study on the dog, Neuroradiology 16:340 (1978).

Meldrum BS, and Nilsson B, Cerebral blood flow and metabolic rate early and late in prolonged epileptic seizures induced in rats by bicuculline, Brain 99:523 (1976).

Mendelow AD, McCalden TA, Hattingh J, Coull A, Rosendorff C, and Eidelman BH, Cerebrovascular reactivity and metabolism after subarachnoid hemorrhage in baboons, Stroke 12:58 (1981).

Miller CL, Lampard DG, Alexander K, and Brown WA, Local cerebral blood flow following transient cerebral ischemia. I. Onset of impaired reperfusion within the first hour following global ischemia, Stroke 11:534 (1980).

Miller CL Alexander K, Lampard DG, Brown WA, and Griffiths R, Local cerebral blood flow following transient cerebral ischemia. II. Effect of arterial pCO_2 on reperfusion following global ischemia, Stroke 11:542 (1980).

Nemoto EM, Snyder JV, Carroll RG, and Morita H, Global ischemia in dogs: Cerebrovascular CO_2 reactivity and autoregulation, Stroke 6:425 (1975).

Pulsinelli WA, Levy DE, and Duffy TE, Regional cerebral blood flow and glucose metabolism following transient forebrain ischemia, Ann Neurol 11:499 (1982).

Schmid-Schönbein H, Microrheology of erythrocytes, blood viscosity and the distribution of blood flow in the microcirculation, in: "Handbuch der allgemeinen Pathologie III/7 Mikrozirkulation", Altmann HW, Büchner F, Cottier H, Grundmann E, Holle G, Letterer E, Masshoff W, Meessen H, Roulet F, Seifert G, Siebert G, eds., Springer, Berlin-Heidelberg-New York (1977).

Siemkowicz E, Cerebrovascular resistance in ischemia, Pflügers Arch

388:243 (1980).

Siesjö BK, Cell damage in the brain: A speculative synthesis, J Cereb Blood Flow Metabol 1:155 (1981).

Siesjö BK, and Abdul-Rahman A, Delayed hypoperfusion in the cerebral cortex of the rat in the recovery period following severe hypoglycemia, Acta Physiol Scand 106:375 (1979).

Snyder JV, Nemoto EM, Carroll RG, and Safar P, Global ischemia in dogs: intracranial pressures, brain blood flow and metabolism, Stroke 6:21 (1975).

Snyder BD, Ramirez-Lassepas M, and Sukhum P, Failure of penthotal to protect from anoxic cerebral injury, Stroke 9:99 (1978).

Snyder BD, Ramirez-Lassepas M, Sukhum P, Fryd D, and Sung JH, Failure of thiopental to modify global anoxic injury, Stroke 10:135 (1979).

Steen PA, Newberg LA, Milde JH, and Michenfelder JD, Nimodipine improves cerebral blood flow and neurologic recovery after complete cerebral ischemia in the dog, J Cereb Blood Flow Metabol 3:38 (1983).

Stullken EH Jr, and Sokoll MD, The effects of heparin on recovery from ischemic brain injuries in cats, Anesth Analg 55:683 (1976).

Takagi S, Cocito L, and Hossmann KA, Blood recirculation and pharmacological responsiveness of the cerebral vasculature following prolonged ischemia of cat brain, Stroke 8:707 (1977).

Tranum-Jensen J, Janse MJ, Fiolet JWT, Krieger WJG, Naumann d'Alnoncourt VC, and Durrer D, Tissue osmolality, cell swelling, and reperfusion in acute regional myocardial ischemia in the isolated porcine heart, Circ Res 49:364 (1981).

Tweed WA, Wade JG, and Davidson WJ, Mechanisms of the "low-flow" state during resuscitation of the totally ischemic brain, Can J Neurol Sci 4:19 (1977).

Wade JG, Amtorp W, and Sorensen SC, The "low-flow" state in cerebral ischemia, Arch Neurol 32:381 (1975).

Welch KMA, Meyer JS, Teraura T, Hashi K, and Shinmaru S, Ischemic anoxia and cerebral serotonin levels, J Neurol Sci 16:85 (1972).

White BC, Gadzinski DS, Hoehner PJ, Krome C, Hoehner T, White JD, and Trombley JH Jr, Effect of flunarizine on canine cerebral cortical blood flow and vascular resistance post cardiac arrest, Ann Emerg Med 11:119 (1982).

White BC, Winegar CP, Henderson O, Jackson RE, Krause G, Ohara T, Goodin T, and Vigor ND, Prolonged hypoperfusion in the cerebral cortex following cardiac arrest and resuscitation in dogs, Ann Emerg Med 12:414 (1983).

POST-ISCHEMIC PATHOPHYSIOLOGY IN THE GERBIL BRAIN -

CHANGES OF EXTRACELLULAR K^+ AND CA^{++}

T. Yamaguchi, H.G. Wagner and I. Klatzo

Lab. Neuropath. Neuroanat. Sci., National Institute of
Neurological and Communicative Disorders and Stroke
National Institutes of Health, Bethesda, Md 20892

INTRODUCTION

It is increasingly recognized that following temporary ischemia cerebral tissue damage may further progress. A protracted development of
ischemic injury has been described as "maturation" phenomenon (Ito et
al., 1975; Klatzo, 1975). One of its features appears to be the direct
relationship between the intensity of the insult and the rate of maturation. A similar relationship has also been observed following recirculation employing biochemical and other parameters (Klatzo, 1975). Irrespective of unequivocal evidence indicating progression of ischemic tissue
damage after recirculation, it remains largely obscure which factors are
essential and which are merely coincidental. It is also apparent that the
selective vulnerability of various brain tissue elements may considerably
influence post-ischemic pathology. In order to obtain further insight into the role of the different factors operative in post-ischemic pathophysiology, a model of short-lasting cerebral ischemia was used in which
a protracted unfolding of various pathophysiological events could be
expected.

MATERIAL AND METHODS

Five minute cerebral ischemia in adult (12-14 weeks old) gerbils
was produced by bilateral clamping of the common carotid arteries under light halothane anesthesia. The following parameters were studied
during ischemia and at various time intervals after recirculation as earlier reported (Suzuki et al., 1983; Suzuki et al., 1983a):

(1) Morphological changes,
(2) Regional cerebral blood flow (rCBF),
(3) Local cerebral glucose utilization (LCGU),
(4) Blood-brain barrier (BBB) function, and
(5) Spontaneous neuronal activity.

Respective observations were also obtained from 3-weeks old animals. Measurements of the extracellular K^+ - and Ca^{++} -concentrations of the brain were carried out in adult gerbils according to the method described below. Ion-selective electrodes (Walker, 1971) were employed to measure continuously and simultaneously K_e^+ and Ca_e^{++}. The procedure required stereotactic immobilization of the animal and exposure and drilling of the cranium for placement of the electrode. As each experiment lasted for one hour or more, the animals were i.p. anesthetized with sodium pentobarbital (40 mg/kg). Otherwise, we followed the same procedure of exposing the common carotid arteries through a midline cervical incision and occluding the vessels by weighting loose ligatures placed around them. Care was taken that the body temperature remained at 37° C, and that there was only minimal injury to the pia and cortex at the site of electrode placement. The ion-selective electrodes were multibarrelled glass pipettes with a combined tip diameter of about 10-20 microns. Two of the individual pipette tips were siliconized with 15% hexamethyldesilane (Fisher); one was filled with the K^+ -ion selective resin (Corning 477317), the other with a Ca^{++} -ion selective resin (WPI IE-202). The third barrel, filled with 154 mM NaCl, was used as a reference electrode.

A silver-silver chlorided lead was inserted into each electrode and connected to high-input impedance-capacitance compensated preamplifiers. The signals were amplified as necessary and displayed on a multichannel chart recorder. The electrode tips were positioned in the cortex or hippocampus by a hydraulic calibrated microdrive. Calibration of the ion-selective electrodes was performed before and after each experiment using test solutions made up with known concentrations of KCl and $CaCl_2$. Local EEG recordings were derived from the reference electrode signal and displayed on the chart recorder.

In frozen sections of gerbil brain the thickness of the cortex was found to be 1.055 ± 0.059 mm ($\pm$ SD) at the site of measurement. The depth of the CA1 sector from the cortical surface was 1.355 ± 0.053 mm. The tip of the ion-selective electrode was placed 0.70 mm below the cortical surface for measurements in cerebral cortex, or 1.35 mm for measurements in the hippocampus. Paraffin oil was then poured on the surface of the dura to prevent dehydration of the exposed cerebral cortex. Before carotid occlusion, the gerbils were observed and electrode signals were recorded for a 20-30 min control period.

RESULTS

1. Morphological Observations

Morphological evidence of ischemic brain damage in adult gerbils subjected to 5 minutes of cerebral ischemia by bilateral carotid occlusion was confined to the hippocampus. A conspicuous destruction of CA1 neurons was found in animals sacrificed after one week (Suzuki et al., 1983). Two days after recirculation the CA1 pyramidal neurons appeared to be well preserved in the light microscope, whereas electron microscopy revealed a progressive accumulation of endoplasmic reticulum lamellae. Severe injury of CA1 neurons was evident on the third day. On the fourth day the CA1 sector showed pronounced neuronal disintegration with marked microglial activity. On the other hand, no noticeable damage of CA1 neurons was seen in 3 weeks old gerbils sacrificed after one week, whereas morphological studies after one day revealed "reactive cell changes" as previously described by Ito et al. (1979). These are presumably reversible. No neuronal destruction in the CA1 sector was observed in adult gerbils, where K^+ - and Ca^{++} -activities were determined in sodium pentobarbital anesthesia.

2. Cerebral Blood Flow

Measurements of the regional cerebral blood flow (rCBF) by 3H-nicotine revealed after 5 minutes ischemia flow values of below 10 ml/100 g min in the frontal and occipital cortex and hippocampus, whereas the cerebellum which is supplied by the vertebral arteries was slightly hyperemic (Suzuki et al., 1983). Evaluation of rCBF by qualitative 14C-iodoantipyrine autoradiography in adult and young gerbils during bilateral carotid occlusion uniformly indicated severe ischemia in both hemispheres, except central brain regions supplied by the vertebral artery system. In adult animals with prompt recirculation, spotty hyperperfusion was seen in cerebral cortex, hippocampus and basal ganglia. Ten minutes after recirculation diffuse hypoperfusion affecting most of the hemispheres became evident (Suzuki et al., 1983). No significant rCBF changes were observed at later time intervals.

3. Local Cerebral Glucose Utilization

Qualitative evaluation of the local cerebral glucose utilization (LCGU) in adult gerbils indicated at 10 minutes after recirculation, i.e. at the time of diffuse postischemic hypoperfusion, a conspicuous increase in hippocampus, whereas it appeared to be reduced in the rest of the brain. No significant changes were observed at other time intervals (Suzuki et al., 1983).

4. Blood-Brain Barrier Function

The blood-brain barrier of adult gerbils showed a biphasic opening, the first occurring shortly after release of occlusion and the second after 3 days, which was associated then with severe neuronal damage in the CA1 sector (Suzuki et al., 1983).

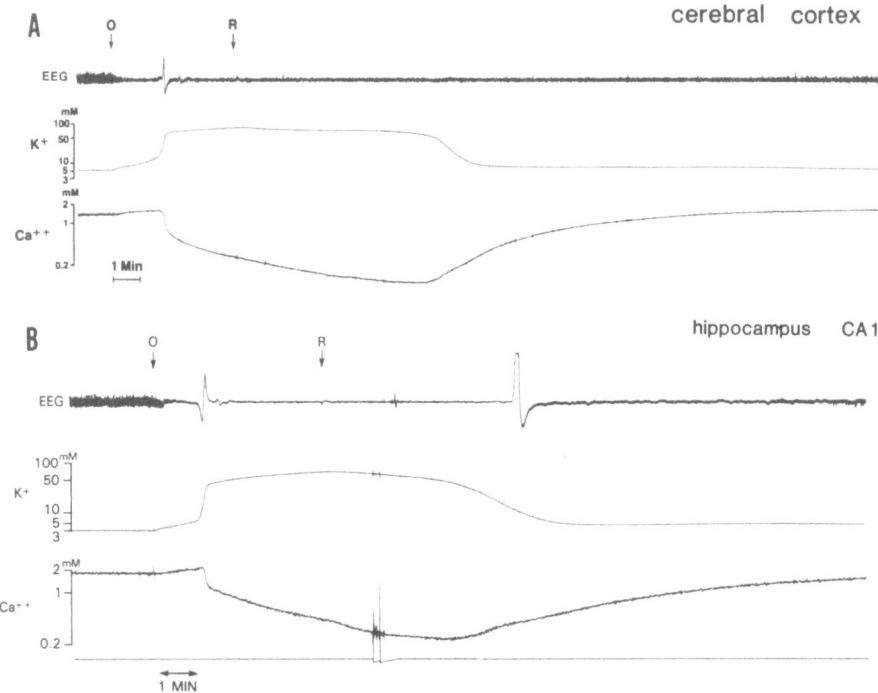

Fig.1: Time course of K^+_e and Ca^{++}_e in **(A)** cerebral cortex, and **(B)** CA1 region of hippocampus before, during and after 5 minutes of cerebral ischemia. **O:** moment of occlusion, **R:** release. The local EEG is shown on top.

5. Neuronal Activity

Studies of the spontaneous neuronal activity demonstrated cessation of firing in both CA1- and cortical neurons within one minute after onset of ischemia and a return of activity 10-20 minutes after recirculation (Suzuki et al., 1983a). Subsequently, CA1 neurons revealed marked hyperactivity during the first day, whereas the activity of cortical neurons was within the normal range. At 48 hours no spontaneous activity could be recorded from CA1 neurons, while neurons of the cerebral cortex behaved normally.

6. Extracellular K^+ and Ca^{++}

Figure 1 shows typical time courses of K^+ and Ca^{++} in cortex and hippocampus of gerbils exposed to 5 min ischemia. The ionic changes in the two structures were basically similar, although exact values of ion-concentrations and duration of each phase were slightly different (Table I). The local EEG became isoelectric within 20 seconds upon occlusion. Initially, K^+_e and Ca^{++}_e increased slowly for about 49 sec both in cortex and hippocampus. Thereafter, K^+_e increased rapidly to 45 mM in the cortex and to 41 mM in the hippocampus. Subsequently, K^+_e continued to increase, but at a much slower rate reaching more than 75 mM in the cortex and 78 mM in the hippocampus. Conversely, Ca^{++}_e decreased to less than 50% of the preocclusion value. The onset of the drop in Ca^{++}_e coincided with the sharp rise of K^+_e. Interestingly, when occlusion was terminated K^+_e responded fairly prompt by starting to decrease slowly, while Ca^{++}_e continued to fall. Ca^{++}_e reached a minimum some minutes after K^+_e had begun to decrease again. Recovery of Ca^{++}_e was appreciably protracted. In cerebral cortex, K^+_e returned to the base-line in about 10 minutes after recirculation - in the hippocampus already after 8.3 min - whereas Ca^{++}_e did not normalize until after 23 min in cortex, or after 20 min in hippocampus, respectively. In some cases, the local EEG showed some activity after K^+_e had recovered to the pre-ischemic level.

DISCUSSION

The described changes resulting from 5 minutes of cerebral ischemia in gerbils strongly support the assumption that the development of ischemic injury takes place mostly after recirculation. Cerebral cortex and hippocampus had a diametrically different outcome from ischemic injury, although both regions were exposed to a similar level of insult. Extracellular measurements of spontaneous action potentials revealed both in cerebral cortex and the CA1 sector a cessation of neuronal activity within one minute and a return within 10-20 minutes after recirculation (Suzuki et al., 1983a). Furthermore, the changes of K^+_e and Ca^{++}_e during and briefly after 5 minutes occlusion did not indicate significant differences between cortex and hippocampus. This is of particular interest, since Ca^{++} has been implicated as a potential trigger of the chain of reactions leading to destruction of neurons in the hippocampus (Siesjö, 1981; Hass, 1981). An excessive entry of Ca^{++} into CA1 and CA3 neurons has also been histochemically demonstrated in status epilepticus at the ultrastructural level (Griffiths et al., 1982).

The rather dramatic changes of K^+_e and Ca^{++}_e reflect major movements of these ions in the extracellular space, where the tip of the

TABLE 1

K^+_e and Ca^{++}_e in cerebral cortex and hippocampus of gerbils exposed to 5 min ischemia

	Baseline	Depolarization phase		At 5 min	Repolarization phase	
		At the start	At the end		At the start	At the end
CORTEX n=5						
K^+	4.2+0.6	9.0+1.6	54.3+5.3	75.3+5.2	49.6+5.8	4.2+0.5
Ca^{++}	1.28+0.11	1.47+0.12	0.88+0.09	0.42+0.11	0.32+0.07	1.28+0.14
HIPPOCAMPUS n=5						
K^+	3.4+0.2	5.7+0.2	41.2+7.1	78.8+0.11	44.4+5.7	3.8+0.4
Ca^{++}	1.37+0.15	1.69+0.16	1.26+0.24	0.70+0.11	0.40+0.07	1.36+0.17

K^+ and Ca^{++} are given in mM; mean + SEM

electrode was located. It is not known in detail how the outpouring of potassium and influx of calcium into cell elements is brought about in ischemia. Other work has indicated similar changes and revealed dramatic shifts of Na^+ and Cl^- in addition (Hossmann et al., 1973). At first, K^+_e and Ca^{++}_e increased gradually and similarly for almost 50 seconds prior to entering the phase of rapid change. Then, Ca^{++}_e reversed and decreased rapidly, whereas K^+_e showed a pronounced rise. It can be assumed that the rapid changes were associated with membrane depolarization. A value of 75 mM K^+_e seems to be extraordinarily high, however, part of this can undoubtedly be attributed to shrinkage of the extracellular space as cells depolarized and imbibed water. The delay in recovery of Ca^{++}_e is of interest in comparison with that of K^+_e which normalized sooner. Since Ca^{++}_e has been suggested as a trigger of cell damage, the delay in recovery may be significant. Neuronal repolarization was clearly evident in EEG recordings from hippocampus, whereas its absence in the cortex could be attributed to a relatively slower repolarization, which was filtered out by the low end of the EEG amplifier band pass.

No significant differences between cerebral cortex and hippocampus at the end of 5 minutes ischemia were noted in our preliminary observations with regard to several basic energy metabolites. It appears then that following a similar impact of ischemia, later developing regional differences in severity of ischemic injury can be attributed to selective vulnerability. Therefore, the observations of selective changes in the hippocampus after exposure to an intensity of ischemia equal to that in other regions strongly corroborate Vogt's (1922) theory of pathoclisis. The theory assumes that different, intrinsic physico-chemical properties of various neuronal cell types account for their different response to various noxious factors. The theory of pathoclisis itself does not explain the nature of mechanisms operative in the development of post-ischemic injury. Therefore, a comparative analysis of the regional differences may provide some insight into the factors involved.

A first, conspicuous difference was observed 10 minutes after recirculation. Autoradiographic studies demonstrate then a strikingly increased uptake of 2-deoxyglucose in the hippocampus in contrast to rCBF which was indicating diffuse hypoperfusion of the hemispheres (Suzuki et al., 1983). Uncoupling of glucose utilization and CBF has been reported by Pulsinelli et al. (1982) in rats, which during post-ischemic hypoperfusion had a brief increase in glucose utilization in the striatum and hippocampus, i.e. in structures which prove to be especially vulnerable. It should be pointed out that the in our studies increased glucose utilization of the hippocampus 10 minutes following recirculation can hardly be explained by an increase of neuronal activity. At that time, only the beginning of a return of spontaneous action potentials could be recorded in the CA1 sector and in the cortex. Similarly, studies of Diemer and Siemkowicz (1980) demonstrate a markedly increased uptake of 2-deoxyglucose in the hippocampus at a time when the EEG had not yet returned. An increased glucose utilization of the hippocampus could be due to an enhancement of anaerobic glycolysis, as suggested by Ginsberg et al.

(1977) and Welsh et al. (1980), or to a release of intracellular potassium demonstrated in our studies. This would stimulate metabolism of hippocampal neurons briefly after recirculation. As to the later post-ischemic
periods, our studies indicate that the greatly increased energy demands
resulting from high frequency neuronal discharge throughout the first day
after recirculation in the CA1 sector was not supported by a corresponding increase in rCBF. This could have led to secondary, relative
hypoxia. Brain damage resulting from a disparity in substrate demand and
supply has been considered by Hossmann et al. (1973), Snyder et al.
(1975) and Levy et al. (1979).

It can be assumed that in cerebral ischemia various, selectively
vulnerable brain structures have individual thresholds for post-ischemic
injury. Thus, though 5 minutes ischemia is sufficient to produce severe,
selective injury in the CA1-sector, the same intensity appears to be
below the threshold of ischemic damage in other structures. Otherwise,
according to our most recent observations, the threshold of ischemic
vulnerability appears to be higher in young animals. This is concluded
from the lack of CA1 destruction following 5 minutes ischemia in 3
weeks old gerbils, and from the generally milder changes of basic energy metabolites, such as ATP, phosphocreatine, glucose and lactate at
the end of ischemic occlusion.

Besides maturation of the severity of neuronal changes, there is
usually a progression in the size of an ischemic lesion as observed by
Ito et al. (1975) and Pulsinelli et al. (1982). The development of ischemic brain edema may play a paramount role in this respect as a common
denominator of effects resulting from accumulation of, e.g. free fatty
acids, kinins, serotonin, and prostaglandins in severely ischemic tissue.
As shown by Fujimoto et al. (1976) an ischemic insult is promptly followed by the intracellular accumulation of water. Opening of the blood-
brain barrier induces vasogenic edema, which may occur shortly after recirculation in association with reactive hyperemia, or, the barrier may
open after some delay in association with severe injury of the tissue as
observed in studies on gerbils (Suzuki et al., 1983) and cats (Wagner et
al., 1983). With progression of ischemic injury there is an enlargement
of the extracellular compartment due to the disruption of membranes of
dying cells, and a pronounced increase in osmolality leading to further
accumulation of water in the cerebral tissue. The resulting increase in
tissue pressure in severely edematous areas is likely to further depress
rCBF which then may fall below the critical threshold of viability in adjacent penumbra areas where the neuronal elements are struggling for survival. This eventually results in further spreading of infarction like a
forest fire engulfing still viable adjacent areas. The elucidation of the
secondary mechanisms operative in ischemic brain injury is especially important as a basis for the design of effective therapeutic measures after
an ischemic insult has occurred. Not depreciating the importance of prevention of stroke, there is no question that progress in the treatment of
stroke post facto is feasible and urgently needed.

CONCLUDING REMARKS

(1) Selective vulnerability of brain tissue is related to differences in individual thresholds to ischemic injury of various brain structures. This accounts for the selective damage confined to certain types of neurons, e.g. the CA1 pyramidal cells, while other regions exposed to similar intensity of ischemia remain intact. The brain of young animals generally seems to be more resistant, i.e. to have higher thresholds against ischemic injury.

(2) Final outcome of ischemic injury is determined by the various processes taking place in the post-ischemic period. Uncoupling of metabolic demand from supply resulting in relative hypoxia, disturbances in protein synthesis, release of free fatty acids, kinins, serotonin, and prostaglandins may play an important role in the progression of injury among those changes.

(3) A common denominator is the development of ischemic brain edema which, if associated with a marked increase in tissue pressure, may lead to a progressive enlargement of the ischemic infarct by engulfing adjacent areas of the penumbra zone.

REFERENCES

1. Diemer NH, and Siemkowicz E, Regional glucose metabolism and nerve cell damage after cerebral ischemia in normo- and hypoglycemic rats, in: "Circulatory and Developmental Aspects of Brain Metabolism", Spatz M, Mrsulja BB, Rakic L, Lust D, eds., Plenum Press, New York, London (1980).
2. Fujimoto T, Walker JT, Spatz M, and Klatzo I, Pathophysiologic aspects of ischemic edema, in: "Dynamics of Brain Edema", Pappius H, Feindel W, eds., Springer, Heidelberg, New York (1976).
3. Ginsberg MD, Reivich M, Giandomenico A, and Greenberg JH, Local glucose utilization in acute focal cerebral ischemia: Local dysmetabolism and diaschisis, Neurology 27:1042 (1977).
4. Griffiths T, Evans MC, and Meldrum BS, Intracellular sites of early calcium accumulation in the rat hippocampus during status epilepticus, Neurosci Let 30:329 (1982).
5. Hass WK, Beyond cerebral blood flow, metabolism and ischemic threshold: Examination of the role of calcium in the initiation of cerebral infarction, in: "Cerebral Vascular Disease", Meyer JS, Lechner H, Reivich M, Ott EO, Aranibar A, eds., Excerpta Medica, Amsterdam (1981).
6. Hossmann KA, Lechtape-Grüther H, and Hossmann V, The role of cerebral blood flow for the recovery of brain after prolonged ischemia, Z Neurol 204:281 (1973).
7. Ito U, Spatz M, Walker JT, and Klatzo I, Experimental cerebral ischemia in Mongolian gerbils. I. Light microscopic observations, Acta Neuropath 32:209 (1975).

8. Klatzo I, Pathophysiologic aspects of cerebral ischemia, in: "The Nervous System", Tower DB, ed., Raven Press, New York (1975).

9. Levy D, Van Uitert R, and Pike C, Delayed post-ischemic hypoperfusion: A potentially damaging consequence of stroke, Neurology 29:1245 (1979).

10. Pulsinelli WA, Levy DE, and Duffy TE, Regional cerebral blood flow and glucose metabolism following transient forebrain ischemia, Ann Neurol 11:499 (1982).

11. Siesjö BK, Cell damage in the brain: A speculative synthesis, J Cereb Blood Flow Metabol 1:155 (1981).

12. Snyder J, Nemoto E, Carroll R and Safar P, Global ischemia in dogs: Intracranial pressure, blood flow and metabolism, Stroke 6:21 (1975).

13. Suzuki R, Yamaguchi T, Kirino T, Orzi F, and Klatzo I, The effects of 5-minute ischemia in Mongolian gerbils: I. Blood-brain barrier, cerebral blood flow, and local cerebral glucose utilization changes, Acta Neuropath 60:207 (1983).

14. Suzuki R, Yamaguchi T, Li CL, and Klatzo I, The effects of 5-minute ischemia in Mongolian gerbils: II. Changes of spontaneous neuronal activity in cerebral cortex and CA1 sector of hippocampus, Acta Neuropath 60:217 (1983a).

15. Vogt C, and Vogt O, Erkrankungen der Großhirnrinde im Lichte der Topistik, PatDhoklise und Pathoarchitektonik, J Psychiatr Neurol 28:9 (1922).

16. Wagner H, Cahn R, Kuroiwa T, Ting P, Yamaguchi T, and Klatzo I, Role of the blood-brain barrier opening to proteins in pathophysiology of cerebral ischemia, J Cereb Blood Flow Metab 3:S416 (1983).

17. Walker JL, Specific liquid exchanger microelectrodes, Anal Chem 43:89A (1971).

18. Welsh FA, Greenberg JH, Jones SC, Ginsberg MD, and Reivich M, Correlation between glucose utilization and metabolite levels during focal ischemia in cat brain, Stroke 11:79 (1980).

ROLE OF SYNAPTIC TRANSMISSION FAILURE IN THE NEUROLOGIC DEFICIT OF ISCHEMIC BRAIN INJURY

Edwin M. Nemoto, M.R. Lin, G.K. Shiu, and E. Ragupathy[*]

Dept. Anesthesiol. and Crit. Care Med., Univ. Pittsburgh
Pittsburgh, PA 15261; Brain-Behavior Res. Ctr., * Sonoma
Developmental Ctr., Eldridge, CA 95431, USA

INTRODUCTION

The successful care, treatment and prognosis of patients suffering cerebral insults depend upon a thorough understanding of the factors and mechanisms leading to the ultimate neurologic deficit. Past studies have revealed several major facets in the pathogenesis of cerebral ischemic-anoxic injury. First, the degree of the neurologic deficit sustained depends upon not only the initial insult but also the pathological processes developing post-insult that aggravate and extend the injury. These are among others the no-reflow phenomenon, delayed postischemic hypoperfusion and hypermetabolism, free-radical generation, and lipid peroxidation. Second, the degree of the neurologic deficit sustained may be attributable not only to irreversible neuronal necrosis, but also to a - possibly reversible - failure of synaptic transmission without failure of oxidative phosphorylation. Finally, most recent studies suggest that the evolution of ischemic brain damage may be related to a release of free fatty acids (FFA) from membrane phospholipids (PL). Membrane PLs are likely to be involved in membrane transduction processes of synaptic transmission, namely channel gating, membrane-bound enzyme activities, such as adenylate cyclase, phosphodiesterases, and receptor sensitivity. Our intent is to present evidence that failure of synaptic transmission could contribute to the neurologic dysfunction of ischemic brain injury.

259

APPARENT DILEMMAS IN THE PATHOGENESIS OF BRAIN INJURY OF ISCHEMIC ANOXIA

Current dilemmas in the pathogenesis of ischemic brain injury provide circumstantial, indirect evidence that failure of synaptic transmission processes without disturbances of energy metabolism may contribute to the neurologic dysfunction of ischemic brain damage. First, especially after hypoxic insults there is a remarkable inconsistency between the absence of noticable neuropathological changes at the light-microscopical level and the degree of neurological dysfunction. The reasons for these discrepancies are unkown, but it could be surmised that synaptic transmission had failed inspite of a functional energy metabolism. Second, after a sufficiently long duration of complete global brain ischemia, e.g. 15-30 min followed by recirculation, cerebral high-energy phosphate concentrations are restored to near-normal levels within one hour while abnormalities of the EEG and of brain catecholamine levels persist (1-3). These findings indicate that the recovery of spontaneous synaptic activity is delayed as compared to that of oxydative metabolism. This appears to be correlated with abnormalities of neurotransmitter turnover.

Third, the fact that transient amelioration of neurologic deficits resulting from cortical or subcortical lesions can be accomplished by neurotransmitter drugs, such as amphetamine, strongly suggests that viable, non-communicating neurons are responsible for the dysfunction (4). Thus function returns, if the neurons are stimulated to release their neurotransmitters or, if the appropriate receptors are activated. Finally, the delayed return of neurologic function weeks or months after a devastating cerebral insult, or the complete return to consciousness after prolonged coma leaves one or two possibilities. It may indicate the development of alternate neuronal pathways to replace irreversibly damaged neuronal circuits, or the recovery of synaptic function of reversibly damaged neurons.

EVIDENCE FOR DEFECTIVE SYNAPTIC TRANSMISSION WITHOUT ENERGY FAILURE IN ISCHEMIC ANOXIC BRAIN INJURY

Perhaps one of the most surprising findings in investigations on ischemic brain injury is the apparent tolerance of cerebral mitochondria to ischemia, at least when tested in vitro (5,6). The respiratory control ratio was depressed significantly only after 30 - 40 min of compression ischemia in rabbits due to a decrease in state 3 and an increase in state 4 respiration. Delayed secondary decreases of the mitochondrial respiratory control ratio were not observed with recirculation and recovery (5). In rats 30 min of complete compression ischemia also decreased the mitochondrial respiratory control ratio, but complete recovery of mitochondrial function occurred after recirculation (6). When ischemia was incomplete, however, a further deterioration of mitochondrial function occurred.

The findings on the resistance of mitochondrial function to ischemia are generally consistent with the recovery of the cerebral high-energy phosphates after recirculation. As repeatedly shown by Siesjö and associates (7) the cerebral energy-charge potential, phosphocreatine and
lactate levels are restored to normal after up to 30 min of ischemia. In
dogs subjected to 30 min of ischemia, brain high-energy phosphates were
completely restored after recirculation (1) but not after 60 min of
ischemia. Despite recovery of the brain energy-charge potential after 30
min of ischemia EEG abnormalities persisted. However, the brain remained
responsive to strychnine stimulation indicating that the potential for
spontaneous electrical activity was preserved. Thus the processes involved in synaptic transmission are apparently more vulnerable to ischemia
than mitochondrial oxidative phosphorylation. In 1971, Williams and
Grossman (8) showed that during hemorrhagic hypotension the excitatory

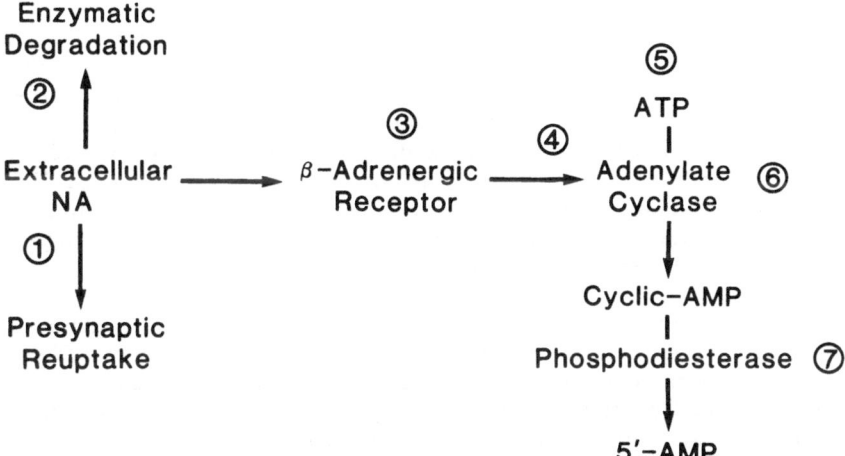

Fig.1: Schematic illustration of the rationale for noradrenalin (NA)-in-
       ducible cAMP accumulation as an indicator of neuronal oxidative
       metabolism and synaptic mechanisms. Exogenous NA added to
       cerebral cortex slices may be taken up into synaptic vesicles (1)
       or degraded enzymatically (2). Extracellular NA stimulates beta-
       adrenergic receptors (3) the sensitivity of which may depend on
       their prior history. The beta-adrenergic receptors are coupled (4)
       to adenylate cyclase (6). Both may be affected by ischemia.
       Adenylate cyclase converts ATP (5) to cAMP, and is therefore
       to some extent dependent upon the cellular oxidative metabolism.
       Cyclic-AMP may be degraded by phosphodiesterase (7) to 5'-AMP.

postsynaptic potentials of pyramidal tract neurons failed before breakdown of the membrane potential occurred. This study also illustrated the
greater sensitivity of synaptic transmission during ischemic anoxia as
compared with the neuronal membrane potential or the energy state.

What evidence is available indicating that synaptic transmission is
altered following ischemia? With energy failure of ischemia, cerebral
neurotransmitters are massively released into the extracellular space.
Accordingly, elevated levels of noradrenalin (NA) and serotonin (5-HT)
are found in the CSF of patients after ischemic and hemorrhagic cerebrovascular disease (9,10). In gerbils subjected to ischemia NA levels of
cerebral cortex declined by 30% after one hour of unilateral carotid
artery occlusion and by 50% after 6-12 hours (11,12). On the other hand,
dopamine (DA) fell by 10% after one hour and by 25% after 6 hours.
In the rat embolic ischemia caused an increase or little change at four
hours post-insult (13). Brain 5-HT levels reportedly declined after
ischemia (12) while it was increased in one study (11). The variability
of neurotransmitter (NT) changes is indicative of the complexity of the
insults, the variability of the tissue response, and the mechanisms involved in synthesis and metabolism of NTs, especially of biogenic amines. Several enzymes involved in NT-synthesis and -degradation, namely,
tyrosine- and tryptophane hydroxylases, dopamine ß-hydroxylase, and
monoamine oxidase are O_2-dependent with a K_m of 5-10 torr (14). Thus
it is very likely that hypoxia and postischemic hypoperfusion markedly
affect tissue levels of biogenic amines (15). Their decline in the cerebral tissue could result from extracellular release, diffusion into the
CSF, decreased synthesis and increased degradation. The energy failure
of ischemia also prevents reuptake of NTs, the primary method of NT-
inactivation. The variable responses of NT-levels in the tissue could
also be due to variable patterns of injury resulting in hypo- and hyperperfusion affecting NT-synthesis and -metabolism.

The massive release of NTs into the extracellular space during
ischemia results in extensive neuronal depolarization which may affect
postsynaptic receptor mechanisms, such as receptor desensitization or
-hypersensitization. To study these changes we considered the NA-inducible cyclic-AMP accumulation as a gross indicator of both neuronal
viability and the efficacy of postsynaptic mechanisms. The rationale for
this approach is as follows (Fig. 1). Exogenous NA added to cerebral cortex slices may be either taken up into synaptic vesicles (1) or enzymatically degraded (2). NA stimulates ß-adrenergic receptors (3) whose sensitivity may be altered depending upon its prior history. The activated receptor is coupled to adenylate cyclase (6). Its activity and receptor
coupling may be affected by various metabolites and neurohormones.
Adenylate cyclase acts on ATP (5) forming cAMP which can be degraded
to 5'-AMP by phosphodiesterase (7). Thus the extent of cAMP accumulation in response to NA stimulation is a function of:

(a) NA-reuptake and -degradation,
(b) number and sensitivity of receptors,

(c) adenylate cyclase activity and its coupling to the receptor,
(d) ATP-levels and availability, and
(e) phosphodiesterase activity.

In general it has been shown, however, that phosphodiesterase plays a minor role in the control of cAMP levels as compared to adenylate cyclase.

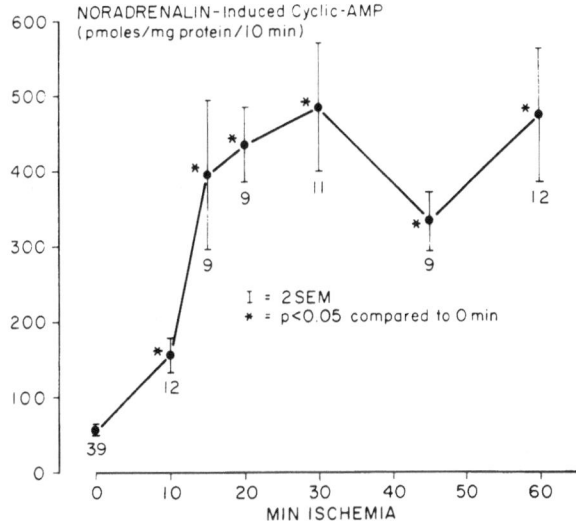

Fig.2: Changes in noradrenalin (NA)-inducible cAMP of cerebral cortical slices from rats subjected to various durations of decapitation ischemia at 37.0^{o} C . NA-inducible cAMP was obtained as the difference in concentration prior to and after 10 min exposure of the cortical slices to 11.2 µM NA. The number of rats studied and the level of signifiance are indicated. Reproduced from Lin et al., J Neurochem (16) with permission from Raven Press, New York.

We hypothesized that the massive release of NTs during ischemia causes receptor desensitization. We studied the effect of variable durations of complete global ischemia without recirculation in vivo on NA-inducible cAMP accumulation in cerebral cortex slices in vitro (16, 17). Surprisingly, a 10-fold increase of the NA-inducible cAMP level was found between 0 and 15-20 min of ischemia without a further increase after longer periods of up to 60 min (Fig. 2). With recirculation after dif-

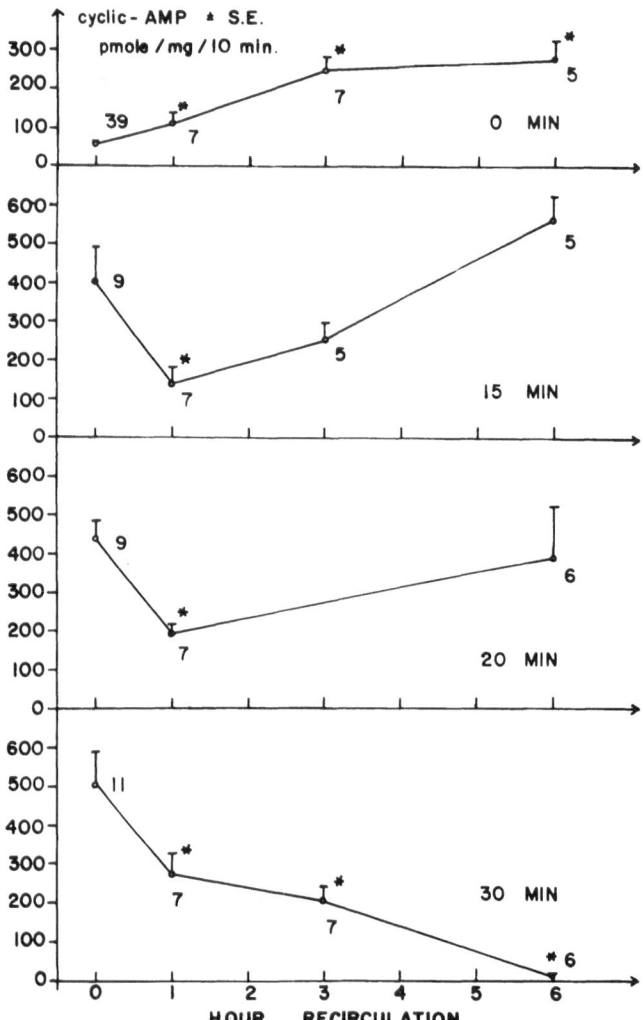

Fig.3: Noradrenalin (NA)-inducible cAMP-concentrations ($\pm$ SEM) in rat
cerebral cortex slices slices after 0, 5, 15, 20 and 30 min ischemia followed by recirculation of up to 6 hrs. Complete global
ischemia was induced by high-pressure neck tourniquet plus arterial
hypotension. The numbers of animals studied are given. $p < 0.05$
vs 0 min of recirculation. Reproduced from Lin et al. (17) with
permission from Springer, New York.

ferent periods of ischemia, the level of NA-inducible cAMP varied with a transient decrease at 3 hours as compared to the level at the end of ischemia without recirculation. This was followed by an increase at 6 hours of recirculation except after 30 min of ischemia (Fig. 3). Thirty min of ischemia and 6 hours of recirculation resulted in a marked decrease of the cAMP response to NA (Fig. 4). The changes in NA-inducible cAMP might be attributable to a number of mechanisms. First, changes of reuptake and degradation of NA may be considered which would result in varying levels of receptor activation in relation to the extracellular NA-concentration. Second, the alterations may affect any receptor mediated event including receptor sensitivity, receptor-adenylate cyclase coupling, adenylate cyclase activity, ATP-availability, and phosphodiesterase activity. However, because of the high, near supramaximal level of NA used in these studies, it is unlikely that the results reflect changes in reuptake or degradation. Furthermore, the findings are unlikely to have resulted from a reduced availability of ATP, since the ATP levels of the cortical slices were elevated in ischemic tissue as compared to that of the controls. These data provide support for the notion that secondary events after recirculation following global cerebral ischemia lead to transmission failure without affecting energy metabolism. Furthermore, Schwartz et al. (18) who studied the effect of ischemia on cAMP levels in the brain did not observe - at least in vitro - changes in activity of adenylate cyclase (AC), or phosphodiesterase (PDE). Yet significant changes of Na^+ -K^+ -ATPase activity were found 20 hours post-insult. These findings would disagree with the conclusion that alterations of AC or PDE are responsible for the changes of NA-inducible cAMP. This leaves receptor-AC coupling or receptor sensitivity as primary mechanisms which, however, has to be confirmed by future studies.

There is some evidence that ischemia affects synaptic membrane transport. In cats subjected to focal ischemia by middle cerebral artery occlusion, uptake of 4-amino butyric acid by cortical synaptosomes was reduced to 50 % of normal (19). However, the reduction in uptake was attributable to widespread neuronal damage rather than to specific changes of the synaptic compartment. In gerbils subjected to unilateral ischemia, no changes of the synaptosomal uptake of DA, GABA or glutamate were found after 8 hours of ligation. After 16 hours, the uptake was markedly decreased with 15, 28, and 47% inhibition for DA, GABA and glutamate, respectively (20). Anoxia of rat cerebral cortex irreversibly decreased the rate of NT-accumulation, but not its release (21). The inhibitory effect of anoxia on NT-accumulation was enhanced at low pH. The authors concluded that NT-uptake systems are more easily damaged by anoxia than energy metabolism or ion transport. Failure of NT-reuptake could render the receptors refractory to stimulation resulting in desensitization. However, at least as NA-inducible cAMP is concerned, initial hypersensitivity against NA developed during ischemia. A marked reduction of the response was observed only 6 hours after recirculation.

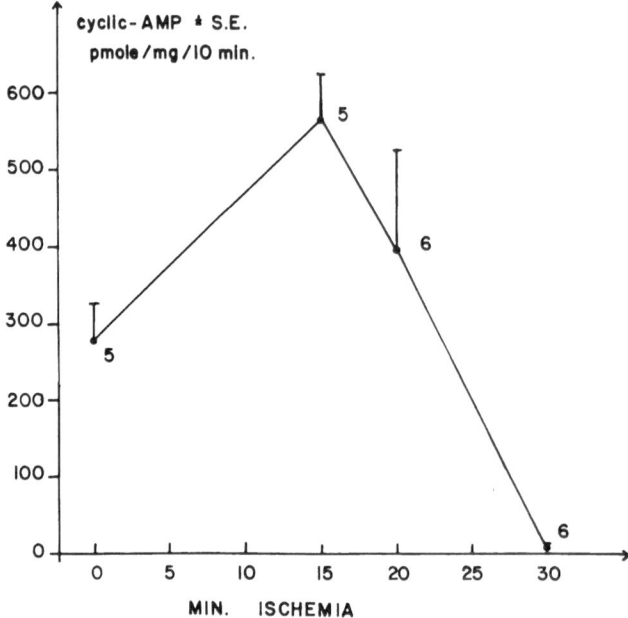

Fig.4: Noradrenalin (NA)-inducible cAMP-concentrations ($\pm$ SEM) in rat cerebral cortex studied in vitro after 0, 15, 20 and 30 min of ischemia and 6 hrs recirculation in vivo. The number of rats studied is indicated. Reproduced from Lin et al. (17) with permission from Springer, New York.

In studies on the effect of ischemia on synaptosomal transport, we used a rat model of complete global brain ischemia combining arterial hypotension and a high pressure neck tourniquet. The rats were subjected to variable durations of ischemia followed by 6 hours of recirculation. Arterial blood pressure and arterial blood gases remained within normal limits. At the end of 6 hours of recirculation the brains were frozen with liquid N_2 in situ, and the frozen material was shipped under dry ice to Dr. E. Ragupathy in Eldridge, California for assay of synaptosomal transport of leucine, GABA and glutamate. Synaptosomal transport of these amino acids was unaffected by 15 or 20 min ischemia followed by 6 hours recirculation (Table 1). However, with 30 min ischemia and 6 hours recirculation transport of GABA and glutamate was attenuated by 27 or 32%, respectively. The time course was similar to that of the NA-inducible cAMP response. Thus it appears again that the secondary detrimental effects developing after 30 min of ischemia and 6 hours of recirculation are unlikely to have resulted from a failure of oxidative metabolism.

TABLE 1

Synaptosomal Uptake of Amino Acids in Percent of Control[+] in Rat

Cerebral Cortex after Global Ischemia

	Duration of Ischemia (min)[++]		
	15	20	30
LEU	104 ± 7	97 ± 5	109 ± 8
GABA	98 ± 2	93 ± 3	73 ± 6*
GLU	103 ± 6	93 ± 7	68 ± 5*

n = 6-10 for each value; [+] control uptake was for: LEU = 7-10 pM/mg prot 10 min; GABA = 36-60 pM/mg prot 10 min; GLU = 77-79 pM/mg prot 10 min; [++] followed by 6 hrs of recirculation; p < 0.001 vs. controls

MECHANISMS OF MEMBRANE FAILURE

Reference to "membrane" under this heading applies especially to the synaptic membrane. However, since we do not know to which extent these processes are restricted to synaptic membranes, we assume that they occur in all membranes with receptors and membrane-bound enzymes involved in activation of second messengers, such as cAMP and Ca^{++}. The accumulation of free fatty acids (FFA) during complete global brain ischemia is unique with regard to the catastrophic biochemical events. It is unique because unlike other biochemical changes noted to date, only accumulation of FFAs continues in a significant manner after ischemia periods longer than 10-15 min. In fact FFA accumulation continues even after 60 min of ischemia (Fig. 5; 22), probably for up to 8 hours (23). Thus, we have postulated that FFA accumulation is the first biochemical indicator for the evolution of ischemic brain injury, at least during complete global brain ischemia (22).

The FFAs released during ischemia may be important mediator substances of ischemic brain damage in general, and of synaptic transmission failure in particular. FFAs in high concentrations have potential detrimental effects per se or by their potent metabolites, such as the pro-

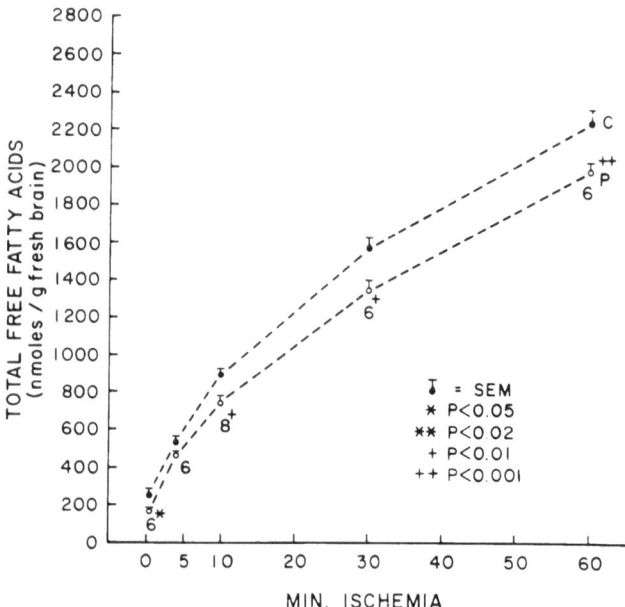

Fig.5: Total free fatty acids of rat brain with increasing duration of decapitation ischemia. The brains were kept at 37° C post decapitation. Rats were either injected with 0.9% NaCl (C) or 60 mg/kg b.w. pentobarbital (P) 10 min prior to decapitation. The number of rats studied is indicated. Total free fatty acids were determined by TLC and gas liquid chromatography. Reproduced from Shiu et al., J Neurochem (22) with permission of Raven Press, New York.

staglandins or leukotrienes. Release of FFAs is apparently a normal consequence of synaptic transmission and membrane channel gating (24). It is currently thought that receptor activation by a neurohormone triggers hydrolysis of phosphatidylinositol-4,5-biphosphate ($PI-P_2$) by phosphodiesterase (PDE) resulting in formation of 1,2-diacylglycerol and inositol triphosphate (IP_3; Fig. 6). The diglyceride activates protein kinase C and phosphorylates receptor proteins, thereby altering receptor sensitivity and membrane proteins and, consequently membrane permeability. IP_3 mobilizes intracellular Ca^{++} resulting in a multiplicity of effects including phospholipase-activation as a mechanism of FFA accumulation. The diglyceride is further degraded to glycerol and FFAs. Stearic and arachidonic acids are the predominant fatty acids in the one- and two po-

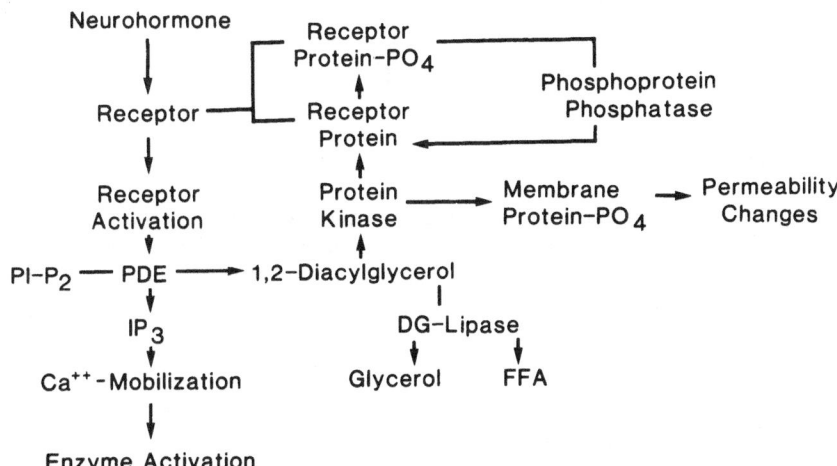

Fig.6: Hypothetical membrane processes following neurohumoral receptor activation. The neurohormone triggers hydrolysis of phosphatidylinositol-biphosphate by phosphodiesterase leading to the formation of 1,2-diacylglycerol and inositol triphosphate (IP_3). Diacylglycerol may then be metabolized by diglyceride lipase resulting in the formation of glycerol and free fatty acids (FFA). Diacylglycerol stimulates protein kinase leading to phosphorylation of receptor- and membrane proteins and, ultimately, to alterations of receptor sensitivity and membrane permeability. Dephosphorylation of the proteins is carried out by phosphoprotein phosphatase. Inositol triphosphate formed by hydrolysis of $Pi-P_2$ induces Ca^{++}-mobilization and activation of phospholipases catalyzing release of free fatty acids from phospholipids.

sition of phosphatidylinositol. Thus FFAs released during ischemia may arise from diglycerides catalyzed by diglyceride lipase or through activation of phospholipases secondary to Ca^{++} -mobilization by IP_3 . A release of FFAs occurring later may be attributed to lysosomal phospholipases in addition. The massive release of FFAs during ischemia is probably associated with the ischemic neuronal depolarization resulting from the discharge of neurotransmitters. The liberation of FFAs from membrane phospholipids, if not restored, may result in permanent alterations of receptor sensitivity in membrane transduction. However, the precise molecular mechanisms involved in synaptic transmission under normal and pathological conditions remain to be elucidated.

ACKNOWLEDGEMENTS

        This work was supported in part, by the American Heart Association, Western Pennsylvania affiliate, grant No. 5-33730 and the U.S. Public Health Service, grant No. NS15659, and halothane was provided by Ayerst Laboratories.

REFERENCES

1. Hinzen DH, Müller U, Sobotka P, Genert E, Lang R, Hirsch H, Metabolism and function of dog's brain recovering from longtime ischemia, Am J Physiol 223:1158 (1972).
2. Brown RM, Carlsson A, Ljunggren B, Siesjö BK, Snider SR, Effect of ischemia on monoamine metabolism in the brain, Acta Physiol Scand 90:789 (1974).
3. Calderini G, Carlsson A, Nordström CH, Influence of transient ischemia on monoamine metabolism in the rat brain during nitrous oxide and phenobarbitone anaesthesia, Brain Res 157:303 (1978).
4. Glick SD, Zimmerberg B, Pharmacological modification of brain lesion syndromes, in: "Recovery from brain damage, research and theory", Finger S, ed., Plenum Press, New York (1978).
5. Schutz H, Silverstein PR, Vapalahti M, Bruce DA, Mela L, Langfitt T, Brain mitochrondrial function after ischemia and hypoxia. I. Ischemia induced by increased intracranial pressure, Arch Neurol 29:408 (1973).
6. Rehncrona S, Mela L, Siesjö BK, Recovery of brain mitochondrial function in the rat after complete and incomplete cerebral ischemia, Stroke 10:437 (1979).
7. Siesjö BK, Brain energy metabolism, John Wiley & Sons, New York (1978).
8. Grossman RG, Williams V, Electrical activity and ultrastructure of cortical neurons and synapses in ischemia, in: "Brain hypoxia", Brierley JB, Meldrum BS, Lippincott JP, Lippincott Co., Philadelphia (1971).

9. Meyer JS, Stoica E, Pascu I, Shimazu K, Hartmann A, Catecholamine concentrations in CSF and plasma of patients with cerebral infarction and haemorrhage, Brain 36:277 (1973).

10. Meyer JS, Okamoto S, Shimazu K, Koto A, Ohuchi T, Sari A, Ericsson AD, Cerebral metabolic changes during treatment of subacute cerebral infarction by alpha and beta adrenergic blockade with phenoxybenzamine and propranolol, Stroke 5:180 (1974).

11. Mrsulja BB, Mrsulja BJ, Spatz M, Ito U, Walker JT Jr., Klatzo I, Experimental cerebral ischemia in mongolian gerbils, IV. Behaviour of biogenic amines, Acta Neuropath 36:1 (1976).

12. Welch KMA, Chabi E, Buckingham J, Bergin B, Achar VS, Meyer JS, Catecholamine and 5-hydroxytryptamine levels in ischemic brain, Stroke 8:341 (1977).

13. Kogure K, Scheinberg P, Kishikawa H, Busto R, The role of monoamines and cyclic AMP in ischemic brain edema, in: "Dynamics of brain edema", Pappius HM, Feindel W, eds., Springer-Verlag, Berlin (1976).

14. Strang RHC, Estimation of Km values of enzymes requiring molecular O_2 as a substrate, Biochem J Letters 193:1033 (1981).

15. Davis JN, Carlsson A, The effect of hypoxia on monoamine synthesis, levels and metabolism in rat brain, J Neurochem 21:783 (1973).

16. Lin MR, Henteleff HB, Nemoto EM, Noradrenalin-inducible cyclic-AMP accumulation in rat cerebral cortex: Changes during complete global ischemia, J Neurochem 40:595 (1983).

17. Lin MR, Nemoto EM, Kessler PD, Alterations in whole brain cyclic-AMP and cerebral cortex NA-inducible cyclic-AMP in rats during and after complete global ischemia, in: "Brain protection-morphological, pathophysiological and clinical aspects", Wiedemann K, Hoyer S, eds., Springer Verlag, New York (1983).

18. Schwartz JP, Mrsulja BB, Mrsulja BJ, Passonneau JV, Klatzo I, Alterations of cyclic nucleotide-related enzymes and ATPase during unilateral ischemia and recirculation in gerbil cerebral cortex, J Neurochem 27:101 (1976).

19. Strong AJ, Tomlinson BE, Venables GS, Gibson G, Hardy JA, The cortical ischaemic penumbra associated with occlusion of the middle cerebral artery in the cat: 2. Studies of histopathology, water content, and in vitro neurotransmitter uptake, J Cerebral Blood Flow Metab 3:97 (1983).

20. Weinberger J, Cohen G, Nerve terminal damage in cerebral ischemia: Greater susceptibility of catecholamine nerve terminals relative to serotonin nerve terminals, Stroke 14:986 (1983).

21. Pastuszko A, Wilson DF, Erecinska M, Neurotransmitter metabolism in rat brain synaptosomes: Effect of anoxia and pH, J Neurochem 38:1657 (1982).

22. Shiu GK, Nemmer JP, Nemoto EM, Reassessment of brain free fatty acid liberation during global ischemia and its attenuation by barbiturate anesthesia, J Neurochem 40:880 (1983).

23. Gercken G, Brauning C, Quantitative determination of hydrolysis products of phospholipids in the ischaemic rat brain, Pflügers Arch 344:207 (1973).

24. Berridge MJ, A novel cellular signaling system based on the integration of phospholipid and calcium metabolism, in: "Calcium and cell function", Cheung WU, ed., Academic Press, New York (1983).

INCREASED VULNERABILITY OF THE TRAUMATIZED BRAIN

TO EARLY ISCHEMIA

L. W. Jenkins, A. Marmarou, W. Lewelt, and D.P. Becker

Division of Neurological Surgery, Medical College of
Virginia, Richmond, Virginia, USA

Hypoxia and arterial hypotension are the most frequent causes of secondary brain damage in head injured patients and contribute significantly to morbidity and mortality (1,2,3). Early ischemia due to a delay in hospital admission is difficult to document and even more difficult to study in humans. However, laboratory studies have recently demonstrated that early secondary hypotension and hypoxia increases both mortality and morbidity following experimental head trauma (4). Yet, it is unclear whether this is the result of a direct effect on the brain or upon peripheral systems of the animal. Therefore, an increased vulnerability of the brain to ischemia or hypoxia following trauma has not been demonstrated to date. An increased vulnerability of the traumatized brain to ischemia may result from a number of factors. First, impairment of the responsiveness of the cerebral circulation as the result of trauma (1,2,5,6) may increase the risk of cerebral ischemia in patients with secondary hypotension and hypoxia. Second, metabolic derangements which result directly from the primary traumatic insult may increase the vulnerability of the brain to hypoxic and ischemic damage. Finally, peripheral reactions to trauma, hemorrhagic hypotension, or apnea may alter the response of the traumatically damaged CNS to hypoxia and ischemia. The present study was undertaken to examine the effect of an early "secondary" ischemic insult upon the function and structure of the traumatized brain. The response of the normal brain to the same level of ischemia was studied for comparison. The purpose of the current study was to test the hypothesis that a traumatic insult renders the CNS more vulnerable to cerebral ischemia.

Although fluid percussion trauma is considered a model of primary brain stem injury on the basis of pathological changes, a major feature is that graded levels of trauma can be produced with different primary and secondary effects depending on the direction and magnitude of impact (cf. Becker, this volume). With a central injury of less than 3 atmospheres directed toward the marginal gyri of the mid-parietal cortex of ventilated animals, there is no evidence for the development of secondary brain ischemia during the first 24 hours following trauma. There are, however, functional, metabolic, and structural changes which have been previously described (5,6,7). The lack of ischemia allows for an examination of the addition of a "controlled ischemic insult" to the primary traumatic injury in a more reproducible and systematic manner.

In this study a level of fluid percussion injury (2.5 - 2.8 atmospheres) was employed which produces a strong sympathoadrenal response (11) without secondary ischemia (5,6). This trauma level did not induce structural alterations in the neocortex except at the contusion site beneath the injury screw, whereas structural alterations of the neuronal perikarya and dendritic and axonal domains were identified in the brain stem (7,8). We chose a level of incomplete cerebral ischemia which causes qualitatively reversible changes of tissue structure and EEG activity. The ischemic or traumatic insult alone thus did not induce irreversible structural damage of the cerebral cortex and forebrain. This provided an optimal opportunity to examine the vulnerability of the traumatized brain to ischemia and, whether ischemia exacerbates damage already incurred during trauma.

METHODS

The fluid percussion model was used in cats as previously described in detail (5,7). Global incomplete cerebral ischemia was induced by intrathoracic arterial ligation to reduce collateral circulation in combination with bilateral carotid occlusion and hemorrhagic hypotension producing a reduction of MABP to 90 mmHg. Adult male cats weighing between 2.5-3.5 kg were subjected to either a central traumatic impact of 2.5-2.8 atmospheres or 7 minutes of incomplete cerebral ischemia, or a combination of both. All animals were anesthetized with pentobarbital (30 mg/kg) and maintained on 70% N_2O and 30% O_2 and supplemental pentobarbital. Muscle relaxation was maintained either with gallamine triethiodide (5-10 mg/kg) or pancuronium bromide (0.05-0.1 mg/kg). Cerebral blood flow was measured prior and sequentially after trauma, ischemia or trauma/ischemia combined using the microspheres method. Right and left hemispheric EEG recordings were made continuously. All animals were subjected to perfusion fixation with aldehydes via the thoracic aorta for light and electron microscopic examination. Additionally, the flow of the fixative to the brain was confirmed and quantified by microspheres in acute animals subjected to a combination of trauma and ischemia. Three experimental groups were investigated:

(1) 3 cats with trauma and 24 hours survival,
(2) 3 cats with 7 minutes of incomplete ischemia - two with 2 hours and one with 24 hours survival,
(3) 4 cats with trauma followed by 7 minutes of ischemia one hour later - three with 2 hours and one with 12 hours of survival.

TABLE 1

ML/100GM/MIN	FIXED WET WEIGHT Control (N = 3)	4hrs (N = 3)	MEAN (SD) 10hrs (n = 2)	18hrs (N = 2)	24hrs (N = 1)
TOTAL CBF	35.0(14.1)	49.5(6.5)	46.9(11.2)	52.0(1.27)	30.0
HEMI CBF	34.8(14.5)	51.4(4.8)	47.7(9.7)	53.0(1.13)	30.6
STEM CBF	33.3(11.4)	38.3(11.8)	39.9(12.4)	46.1(1.13)	27.3
MABP	135(6)	123(19)	126(17)	96(38)	128
CPP	130(6)	115(18)	121(15)	88(38)	116
CVR	4.3(1.4)	2.4(.59)	2.0(1.0)	1.9(.59)	3.9
pCO_2	28.5(2.4)	31(1.2)	28(1.2)	29.4(1.5)	30
pH_a	7.39(.03)	7.43(.02)	7.41(.01)	7.49(.04)	7.4

Cerebral blood flow (microspheres method) and systemic variables (MABP: mean arterial blood pressure; CPP: cerebral perfusion pressure; CVR: cerebrovascular resistance; arterial pCO_2 , and pH) measured prior to and at various intervals after fluid percussion in cats.

RESULTS

Three animals subjected to 2.5 - 2.8 atmospheres of fluid percussion injury and 24 hours survival had transient changes of the EEG at impact but progressive recovery over the next hour. The acute CBF changes following injury were as previously reported (6). There was marked hyperemia followed within one hour by a normal to slightly decreased flow. Chronic CBF studies over 18 to 24 hours post trauma revealed persistent hyperemia (Table 1). Except at the contusion site, no structural changes were observed in the cerebral cortex, hippocampus, basal ganglia or other forebrain structures (Fig. 1). However, some swelling of the subcortical white matter and neocortex were seen even away from the site of impact. This is consistent with specific gravity measurements of animals with chronic survival sustaining the same level and type of injury (cf. Becker, this volume). In the brain stem petechial hemorrhages were observed in some animals, while in all animals selective axonal damage as previously reported (5,8).

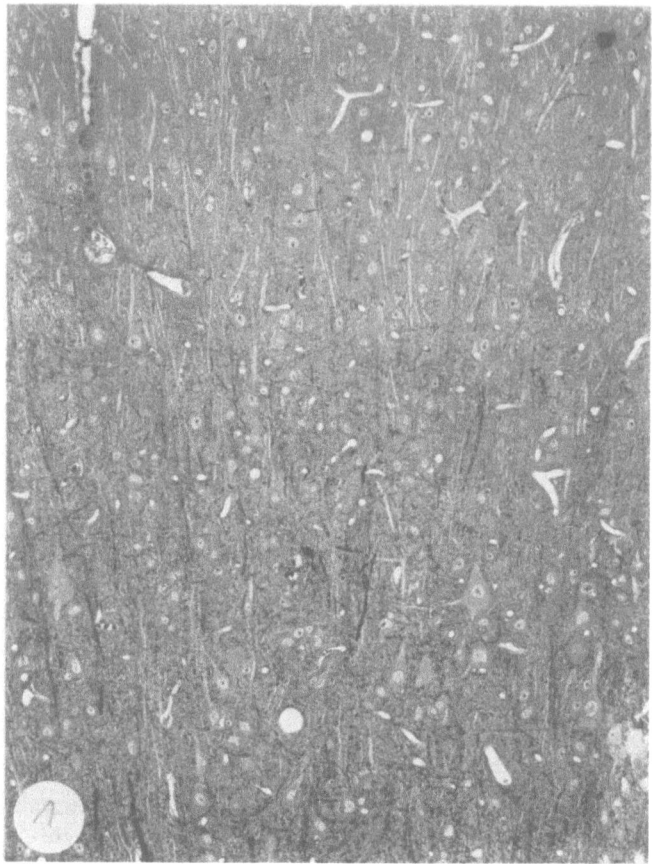

Fig.1: Upper four layers of cerebral cortex obtained from the suprasylvian gyrus 24 hrs after a 2.8 atm fluid percussion injury. Note the
structural preservation of the parenchyma indicating the lack of
structural injury after fluid percussion. 125x

Ischemia

The EEG of all animals subjected to 7 minutes of incomplete ischemia became isoelectric within 30 seconds of carotid ligation and induction of hypotension. However, the spontaneous electrical activity returned within several minutes after release of the ligatures and blood
reinfusion. In one animal surviving 24 hours following ischemia the EEG
remained robust until sacrifice. The CBF measurements revealed in all
cats pronounced cerebral ischemia during the insult followed by acute
hyperemia. CBF had a tendency to normalize over time in one chronic
survivor but was still above the control level (Table 2). The cerebral
structures appeared normal in all regions except for slight chromatin
clumping of neuronal nuclei which, however, was the result of the perfusion method (Fig. 2).

TABLE 2

ML/100GM/MIN	FIXED WET WEIGHT	MEAN (SD)		
	CONTROL (n = 3)	ISCHEMIC INSULT (n = 3)	1 MINUTE POST ISCHEMIA (n = 1)	60 MINUTE POST ISCHEMIA (n = 3)
TOTAL	30.7(13.1)	6.2(3.6)	31.6	47.7(24.9)
HEMI	29.1(11.8)	5.1(3.2)	27.8	44.5(19.7)
STEM	29.5(11.9)	8.5(3.6)	43.0	48.3(20.1)
pCO_2	28.5(4.3)	25.8(6.5)	26.0	27.3(7.4)
pH_a	7.39(.04)	7.33(.2)	7.4	7.42(.05)
MABP	128.3(30.1)	86.7(11.6)	125	127.5(3.5)

Cerebral blood flow and systemic variables (cf. Table 1) of experimental animals with 7 minutes of incomplete cerebral ischemia obtained prior, during the insult and 1 or 60 minutes thereafter.

Trauma Plus Ischemia

Four animals were studied in this group, three sacrificed at 2 hours post ischemia and one surviving for 12 hours post ischemia. All animals were subjected to the same levels of trauma as in group 1 and ischemia as in group 2. Similar EEG changes following trauma were seen as in group 1. However, the EEG became isoelectric in all animals with induction of ischemia one hour after trauma. Following ischemia, only one animal showed some electrical recovery which was minimal. There was no recovery of any EEG activity in the chronic animal, although CBF was above normal at sacrifice. The CBF history was similar in all animals with comparable levels of incomplete ischemia as in the group 2 animals. Following the ischemic insult, however, CBF was not increased but slightly depressed or normal (Table 3).

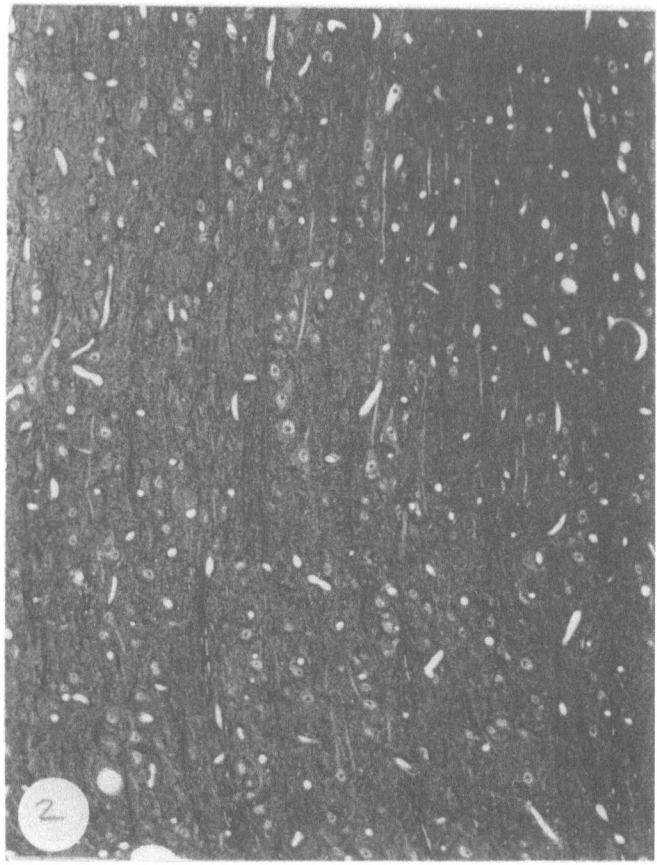

Fig.2: Upper four layers of cerebral cortex from the suprasylvian gyrus
studied 24 hrs after 7 min of incomplete cerebral ischemia. Note
normal cellular structure except for slight chromatin clumping.
125x

        The structural changes seen in all animals were most pronounced in
the experiment with 12 hours survival. The changes consisted of ischemic
neuronal damage and perineuronal and perivascular swelling as previously
described (9). In acute animals, structural changes were only found in
neocortical areas of the arterial boundary zones, while in the chronic
animal other brain regions, such as the CA 1, CA 3 and CA 4 sector of
the hippocampus and cerebellar Purkinje neurons were involved as well.
Additionally, cortical gray and white matter appeared much more edematous as concluded from the presence of swollen astrocytes and an enlargement of the extracellular compartment (Fig. 3).

TABLE 3

ML/100GM/MIN	FIXED WET WEIGHT		MEAN (SD)		
	45 MINUTE POST TRAUMA CONTROL (N = 4)	ISCHEMIC INSULT (N = 4)	30 MINUTE POST ISCHEMIA (N = 1)	60 MINUTE POST ISCHEMIA (N = 4)	FIX
TOTAL	39.8(11.3)	8.8(2.2)	44.6	31.3(5.4)	171
HEMI	41.8(13.6)	7.6(2.4)	48.2	30.6(5.5)	177
STEM	33.4(3.8)	13.3(2.7)	36.0	33.8(3.1)	154
pCO_2	29.2(2.0)	29.6(3.8)	33.0	30(3.3)	
pH_a	7.39(.05)	7.35(.12)	7.4	7.38(.03)	
MABP	112(18.0)	93.7(15.3)	110	97(28.8)	128.5

Cerebral blood flow and systemic variables of experimental animals subjected to both trauma and ischemia. Control CBF was obtained 45 minutes after fluid percussion injury. CBF was also measured during the ischemic insult and 30 and 60 minutes post ischemia. The flow of fixative (fix) was also measured with microspheres during perfusion fixations.

DISCUSSION

Based upon our endpoints the results suggest that the brain is more sensitive to ischemia after head injury. To our knowledge, this data provides the first direct evidence that head injury enhances the vulnerability of cerebral tissue to ischemia. In addition, since the forebrain was minimally damaged by the currently employed type of central fluid percussion injury and capable of considerable recovery, our results indicate that even mild traumatic injury renders the brain more vulnerable to ischemia. The mechanisms of the increased vulnerability remain unknown but some speculation seems warranted. It must be emphasized that this study focussed on cortical loci, especially the suprasylvian gyri which are arterial boundary zones. The EEG electrodes were placed over these regions to correlate functional activity with the structural alterations. It is noteworthy that the morphological evaluation of animals subjected to trauma and ischemia revealed traditional ischemic selective vulnerability. The cortical layers 3, 5, and 6 of the arterial boundary zones were the most vulnerable while other, traditionally selective vulnerable areas were affected to a variable extent only. This may reflect undetected differences in magnitude of the ischemic insult due to the averaging

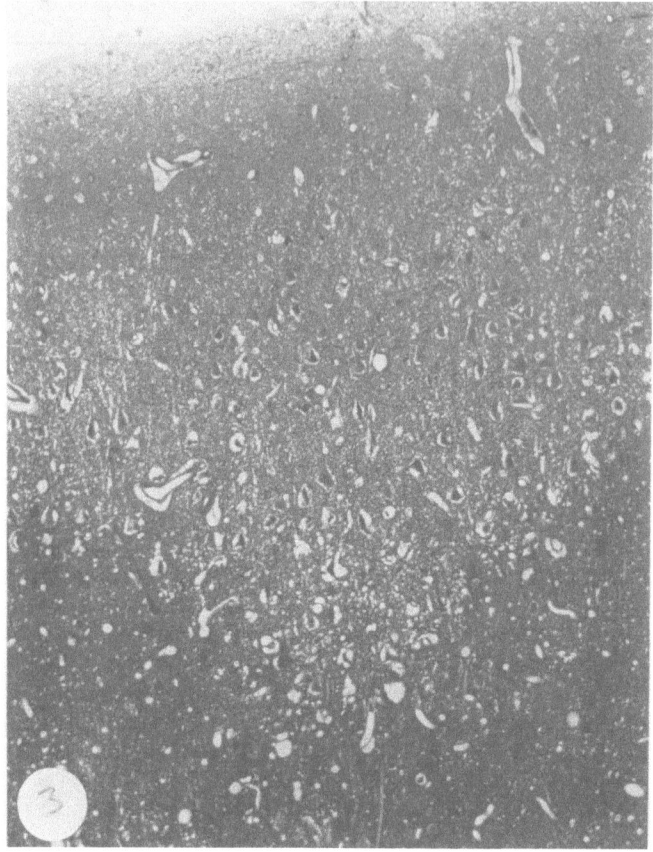

Fig.3: Cerebral cortex of an experimental animal subjected to trauma and
7 min incomplete cerebral ischemia. The specimen was obtained
from the suprasylvian gyrus as in the other experimental groups.
Note dark and shrunken neurons as well as swollen perineuronal
and perivascular astrocytic processes. 125x

effect of the microspheres method in large tissue samples. Such differences were probably most pronounced in the arterial boundary zones because
of the hypotensive component. One would expect a combined pattern of
alterations, i.e. selective vulnerability resulting from ischemia plus an
exacerbation of primary brain stem pathology due to fluid percussion, if
the two insults were indeed cumulative. However, this was not the case.
Therefore, the question remains whether the cells which succumbed to the
secondary ischemic insult were those most severely injured from trauma,
or whether other factors were responsible. It is always dangerous and
usually incorrect to single out one factor as the cause of a complex
pathophysiological process, such as trauma or ischemia. Yet, one possible
explanation for the above results may be proposed, nevertheless. It is

well recognized that hyperglycemia exacerbates cerebral damage in a number of laboratory models of cerebral ischemia (10). Although the responsible mechanisms remain unknown, most speculation has centered around
lactic acidosis. It is well documented that fluid percussion injury of
the level currently examined causes hyperglycemia due to the sympatho-
adrenal surge which persists for hours (11). Moreover, it has recently
been shown that even with mild trauma lactic acid levels increased in cerebral tissue together with a decrease of phosphocreatine, although blood
flow was not limited (Yang, person. commun.). Therefore, hyperglycemia
leading to increased cerebral acidosis in damaged brain may be responsible for the increased ischemic vulnerability observed in the present
study. Control of the serum glucose level may prove an effective treatment to influence the CNS vulnerability to ischemia after trauma. Such
studies are presently underway.

ACKNOWLEDGEMENTS

This work was supported by NIH grants NS19550 and NS12587 as well
as a grant from the Virginia Affiliate of the American Heart Association.

REFERENCES

1. Miller JD, Sweet, RC, Narayan R, Becker DP, Early insults to the
      injured brain, JAMA 240:439 (1978).
2. Langfitt TW, Gennarelli, TA, Can the outcome from head injured be
      improved? J Neurosurg 56:19 (1982).
3. Graham DI, Adams JH, Doyle D, Ischemic brain damage in fatal non-
      missile head injuries, J Neurol Sci 39:213 (1978).
4. Nelson LR, Auen EL, Bourke RS, Barron KD, Malik AB, Crago EJ,
      Popp AJ, Waldman JB, Kimelberg HK, Foster VV, Creel W,
      Schuster L, A comparison of animal head injury models devel-
      oped for treatment modality evaluation, in: "Head injury: basic
      and clinical aspects", Grossman RG, Gildenberg PL, eds., Raven
      Press, New York (1982).
5. Lewelt W, Jenkins LW, Miller JD, Autoregulation of cerebral blood
      flow after experimental fluid percussion injury of the brain,
      J Neurosurg 53:500 (1980).
6. DeWitt DS, Jenkins LW, Lutz H, Wei EP, Kontos HA, Miller JD,
      Becker DP, Regional cerebral blood flow following fluid percus-
      sion injury, 10th International Symposium on Cerebral Blood
      Flow and Metabolism 1(Suppl. 1):579 (1981).
7. Povlishock JT, Becker DP, Miller JD, Jenkins LW, Dietrich WD, The
      morphopathological substrate of concussion? Acta Neuropathol
      47:1 (1979).
9. Jenkins LW, Povlishock JT, Lewelt W, Miller JD, Becker DP, The role
      of post-ischemic recirculation in the development of ischemic
      neuronal injury following complete cerebral ischemia,
      Acta Neuropathol 55:205 (1981).

10. Siesjö BK, Cell damage in the brain: A speculative synthesis, J Cerebr Blood Flow Metabol 1:155 (1981).

11. Rosner MJ, Newsome HH, Becker DP, Mechanical brain injury: The sympathoadrenal response, J Neurosurg 61:76 (1984).

RECENT ADVANCES IN THE STUDY OF CEREBROVASCULAR RECEPTORS

Maria Spatz

Laboratory of Neuropathology and Neuroanatomical Sciences,
NINCDS, National Institutes of Health, Bethesda, Maryland,
20892, USA

The possibility of monoamines interaction with vascular elements in the brain under pathological conditions leading to an altered cerebrovascular function and secondary brain injury was postulated at the time of limited knowledge about their respective receptors (33). The confirmation of this concept has been hindered by the difficulties of studying the regulatory mechanisms of the noticeably inaccessible vascular bed. Needless to say that direct proof of receptors present in the vessels and particularly in the microvessels of the brain has lingered beyond those described in the peripheral vasculature. For a long time it was thought that the cerebral arteries lack reactivities to nervous and pharmacological stimuli, even though the presence of nerve fibers in cerebral arteries was already described centuries ago by Willis. Subsequently, this opinion was overturned by electronmicroscopic, histochemical and specific immunocytochemical evidence of an existing perivascular innervation (4,6,9,11,16,17,29,36,39,43).

The rich network of nerve fibers and/or varicosities in the proximity of cerebral vessels was shown to originate from sympathetic cervical ganglia and/or cerebral nuclei. They comprise (uncontested) noradrenergic, dopaminergic, cholinergic, tryptaminergic and peptidergic innervation. These findings and the biochemical evidence of neuro-hormones in the vessel wall preconceived the vessels as the effector site on which the neuro-transmitters released from nerve endings could act and induce response. Based on these observations and assumptions the notion of cerebrovascular receptors was born and provided an impetus for their exploration. Moreover, a neuronal involvement in the regulation of cerebral blood flow (CBF), systemic blood pressure (SBP) and blood-brain barrier (BBB) permeability stressed the urgency and importance of investigating cerebral vascular receptors especially those of microvessels (3,8,11,17, 31,42,44).

Neither an isolation nor complete chemical characterization of the vascular receptors has been yet achieved, irrespective of their origin. Until recently most of the available information has been derived from pharmacological investigations in vivo and in vitro . They will be only briefly summarized here, since a number of excellent and detailed reviews were published in the last few years (cf. 2,10,31,39).

PHARMACOLOGICAL ARTERIAL PROPERTIES

Based on studies performed in animals and/or man it was shown that the cerebral vasculature is reactive to various adrenergic, dopaminergic, muscarinic and peptidergic agents, as well as to 5-hydroxy-tryptamine and histamine (7,9,11,15,44). A general agreement exists on the complexity of their effects in the vascular bed, since the response of the vessels rests on many factors. It may depend on their direct or indirect action on the vessels, such as a secondary reaction to the systemic blood pressure and heart-rate which could affect the vascular tone. The route of administration also plays an important role for certain substances which can be metabolized before reaching the receptor. Thus, the cerebral vessel could not be pharmacologically characterized with certainty by these compounds when given systemically. On the other hand, topical application of vasoactive agents to pial vessels or analyzing of cylindrical vascular segments aided in elucidation of their pharmacological properties.

These investigations comprised the evaluations of dose-dependent vascular responses to a variety of substances in the presence or absence of their respective blockers. The most important conclusions were as follows:

(1) Cerebral arteries contain contractile α-type and dilatory β_1-type adrenergic receptors

(2) The dose-dependent vasodilatation by dopamine is apparently mediated by dopamine receptors, while vasoconstriction is mediated by α-adrenergic and/or 5-HT receptors

(3) Acetylcholine vasodilatory (low dose) and vasoconstrictory (high dose) arterial reactivities occur through muscarinic type of cholinergic receptors

(4) 5-HT-induced cerebrovascular effects are complex. A marked vasoconstriction is mediated directly by serotoninergic and indirectly by α-adrenergic receptors. Conversely, vascular relaxation - ascribed to differences of animal species or vascular tone - is possibly related to β-adrenergic receptors

(5) Vasodilatation by Substance P or VIP is mediated by peptidergic receptors

(6) Histamine induced contraction of cerebral arteries is mediated by H_1-type-, while dilatation by H_2-type receptors.

Thus, the reactivity of the cerebral arteries to vasoactive substances is consistent with their implied participation in the modulation

of CBF, SBP and blood-brain barrier function. Additional support for this
hypothesis was obtained by (I) observations of specific immunofluores-
cence on microvessels of (a) phenylethanolamine-N-methyl-transferase
(46) - the catecholamine synthetizing enzyme - (b) 5-HT (48), and (II) in
particular by the detection of receptors linked to a biochemical function
(19-21, 25-27, 35,40,45,48,51).

MICROVASCULAR RECEPTOR PROPERTIES

The characterization of these receptors became possible only with
the availability of techniques for the isolation of cerebral microvessels. Moreover, the recent development of separate cultures of cerebrovascular endothelial and smooth muscle cells derived from dissociated
microvascular fraction of the brain permits studies of their individual
properties (45,51). Until now most of the information has been obtained
through investigations of adenylate cyclase responsiveness to neurohormones, since cAMP was suggested to modulate vascular functions, especially BBB-permeability (22). Binding studies have been also useful in
examining microvascular receptors. As a matter of fact, the presence of
β_2 - and weak β_1 -adrenergic receptors was detected by both methods
(18,19,21, 35,40,41). However, α-adrenergic, as well as GABA- and
cholinergic receptors were observed in microvascular fractions with the
use of binding ligands (12-14,18,41). The specific binding of tritiated
ligands to α_1 -adrenergic receptors was found higher in bovine than in
pig vessels but was undetectable in microvessels of rats (18,41). It is
of special interest that the cholinergic receptors were localized in the
endothelium, since enzymatically isolated endothelium from bovine cerebral capillaries bound ^{3}H-quinuclidinyl benzilate (QNB) to the same
extent as the intact capillary fraction. On the other hand, choline ace-
tyltransferase activity - a marker for cholinergic neurons - was higher
in intact capillaries than in the endothelium. This suggests a perien-
dothelial localization at the site of attached nerve terminals (13).

An enhancement of adenylate cyclase (AC) activity was also observed to other than adrenergic agents. These were dopamine, prostaglandins E_1 and E_2 , histamine, 2-chloroadenosine and - among the twen-
ty-three peptides tested - VIP and parathyroid hormone, but not the F-
prostaglandins and 5-HT (20,21,40). Based on various investigations including our own observations concerning the microvascular AC-sensitivity
to adrenergic agents, dopamine, prostaglandins, histamine and serotonin,
it appears that the absence and presence as well as the degree of induced
formation of cAMP rests not only on the animal species but also on the
preparation of the fraction and its functional state. Besides it has to
be kept in mind that isolated microvessels contain not only endothelial
but also smooth muscle cells, pericytes and possible glial cells as well
as nerve endings (45,49).

The novel approach for studying the function of microvessels and
their receptors in pure cultures of vascular elements avoids this pro-

blem. It provides a living cell model for elucidating microvascular regulatory mechanisms which are involved in the processes occurring at the
blood-brain barrier. It allows to stimulate some of the conditions observed in vivo in a normal or abnormal state. The establishment of this
model belongs to the latest development and there has not been sufficient time to explore many experimental avenues. However, the results obtained already gave additional insight of actual and/or implied functions
of cerebral capillaries. The cultured cerebrovascular endothelial cells
display the same enzymatic characteristics as those described in the
microvessels studied in situ or in isolated capillaries (45). These cells
synthetize prostaglandins which can be stimulated by adrenergic agents,
5-HT and Ca^{++} ionophore (50). They also contain PNMT (45), the enzyme
which converts norepinephrine to epinephrine. Besides the cultured endothelial cells are capable of metabolizing and synthetizing 5-HT (32). Indomethacin (the inhibitor of prostaglandin synthesis) decreases formation
of 5-HT. This can be reversed by PGI_2 (48).

These findings indicate an interaction between two known vasoactive substances which could be responsible for controlling the intrinsic
vascular tone. Both the endothelial and the smooth muscle cell cultures
possess adenylate cyclase which is responsive to neurohormones (26, 27,
48, 51). The cultured endothelium contains β_2 - and α_2 -adrenergic receptors linked to adenyl cyclase, whereas the smooth muscle possesses
β_2 - and α_1 -adrenergic receptors coupled to AC (see Table). On the
other hand, β_1 -adrenergic receptors observed in cerebral arteries were
not detected in cultured endothelium and smooth muscle cells. This might
be related to the vessel size and preparation of vessels, but also could
depend on the animal species. The presence of β_2 -adrenergic receptors
coupled to AC is in agreement with findings reported by others in
cerebro-microvascular fractions. Moreover, the existence of β_2 - rather
than β_1 -adrenergic receptors in the microvascular bed is supported by an
attenuation of amphetamine-induced enhancement of blood-brain barrier
permeability to proteins by β_2 and β_1 / β_2 - but not β_1 -adrenergic antagonists (23). The presence of α_1 - and absence of α_2 -adrenergic receptors linked to AC in the smooth muscle cells was not only concluded
from studies of their respective antagonists (prazosine-α_1 , yohimbine-
α_2) but also from an AC reactivity to α_1 - and α_2 -agonists (51).

Phenylephrine (α_1 -agonist) stimulated AC-activity was dose-
dependent and showed a similar pattern but a lower affinity than those
observed with epinephrine and norepinephrine (EC_{50} = 6.5 x 10^{-6} M, 7.5 x
10^{-7} M and 4 x 10^{-6} M, respectively). The α-adrenergic response was not
mediated by adenosine in contrast to that described in some brain regions. Moreover, an additive stimulation of cAMP formation was seen by
simultaneous exposure of the smooth muscle AC system to low doses of
phenylephrine together with either norepinephrine or epinephrine. Clonidine (α_2 -agonist) failed to enhance the AC-activity of broken or intact smooth muscle cells. Nonetheless, α_2 -subtypes of adrenergic receptors were detected with labeled clonidine or yohimbine in cultured
smooth muscle cell preparations using binding techniques (unpublished

TABLE 1

Effect of ß- and α-Adrenergic Antagonists on AC Activity

	Endothelium		Smooth Muscle Cell	
	+Isoproterenol		+Norepinephrine	Epinephrine
	$(10^{-5}$ M$)$	(5×10^{-6})	$(10^{-5}$ M$)$	$(10^{-5}$ M$)$
ß-adrenergic				
1. None	100.0	100.0	100.0	100.0
2. Atenolol (β_1)	114.0	–	–	–
3. Practolol(β_1)	110.0	–	82.0	84.0
4. IPS-339 (β_2)	11.0*	–	–	–
5. Butoxamine (β_2)	–	–	25.0*	21.0*
6. Propranolol (β_1/β_2)	11.0*		49.0*	38.0*
α-adrenergic				
1. None	100.0	100.0	100.0	100.0
2. Prazosine (α_1)	–	100.0	65.0*	60.0*
3. Yohimbine (α_2)	99.0	168.0*	85.0	96.0
4. Phentolamine (α_1/α_2)	99.0	227.0*	103.0	104.0

The values are expressed as percent of basal activity with and without the addition of antagonists. Antagonist-concentrations given were 10^{-5} M except of propranolol (i.e. 10^{-6}).

* Significantly different as compared to the values of agonist-stimulated AC activity. They are based on two separate experiments in the endothelium and 13-16 experiments in the muscle assayed in triplicate. Atenolol and practolol are β_1 -antagonists; IPS-339, butoxamine β_2 -antagonists; propranolol β_1 / β_2 -antagonist. Prazosine and yohimbine are α_1 and α_2 -antagonists, respectively, phentolamine is a α_1 / α_2 -antagonist.

observations). On the whole, the molecular mechanism of α-adrenergic reactivity is not as clear as that of ß-adrenergic AC stimulation (30).

The enhancement of the α-adrenergic AC system reportedly depends on extracellular Ca^{++} and appears to be associated with calcium fluxes and changes of the phosphatidylinositol turnover. However, there is no argument concerning the relationship of α-adrenergic receptors to either extra- or intracellular calcium ions and vascular contractions. Based on the detected presence of α_2-adrenergic receptors coupled to the AC system of cerebrovascular smooth muscle, we suggested that the control of vascular calcium levels and of cerebrovascular activity could be modulated by (1) cerebral and (2) blood-borne amines. Cerebral amines could modulate through the α_1-adrenergic receptors linked to AC and/or α_2-adrenergic receptors not coupled to AC, while blood-born amines via endothelial α_2-adrenergic receptors linked to AC (51). Notwithstanding and most importantly, the additive response of smooth muscle AC system to phenylephrine with norepinephrine or epinephrine strongly supports the contention of α_1-receptors participation in mediating the vasoconstrictive effect of catecholamines involved in the regulation of CBF and SBP.

Prostaglandins were also found to stimulate both the endothelial and the smooth muscle AC systems (26,27,48,51). Endothelial cAMP formation was more stimulated by the E-type prostaglandins and PGI_2 (EC_{50} = $3x10^{-7}$ M and $3x10^{-6}$ M, respectively) than by prostaglandins of the F-series and PGD_2. The latter activates AC at high doses only (27). Moreover, an interaction between PGE_2 and norepinephrine was observed on endothelial AC activity. The dose dependent norepinephrine inhibition of PGE_2-enhanced formation of cAMP could be partially restored by phentolamine (the α_1 / α_2-adrenergic antagonist). Clonidine and 6-fluoro-norepinephrine (α-agonists) and isoproterenol (ß-agonist) blocked also partly, whereas forskolin - an activator of AC (most likely of the catalytic unit) - and PGE_2, synergetically stimulated the endothelial AC-system. Although a detailed evaluation of the data indicates that PGE_2 and norepinephrine interact at the α-adrenergic receptor, the limited degree of inhibition by the α-agonists suggests that the effect of PGE_2 on the endothelial AC system has two components: one is sensitive and the other insensitive to α-adrenergic substances. These observations are not only of scientific interest but have clinical ramifications. The dose-dependent inhibitory effect of norepinephrine on PGE_2-enhanced formation of cAMP in particular could be related to the diversity of the vascular responses reported in CBF studies in-vivo. The evaluation of prostaglandin effects on cerebrovascular muscle cells is still incomplete. However, it has been observed that contrary to endothelium the AC-system of smooth muscles cells is responsive to PGE and F. Formation of cAMP in smooth muscle was greater in the presence of PGE_1 and PGE_2 than in the presence of $PGF_{1\alpha}$ and $PGF_{2\alpha}$ (48).

Up to now the responsiveness of the AC system to vasoactive substances other than adrenergic agents and prostaglandins was only studied in cultured cerebrovascular endothelium but not in smooth muscle cells.

The results were as follows: high doses of dopamine stimulate, while adenosine, angiotensin I and II, GABA and VIP inhibit formation of cAMP. Acetylcholine, histamine, serotonin, glycine, glutamine, bradykinin, neurotensin and vasopressin were ineffective in disrupted cells (26). However, these results are still inconclusive, since some of the hormones studied might have a direct or indirect effect on intact cells of the cerebral endothelium. It is noteworthy that our findings on endothelial AC inhibition by adenosine and VIP is in contrast with the reported stimulation of microvascular AC by the same substances (cf. above). Therefore, the possibility exists that the stimulation of AC by adenosine occurs in other cells than endothelium (21). Future studies should not only establish their receptor sites and properties but their cerebrovascular function. This is as yet unknown for either adenosine, VIP or GABA. Adenosine and VIP have been implicated in the regulation of CBF while GABA was reported to modulate heart-rate, blood pressure, and chloride transport.

ALTERATION OF RECEPTOR FUNCTION

It is generally known that exposure of cells to high levels of a hormone leads to a desensitization while depletion results in supersensitization of the cell or tissues (24). These events may perturb cellular processes which are modulated by hormonal induced changes in cAMP formation. It should be added that another possibility exists namely an attenuation of cAMP-mediated cellular events associated with changes of the AC-activity, which can be produced by increasing the level of proteases (1).

So far little information is available on modification of the properties of vascular receptors. The decrease of catecholamine-stimulated cAMP formation of microvessels in spontaneous hypertensive rats (SHR) was related to high dopamine levels (40). Similar observations were made on cerebral microvessels obtained from aged rats showing decreased responsiveness of AC to norepinephrine and ^{125}I-iodohydroxybenzylpindolol-binding sites (28). These findings were consistent with a reduced microvascular content of norepinephrine and an impaired capability to raise the density of ß-adrenergic receptors due to a deficient adrenergic input. Our preliminary investigations showed also that desensitization of cultured cerebrovascular endothelial and smooth muscle cells may depend on the medium content of neurohormones and on the duration of exposure.

COMMENTS

The physiological control of the cerebrovascular bed has been centered on the precapillaries without giving any consideration to the capillaries for a long time. The reason was that precapillaries do not need nervous or hormonal stimuli for maintaining a vascular tone since they have an inherent myogenic activity (39). However, recent observations of contractile elements (38) and perivascular innervation draws attention to a possibly active rather than passive function of these vascular elements (39). The endothelial capability of synthetizing vasoactive substances which are considered to regulate CBF, SBP and blood-brain barrier permeability greatly strengthens this idea. We would also like to suggest that a disturbed capillary synthesis and/or release of prostaglandins and 5-HT might play an important role in the development of migraine and the induction of vasospasm in damaged brain. The demonstration of endothelial adrenergic, prostaglandin- and other receptors linked to adenylate cyclase of cerebral microvessels (26,27,51) supports the contention that many of these processes could be mediated by adenyl cyclase responses to either blood-borne and/or central neurohormones. However, it was also shown that not all α-adrenergic receptors are coupled to AC. Their activity may be mediated by Ca^{++}-ions or phosphoinositol metabolism. Most importantly, the demonstration of these receptors on endothelium or smooth muscle cells supports the contention of their participation in the regulation of cerebrovascular and blood-brain barrier function under normal conditions.

The existence of microvascular receptors together with alterations of the levels of cerebral or spinal cord metabolites including monoamines (5,34,37,47) in ischemia or trauma provides a scenario for a disturbance in their interaction and therefore in vascular reactivity. An alteration of receptor-mediated microvascular functions may also result from changes of capillary synthesis and metabolism of vasoactive substances in ischemia and trauma. Disturbances of blood-brain permeability may allow for an increased influx of these hormones from the periphery (47). Thus, the resultant impairment of vascular activity irrespective of its primary or secondary causitive factors may adversely affect the microcirculation and brain function. Future studies may hopefully bring us closer to an understanding of the function of vascular receptors and their involvement in normal and abnormal conditions. This knowledge would be of value in the prevention and treatment of central and peripheral nervous diseases.

REFERENCES

1. d'Alayer J, Berthillier G, Monneron A, Structure of brain adenylate cyclase: Proteolysis-dependent modifications, Biochem 22:3948 (1983).
2. Bevan JA, The human adrenergic neurovascular mechanism, Gen Pharmac 14:21 (1983).
3. Chan-Palay V, Innervation of cerebral blood vessels by norepinephrine, indoleamine, substance P and neurotensin fibers and leptomeningeal indoleamine axons: Their roles in vasomotor activity and local alterations of brain blood composition, in: "Neurogenic control of the brain circulation", Owman C, and Edvinsson L, eds., Pergamon Press, Oxford (1977).
4. Cervos-Navarro J, The structural basis of an innervatory system of brain vessels, in: "Neurogenic control of the brain circula-

tion", Owman C, Edvinsson L, eds., Pergamon Press, Oxford (1977).

5. Cvejic V, Micic DV, Djuricic BM, Mrsulja BJ, Mrsulja BB, Monoamines and related enzymes in cerebral cortex and basal ganglia following transient ischemia in gerbils, Acta Neuropathol 51:71 (1980).

6. Dahl E, Innervation of the cerebral arteries, J Anat 115:53 (1973).

7. Edvinsson L, Owman C, Pharmacological characterization of adrenergic alpha and beta receptors mediating the vasomotor responses of cerebral arteries in vitro, Circulation Res 35:835 (1974).

8. Edvinsson L, Neurogenic mechanisms in the cerebrovascular bed. Autonomic nerves, amine receptors and their effect on cerebral blood flow, Acta Physiol Scand (Suppl.) 427:1 (1975).

9. Edvinsson L, Owman C, Sjöberg NO, Autonomic nerves, mast cells, and amine receptors in human brain vessels. A histochemical and pharmacological study, Brain Res 115:377 (1976).

10. Edvinsson L, Owman C, Pharmacological characterization of postsynaptic vasomotor receptors in brain vessels, in: "Neurogenic control of the brain circulation", Owman C, Edvinsson L, eds., Pergamon Press, Oxford (1977).

11. Edvinsson L, Fahrenkrug J, Hanko J, McCulloch J, Owman C, Uddman R, Vasoactive intestinal polypeptide: Distribution and effects on cerebral blood flow and metabolism, in: "Cerebral microcirculation and metabolism", Cervos-Navarro J, Fritschka E, eds., Raven Press, New York (1981).

12. Estrada C, Krause DN, Muscarinic cholinergic receptor sites in cerebral blood vessels, J Pharm Exp Ther 221:85 (1982).

13. Estrada C, Hamel E, Krause DN, Biochemical evidence for cholinergic innervation of intracerebral blood vessels, Brain Res 266:261 (1983)

14. Grammas P, Diglio CA, Marks BH, Giacomelli F, Wiener J, Muscarinic receptors in rat cerebral cortical microvessels, Fed Proc 41:451 (1982).

15. Gross PM, Harper AM, Teasdale GM, Cerebral circulation and histamine: 1. Participation of vascular H_1 - and H_2 -receptors in vasodilatatory responses to carotid arterial infusion, J Cereb Blood Flow Metabol 1:97 (1981).

16. Falck B, Mchedlishvili GI, Owman C, Histochemical demonstration of adrenergic nerves in cortex-pia of rabbit, Acta Pharmac Tox 23:133 (1965).

17. Hartman BK, Swanson LW, Raichle ME, Preskorn SH, Clark HB, Central adrenergic regulation of cerebral microvascular permeability and blood flow; anatomic and physiologic evidence, Adv Exp Med Biol 131:113 (1980).

18. Harik SI, Sharma VK, Wetherbee JR, Warren RH, Banerjee SP, Adrenergic and cholinergic receptors of cerebral microvessels, J Cereb Blood Flow Metabol 1:329 (1981).

19. Herbst TJ, Raichle ME, Ferrendelli JA, ß-adrenergic regulation of adenosine 3',5'-monophosphate concentration in brain microvessels, Science 204:330 (1979).

20. Huang M, Hanley DA, Rurstad OP, Parathyroid hormone stimulates adenylate cyclase in rat cerebral microvessels, Life Sci 32:1009 (1983).

21. Huang M, Rorstad OP, Effects of vasoactive intestinal polypeptide, monoamines, prostaglandins, and 2-chloroadenosine on adenylate cyclase in rat cerebral microvessels, J Neurochem 40:719 (1983).

22. Joo F, Rakonczay Z, Wollemann M, cAMP-mediated regulation of the permeability in the brain capillaries, Experimentia 31:582 (1975).

23. Johansson BB, Effect of beta-adrenoreceptor antagonists on the increased cerebrovascular permeability to protein induced by amphetamine, Progr Neuro-Psychopharmacol 2:529 (1978).

24. Iyengar R, Birnbaumer L, Agonist-specific desensitization: Molecular locus and possible mechanism, in: "Advances in cyclic nucleotide research, Vol. 14", Dumont JE, Greengard P, Robison GA, eds., Raven Press, New York (1981).

25. Karnushina IL, Palacios JM, Barbin G, Dux E, Joo F, Schwartz JC, Studies on a capillary-rich fraction isolated from brain: Histaminic components and characterization of the histamine receptors linked to adenylate cyclase, J Neurochem 34:1201 (1980).

26. Karnushina IL, Spatz M, Bembry J, Cerebral endothelial cell culture. I. The presence of β_2 and α_2-adrenergic receptors linked to adenylate cyclase activity, Life Sci 30:849 (1982).

27. Karnushina IL, Spatz M, Bembry J, Cerebral endothelial cell culture. II. Adenylate cyclase response to prostaglandins and their interaction with the adrenergic system, Life Sci 32:1427 (1983).

28. Kobayashi H, Maoret T, Spano PF, Trabucchi M, Effect of age on β-adrenergic receptors on cerebral microvessels, Brain Res 244:374 (1982).

29. Larsson LI, Edvinsson L, Fahrenkrug J, Hakanson R, Owman C, Schaffalitzky de Muckadell O, Sundler F, Immunohistochemical localization of a vasodilatory polypeptide (VIP) in cerebrovascular nerves, Brain Res 113:400 (1976).

30. Lefkowitz RJ, De Lean A, Hoffman BB, Stadel JM, Kent R, Michel T, and Limbird L, Molecular pharmacology of adenylate cyclase-coupled α- and β-adrenergic receptors, in: "Advances in cyclic nucleotide research, Vol. 14", Dumont JE, Greengard P, Robison GA, eds., Raven Press, New York (1981).

31. Marin J, Rivilla F, Nerve endings and pharmacological receptors in cerebral vessels, Gen Pharmac 13:361 (1982).

32. Maruki C, Spatz M, Ueki Y, Nagatsu I, Bembry J, Cerebrovascular endothelial cell culture: Metabolism and synthesis of 5-hydroxytryptamine, J Neurochem 43:316 (1984).

33. Moskowitz MA, Wurtman RJ, Catecholamines and neurologic diseases, N Engl J Med 293:332 (1975).

34. Mrsulja BB, Mrsulja BJ, Spatz M, Klatzo I, Action of cerebral ischemia on decreased levels of 3-methoxy-4-hydroxyphenethylglycol sulphate, homovanillic acid and 5-hydroxyindolacetic acid produced by pargyline, Brain Res 98:388 (1975).

35. Nathanson JA, Glaser GH, Identification of ß-adrenergic-sensitive adenylate cyclase in intracranial blood vessels, Nature 278:567 (1979).
36. Nelson E, Rennels M, Innervation of intracranial arteries, Brain 93:475 (1970).
37. Osterholm JL, Bell J, Meyer R, Pyenson J, Experimental effects of free serotonin on the brain and its relation to brain injury, J Neurosurg 31:408 (1969).
38. Owman C, Edvinsson L, Hardebo JE, Gröschel-Stewart U, Unsicker K, and Walles B, Immunohistochemical demonstration of actin and myosin in brain capillaries, Acta Physiol Scand (Suppl.) 452:69 (1977).
39. Owman C, Autonomic innervation of blood vessels with special emphasis on human cerebrovascular nerves and corresponding amine receptors, Gen Pharmac 14:17 (1983).
40. Palmer GC, Palmer SJ, Chronister RB, Cyclic nucleotide systems in the microcirculation of mammalian brain, in: "The cerebral microvasculature", Eisenberg HM, Suddith RL, eds., Academic Press, New York (1980).
41. Peroutka SJ, Moskowitz MA, Reinhard JF, Snyder SH, Neurotransmitter receptor binding in bovine cerebral microvessels, Science 208:610 (1980).
42. Raichle ME, Grubb Jr RL, Eichling JO, Neural and hormonal regulation of brain water permeability, in: "Neurogenic control of the brain circulation", Owman C, Edvinsson L, eds., Pergamon Press Oxford (1977).
43. Reinhard JF, Liebmann JE, Schlossberg AJ, Moskowitz MA, Serotonin neurons project to small blood vessels in the brain, Science 106:85 (1979).
44. Sercombe R, Aubineau P, Edvinsson L, Mamo H, Owman C, Pinard E, and Seylaz J, Neurogenic influence on local cerebral blood flow. Effect of catecholamines or sympathetic nerve stimulation as correlated with the sympathetic innervation, Neurology 25:954 (1975).
45. Spatz M, Mrsulja BB, Progress in cerebral microvascular studies related to the function of the blood-brain barrier, in: "Advances in cellular neurobiology, Vol. 3", Federoff S, Hertz L, eds., Academic Press, New York (1982).
46. Spatz M, Nagatsu I, Maruki C, Yoshida M, Kondo Y, Bembry J, The presence of phenylethanolamine-N-methyltransferase in cerebral microvessels and endothelial cultures, Brain Res 240:191 (1982).
47. Spatz M, Maruki C, Karnushina I, Nagatsu I, Bembry J, Merkel N, The relationship of monoamines to the blood-brain barrier, in: "Advances in the biosciences, Vol. 43, Stroke: Animal models", Stefanovich V, ed., Pergamon Press, New York (1983).
48. Spatz M, Maruki C, Nagatsu I, Ueki Y, Wroblewska B, Mocarski E, Merkel N, Bembry J, Recent progress in the studies related to cerebral microvascular function, J Cereb Blood Flow Metabol 3:S311 (1983).

49. Suddith RL, Savage KE, Eisenberg HM, Ultrastructural and histo-
        chemical studies of cerebral capillary synapse, in: The cerebral
        microvasculature", Eisenberg HM, Suddith RL, eds., Plenum,
        New York (1980).
50. Wolfe LS, Ng Ying Kin NMK, Spatz M, Metabolites of arachidonic
        acid after calcium ionophore stimulation of cultured cerebral
        capillary endothelial cells and brain tissue: Identification of
        lipoxygenase products, J Neurochem 41:S40 (1983).
51. Wroblewska B, Spatz M, Merkel N, Bembry J, Cerebrovascular
        smooth muscle culture. II. Characterization of adrenergic re-
        ceptors linked to adenylate cyclase, Life Sci 34:783 (1984).

ROLE OF CEREBRAL MICROCIRCULATION IN SECONDARY BRAIN DAMAGE

George Mchedlishvili

Beritashvili Institute of Physiology, Georgian Academy
of Sciences, Gotua Street 14, Tbilisi 380060, USSR

The cerebral circulation plays a significant role in secondary brain damage which may result from microcirculatory disorders provoking either delayed brain edema or secondary cerebral ischemia.

DELAYED BRAIN EDEMA

Delayed brain edema may develop due to changes of the systemic and cerebral circulation causing an elevation of blood pressure in the cerebral microvessels. The significance of the pressure level is quite obvious, since it determines the hydrostatic pressure gradient between the blood and parenchyma. The pressure gradient is the principal driving force determining passage of excessive water across the microvascular walls into the parenchyma. If the tissue has previously been damaged, retention of additional fluid in the cerebral parenchyma ensues, hence, formation of delayed brain edema. Leakage of serum proteins and of other blood constituents, e.g. free fatty acids into the cerebral parenchyma occurs when the blood-brain barrier has been damaged (Klatzo, 1981; Maier-Hauff et al., 1983).

The principal circulatory parameters which determine blood pressure in the cerebral microvessels are:

(a) Systemic (central) arterial pressure,
(b) Systemic (central) venous pressure, and
(c) Resistance of the cerebral arterial branching sequence.

These circulatory parameters may change in the period following primary brain damage resulting in a pressure elevation in the cerebral microvessels as a cause of delayed brain edema (Fig. 1).

The systemic arterial pressure determines the microvascular blood pressure. Under conditions of an unchanged venous pressure, the latter depends on the pressure level in the aortic branches minus the pressure which is expended for overcoming the resistance in the arterial branching sequence carrying blood to the microvessels. Therefore, any significant rise of the systemic arterial pressure will result in a proportional increase of blood pressure in cerebral vessels, provided it is not compensated by appropriate constriction of the cerebral arterioles (autoregulation). Experimental results have shown that the development of brain edema correlates with the level and rate of elevation of the systemic arterial pressure (Häggendal, Johansson, 1972).

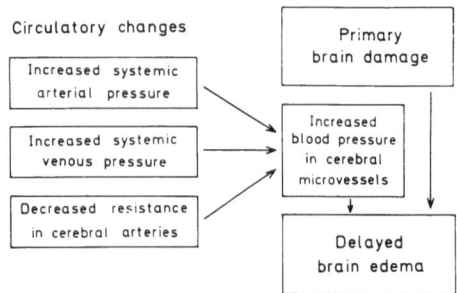

Fig.1: Sequence of events resulting in formation of delayed brain edema and secondary brain damage.

During elevation of the systemic arterial pressure compensatory vasoconstriction occurs primarily in the large cerebral arteries, i.e. in the internal carotid and vertebral arteries (Mchedlishvili, 1972; Mchedlishvili et al, 1973). If, however, the vascular response proves insufficient to maintain the cerebral blood pressure at a normal level, secondary compensatory constriction occurs in the pial arterial ramifications (Mchedlishvili et al, 1973; 1976b). This compensatory response might not uniformely affect all pial arterial ramifications. Vessels with a straight course and poor adrenergic innervation were found to fail in the compensation of an increased arterial pressure resulting in the development of edema in brain areas supplied by those arteries (Gannushkina, 1975; Sakharova, 1980). In the course of the development of brain edema, the systemic arterial pressure has a tendency to decrease (Mchedlishvili et al, 1976a). This response is considered as a compensatory mechanism to restrict development of brain edema.

The systemic venous pressure might also affect development of brain edema (Mchedlishvili et al, 1983), because it is approximately 7-10 times as effective as the systemic arterial pressure to modify blood pressure in the cerebral microvessels (Mchedlishvili et al, 1976a; 1982). This is attributable to the significantly smaller resistance, and to absence of a specific regulatory response of the cerebral venous bed as compared to the arterial system.

The resistance of the cerebral arterial ramifications can exert a marked effect on the blood pressure level in the cerebral microvessels. The major cerebral and pial arteries have been shown to constrict in the course of traumatic and ischemic brain edema (Mchedlishvili, Akhobadze, 1961a; Mchedlishvili et al, 1976a). The vascular response is associated with a reduction of cerebral blood flow (Pappius, 1981; Mchedlishvili, Sikharulidze, in preparation) and should reduce blood pressure in cerebral vessels. Cerebral vasoconstriction is undoubtedly an active vascular response and might be considered as a protective mechanism restricting development of brain edema. Arterial dilation may provoke brain edema, especially when the systemic arterial pressure is elevated (Gannushkina, 1976).

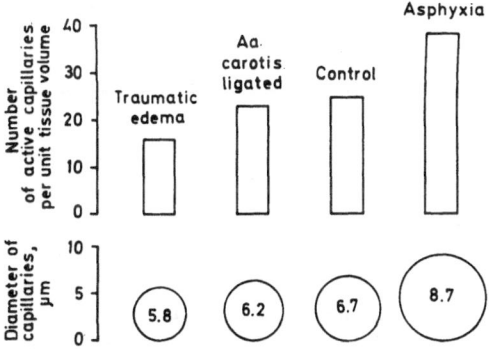

Fig.2: Number of "active" capillaries containing red blood cells and plasma and the corresponding diameter (mean) in cerebral neocortex of rabbits with traumatic edema. The data are compared with the findings on ligation of the carotid arteries, or asphyxia, respectively (from: Mchedlishvili and Akhobadze, 1961b).

SECONDARY MICROVASCULAR DISTURBANCES

Brain tissue may be secondarily damaged, if blood flow in micro-vessels is disturbed. Under a normal arterio-venous pressure difference, blood flow in capillaries can be retarded by an increase in resistance of the microvessels. This may be accomplished either by reduction of the capillary lumen to a certain degree, or by a decrease of blood fluidity in the capillaries. Such microvasculatory changes may follow primary brain damage, e.g. resulting from trauma, ischemia, etc. leading to sec-ondary ischemia and respective brain tissue damage. However, available experimental results are still scant. Normally, the diameter of brain capillaries is 5 μm which is smaller than that of red blood cells, i.e. 7 μm. Therefore, only markedly deformed, and elongated red blood cells pass through the capillary lumen. However, due to the high deformability of red blood cells (Schmid-Schönbein et al, 1969) resistance to blood flow is comparatively small in a capillary. Yet, fluidity is considerably decreased if the lumen diameter is reduced, e.g. to about 4 μm (Chien, 1972). Thus, narrowing of brain capillaries may become a cause of blood flow disturbances and, hence, of secondary brain damage.

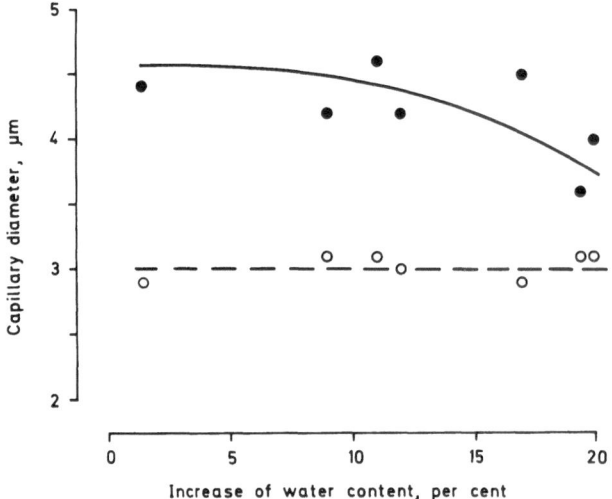

Fig.3: Diameter of "active" capillaries (closed circles) and plasmatic capillaries (open circles) in cerebral neocortex of rabbits in re-lation with the increase of the cerebral water content in post-ischemic brain edema (Mchedlishvili, Sikharulidze, and Varazash- · vili, in preparation).

It has been generally supposed that reduction of cerebral blood flow by edema is brought about by narrowing of capillary diameters, although only few data support this assumption. In earlier studies (Mchedlishvili, Akhobadze, 1961b) a significant decrease of the capillary lumen was not found in traumatic brain edema in rabbits (Fig. 2). Recent studies on postischemic brain edema, however, demonstrate a slight decrease in diameter of so-called "active" capillaries, i.e. which contain red blood cells and plasma. The decrease was considerable in cases with severe edema (Fig. 3). Moreover, quite a number of capillaries were found transformed into plasmatic microvessels with a comparatively small luminal diameter (Mchedlishvili et al., in prepraration). The underlying mechanism of luminal contraction of cerebral capillaries in edematous tissue remains to be explored. An increase of interstitial pressure in the cerebral parenchyma, or swelling of astrocytic endfeet and of endothelial cells may be considered (Reulen et al., 1972).

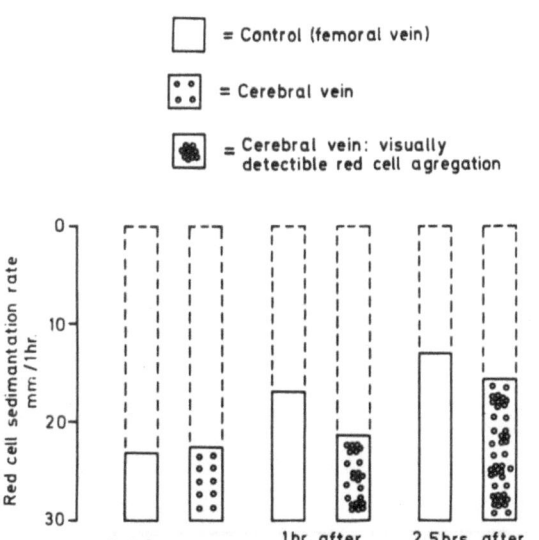

Fig.4: Red blood cell sedimentation rate and incidence of red blood cell aggregates in blood samples drawn from the <u>femoral</u> vein (control), or from <u>cerebral</u> veins (mean) before and 1, or 2.5 hours after mechanical trauma to the brain. Note the increase in red blood cell aggregation in the cerebral vessels (from: Mchedlishvili and Akhobadze, 1961b).

Another mechanism of secondary brain damage may be associated with the slowing of blood flow in cerebral microvessels due to an increase of blood viscosity. Earlier studies of ourselves provided evidence for red blood cell aggregation in the brain vasculature of dogs during development of traumatic brain edema (Mchedlishvili, Akhobadze, 1961b). The degree of aggregation was estimated then by the sedimentation rate of blood samples taken simultaneously from the femoral vein (control) and from the sagittal venous sinus before and 1 or 2.5 hours after trauma to the cerebral cortex. The results are schematically shown in Fig. 4. There it is seen that cerebral trauma is followed not only by an increase in red cell sedimentation in blood from brain vessels, but also by visible aggregation in the test tube. This was never observed in blood samples taken from the femoral vein. The results suggest therefore that the rheological properties of blood passing through the cerebral vasculature undergo specific changes. Red cell aggregation is considered as a direct cause of blood flow retardation and primary stasis in capillaries (Mchedlishvili, 1958). Recent experiments have provided quantitative data that stasis steadily develops in capillaries of cerebral cortex during postischemic brain edema. By studying thick histological sections of cer-

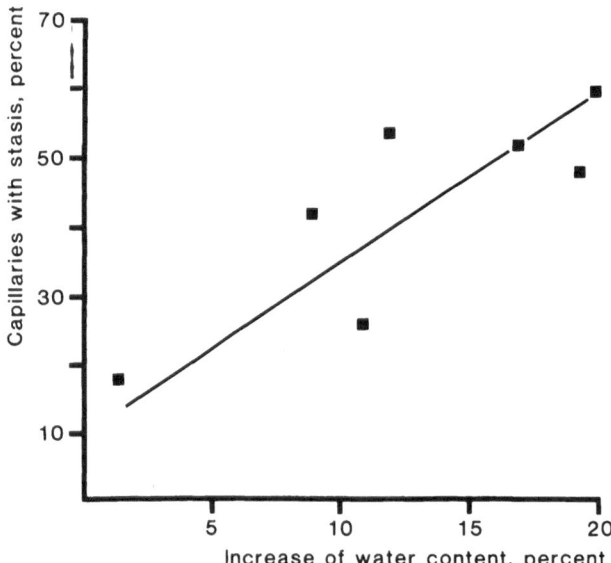

Fig.5: Regression analysis between the number of capillaries with stasis in cerebral cortex and the increase of the tissue water content in postischemic brain edema in rabbits (from: Mchedlishvili, Sikharulidze, and Varazashvili, in preparation).

ebral cortex we found the number of capillaries with stasis to increase in proportion to the increase of the cerebral water content (Fig. 5). Release of gangliosides from damaged brain tissue may be considered as a direct cause of intravascular red cell aggregation together with production of arachidonic acid leading to synthesis of prostaglandins (Mkhejan et al., 1981; Pappius, Wolfe, 1983). A marked narrowing of the lumina of cerebral capillaries, transformation of microvessels into plasmatic capillaries and development of stasis should considerably affect the cerebral microcirculation as a pathophysiological mechanism enhancing secondary brain damage.

REFERENCES

Chien S, Present state of blood rheology, in: "Hemodilution. Theoretical basis and clinical application", Meßmer K, Schmid-Schönbein H, eds., Karger Basel (1972).

Gannushkina IV, Shafranova VP, Morphological evidence and mechanism of spotty character of brain tissue damage in experimental hypertension, in: "Cerebral Vascular Disease", Meyer JS, Lechner H, Reivich M, eds., VII Int. Conference Salzburg 1974, Stuttgart (1975).

Häggendal E, Johansson B, Pathophysiology of the increased cerebrovascular permeability in acute arterial hypertension, Acta Neurol Scand 48:265 (1972).

Klatzo I, Pathological opening of the BBB, in: "Cardiovascular physiology. Microcirculation and capillary exchange", Kovach ACB, Hamar J, Szabo L, eds., Pergamon Press and Akademiai Kiado, Budapest (1981).

Maier-Hauff K, Lange M, Schürer L, Guggenbichler C, Vogt W, Jacob K, Baethmann A, Glutamate and free fatty acid concentrations in extracellular vasogenic edema fluid, in: "Recent progress in the study and therapy of brain edema", Go KG, Baethmann A, eds., Plenum Press, New York (1983).

Mchedlishvili GI, "The capillary circulation", ed., Georgian Academy of Sciences, Tbilisi (1958).

Mchedlishvili GI, "Vascular mechanisms of the brain", ed., Plenum Press, New York (1972).

Mchedlishvili GI, Akhobadze VA, The cerebral arterial system in brain injury and during traumatic edema, Physiol Bohemoslov 10:8 (1961a).

Mchedlishvili GI, Akhobadze VA, The functional state of the capillary and venous systems of the brain in cerebral traumatic edema, Physiol Bohemoslov 10:15 (1969b).

Mchedlishvili GI, Kapuscinski A, Nikolaishvili LS, Mechanisms of postischemic brain edema: contribution of circulatory factors, Stroke 7:410 (1976a).

Mchedlishvili GI, Mitagvaria NP, Ormotsadze LG, Vascular mechanisms controlling a constant blood supply to the brain ("autoregulation"), Stroke 4:742 (1973).

Mchedlishvili GI, Mossakowski M, Itkis M, Sikharulidze N, Januszewski S, Cerebral blood volume changes during the development of brain edema, in: "Recent progress in the study and therapy of brain edema", Go KG, Baethmann A, eds., Plenum Press, New York (1983).

Mchedlishvili GI, Nikolaishvili LS, Antia RV, Are the pial arterial responses dependent on the direct effect of intravascular pressure and extravascular and intravascular pO_2 ,pCO_2 and pH?, Microvasc Res 10:298 (1976b).

Mchedlishvili GI, Sikharulidze NV, Itkis ML, Januszewski S, Effect of systemic arterial and venous pressures on cerebral blood volume, J Physiol USSR 68:64 (1982).

Mkhejan EE, Akopov SE, Sotsky OP, Role of glycolipids in process of aggregation of erythrocytes and thrombocytes, J Exper Clin Med 21:20 (1982).

Pappius HM, Local cerebral glucose utilization in thermally traumatized rat brain, Ann Neurol 12:157 (1976).

Pappius HM, Wolfe LF, Functional disturbances in brain following injury: Search for underlying mechanisms, Neurochem Res 8:63 (1983).

Sakharova AV, Regional peculiarities of adrenergic and cholinenergic innervation of cerebral surface vessels, Bull Exper Biol Med 88:141 (1980).

Schmid-Schönbein H, Wells RE, Goldstone J, Influence of deformability of human red cells upon blood viscosity, Circ Res 25:131 (1969).

INFLUENCE OF EXPERIMENTAL HYDROCEPHALUS ON CEREBRAL VASCULARIZATION

Christian Plets

Laboratory of Experimental Neurosurgery
Catholic University of Louvain (KUL), Belgium

INTRODUCTION

Obstructive hydrocephalus has been produced in dogs by intracisternal injection of kaolin. Kaolin provokes inflammation and fibrosis of the leptomeninges preventing the flow of cerebrospinal fluid from the ventricular system to the subarachnoid space. Morphologically this type of hydrocephalus is mainly characterized by increasing ventricular dilatation together with atrophy of the periventricular white matter. However, the volume of basal ganglia and cortex remains preserved. The objective of our study was to investigate the influence of hydrocephalus on the cerebral blood supply, the reversibility of the angioarchitectonic changes and the possible role of the affected cerebral vessels in the mechanism of ventricular dilatation. To understand the transformation of the cerebral vascular network it was required to record a number of physical parameters, such as CSF pressure, intracranial compliance, and the outflow-resistance of the CSF resorptive system.

MATERIAL AND METHODS

In 44 adult dogs hydrocephalus was induced by intracisternal injection of 200mg of kaolin. After hydrocephalus had developed the hydrodynamic parameters, particularly intraventricular pressure, compliance, pressure-volume index, and outflow-resistance were determined at various times using the mathematical relations developed by MARMAROU (1975). For this purpose a small burr hole was made on both sides at 0,5 cm from the mid-line. A thin catheter was introduced into each lateral ventricle. One of the catheters was connected to a Statham transducer and a monitor for recording of intraventricular pressure. The other was used for the infusion of serum into the CSF system. The pressure in the lumbar

303

subarachnoid space was also measured. After the last pressure measurement, the carotid and vertebrobasilar systems were perfused with Indian ink. The brain was removed and fixed. Slices were made then for histological examination employing the VAN GIESON technique for staining of the vascular structures.

In 6 other dogs, in which hydrocephalus had been induced few months earlier, plexectomy (in four dogs unilaterally, in two dogs bilaterally) was performed. 6 hydrocephalic dogs were treated with a ventriculo-peritoneal shunt. The same parameters were studied together with the angioarchitecture after various time intervals. The hydrodynamic parameters and angioarchitecture were also examined in 12 normal dogs.

RESULTS

1. Untreated Hydrocephalic Dogs

From the morphological and hydrodynamic findings it appeared that two periods can be distinguished in the evolution of this kind of hydrocephalus. The first period of one month duration constitutes an acute phase of hypertensive, obstructive hydrocephalus with rapid dilatation of the ventricular system. The compliance is decreased while the outflow-resistance markedly increased. Already during the acute stage of hydrocephalus remarkable angioarchitectonic changes occur. In the telencephalon, especially the long rami medullares should be mentioned, which no longer follow a stretched ventriculo-petal course but bend in the horizontal direction when entering the white matter. This causes them to adopt a position parallel to the ventricular wall. In spite of the atrophy of the periventricular white matter, their number and length does not decrease. This results in a relative hypervascularization in the subependymal region. In the medulla oblongata the paramedian arterioles appear elongated due to the shrinking of the ventro-dorsal diameter of the bulbus. This causes the arterioles to become sinuous and, past the sulcus limitans, to adopt a recurrent ventriculo-fugal course. A relative hypervascularization was also found in the subependymal zone at the level of the fourth ventricle. In the entire brainstem all arterioles remain directed towards the sulcus limitans in spite of a few changes of their trajectory. In the cerebellum the ventriculo-petal vessels maintain their direction in contrast to those of the telencephalon. Most probably, this is due to the fact that these arterioles are directed towards the nuclei around the roof of the fourth ventricle. These nuclei are not affected by atrophy or deformation of the periventricular white matter found in the telencephalon.

After the acute phase a second, chronic phase follows with normotensive, communicating hydrocephalus. Complete recanalization of the central canal along the whole length of the spinal cord provides for communication between the intraventricular and the spinal subarachnoid

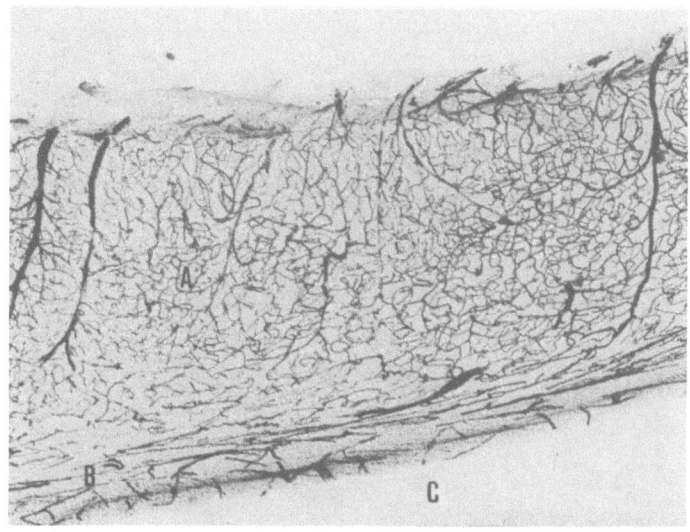

Fig.1: Sagittal section through the parietal area of a dog brain. Hydrocephalus was induced six months earlier. The periventricular white
matter is atrophic. The terminal branches of the ventriculo-petal
arterioles turn into a direction parallel to the ventricular wall
on entering the white matter. In spite of atrophy the periventricular area is not ischemic, A : cortex, B : periventricular white
matter, C : lateral ventricle.

space. The morphology changes more slowly now, but the ventricles dilate
further in spite of a normal intraventricular pressures (Fig. 1).
Large fissures and cysts develop in the bulbus, leading to communicating
hydrosyringobulbia. Although outflow-resistance becomes almost normal,
compliance remains low. Expressing compliance (C) as a function of the
CSF pressure (P), the following relationship holds:

$$C = 0.157 \ e^{-0.127 \ P}$$

The compliance-pressure curve was found to be exponential. It is a
straight line on a logarithmic axis. The compliance in normal dogs was:

$$C = 0.285 \ e^{-0.1268 \ P}$$

Hence, the curves of hydrocephalic and normal dogs are parallel. It follows from these equations that the compliance in hydrocephalic dogs was
reduced by 1.8 times.

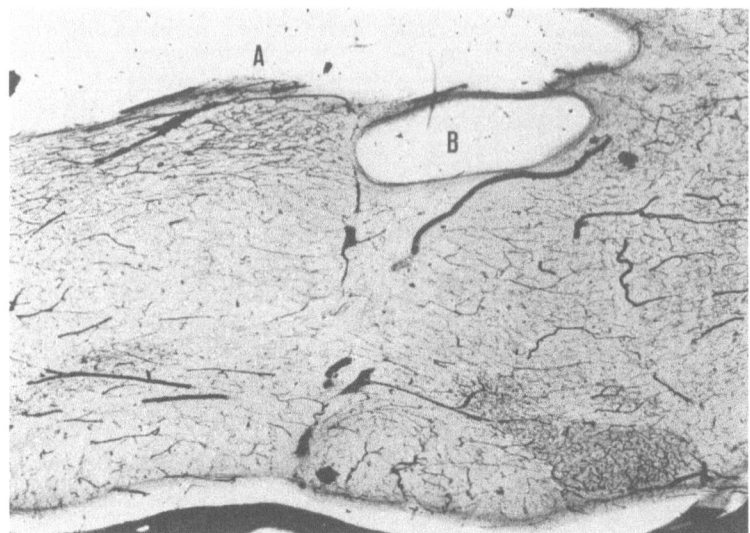

Fig.2: Coronal section through the bulbus 22 months after the onset of
hydrocephalus. The midline and the paramedian arterioles are deviated and displaced by the presence of large intraparenchymatous
cysts (A and B). The areas around the cysts have a dense vascularization.

In this stage the angioarchitecture is mainly characterized by an
accentuation of the changes already found initially. It is particularly
striking that the terminal branches of the rami medullares in the tele-
cephalon ended in a ventriculo-fugal terminal segment after they follow a
course parallel to the wall of the lateral ventricle. The angle formed by
the corpus callosum becomes smaller and smaller. Its larger bloodvessels
are concomitantly more and more bent until they are fully turned in the
direction of the commissural fibres. In the thalamus there is now a
notable reduction of the angle formed by the arteriae thalamoperforantes,
which decreased progressively from 135° to 90°. The most dramatic transformation of the vascular bed occurs in the medulla oblongata. The total
vascular pattern is fundamentally disturbed by the syrinx or cysts.
Moreover, the zones around these cavities are very densely vascularized
with the arterioles and venules running parallel to the wall of the cysts
(Fig. 2). A similar vascular pattern can be observed around the cysts of
the cervical spinal cord. Because of the sagittal bend of the pons, the
rami ad pontem converge on sagittal sections towards the bottom of the
fourth ventricle. In the mesencephalon it is particularly noteworthy that
the ventral and dorsal arteries have a longer and more stretched centri-
fugal terminal segment than normally. Due to the elongation and atrophy
of the superior cerebellar peduncle the intraparenchymatous branches of
the superior cerebellar arteries and of the quadrigeminal arteries approach each other and take a direction parallel to the course of the
fibres.

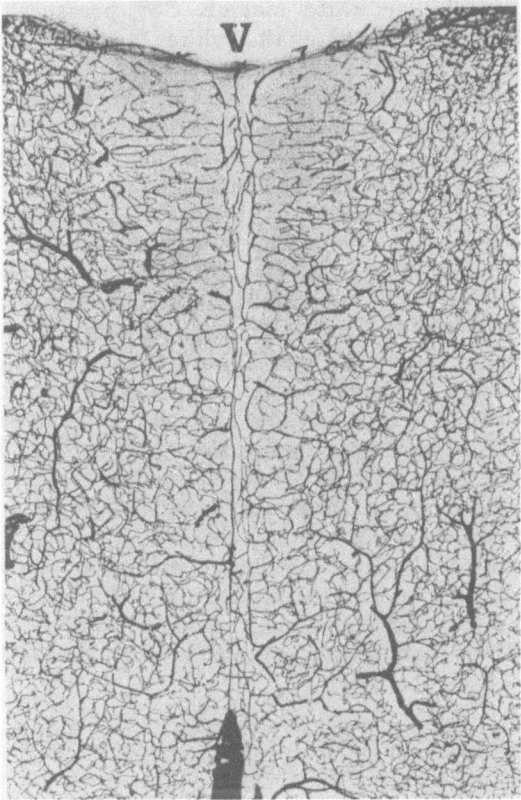

Fig.3: Coronal section through the bulbus of a dog treated with a ventriculo-peritoneal shunt two months after onset of hydrocephalus. Note restoration of the normal course of the paramedian arterioles and of their side-branches. V : bottom of the fourth ventricle.

2. Treated Hydrocephalic Dogs

We examined the effect of ventricular shunting and unilateral or bilateral plexectomy on the blood supply to the hydrocephalic brains. After both types of surgery complete or incomplete morphological recovery of the brain was found only, if surgery was carried out before the end of the eighth month of hydrocephalus. Ventricular shunting before the end of the eighth month resulted in normal diameters of the ventricles and an almost complete recovery of the periventricular structures. The CSF hydrodynamics returned to normal values. The angioarchitecture lost virtually all the typical features of hydrocephalus. Within a few weeks a normal vascular pattern was observed in the entire brain (Fig. 3).

If the shunting procedure was performed later than eight months after induction of hydrocephalus, the ventricles remained dilated, but dilatation did not progress (Fig. 4). There was, however, no increase in the amount of periventricular white matter. CSF pressure, compliance, and outflow-resistance became normal again, while the angioarchitecture retained the most typical features of hydrocephalus. After unilateral plexectomy before the end of the eighth month of hydrocephalus, a collapse of the plexectomized ventricle could be observed, whereas the other ventricle remained dilated (Fig. 5). In the telencephalon the angioarchitecture returned to normal in the treated hemisphere, while the hydrocephalic vascular pattern was maintained in the other. On the side of surgery the dorsolateral thalamus became atrophic and ischemic. The angioarchitectonic characteristics of hydrocephalus continued to develop in the brainstem, cerebellum and in the non-operated hemisphere, although at a slower rate than without plexectomy. Unilateral plexectomy after the eighth month of hydrocephalus did not affect the morphology of the periventricular white matter. Hydrocephalus progressed at a slower rate, however. The angioarchitecture retained its abnormal characteristics in the entire brain. Bilateral plexectomy during the first eight months of hydrocephalus reduced symmetrically both lateral ventricles and allowed for nearly complete recovery of the morphology. Compliance increased, however, without attaining its normal value. The vascular pattern of both hemispheres was equal after the intervention. It was identical to that of a hemisphere with unilateral plexectomy. If the choroid plexus was removed bilaterally after the eighth month of hydrocephalus, the angioarchitecture remained unchanged, except in the dorsal thalamus where ischemia developed.

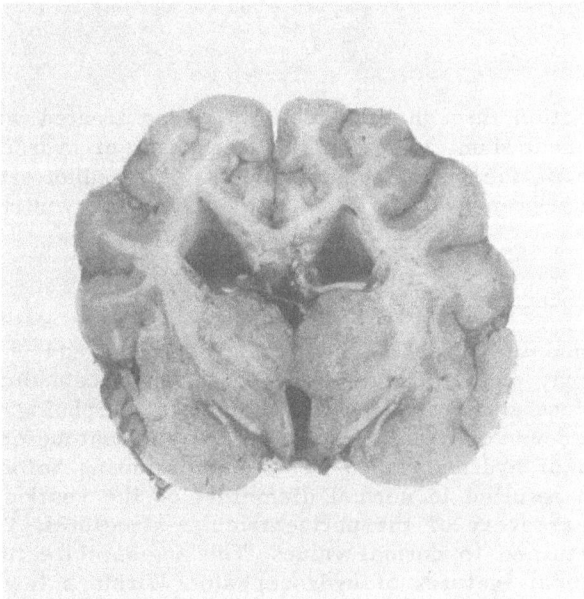

Fig.4: Coronal section through the hemisphere of a dog treated with a ventriculo-peritoneal shunt ten months after onset of hydrocephalus. The lateral ventricles remain dilated in spite of treatment.

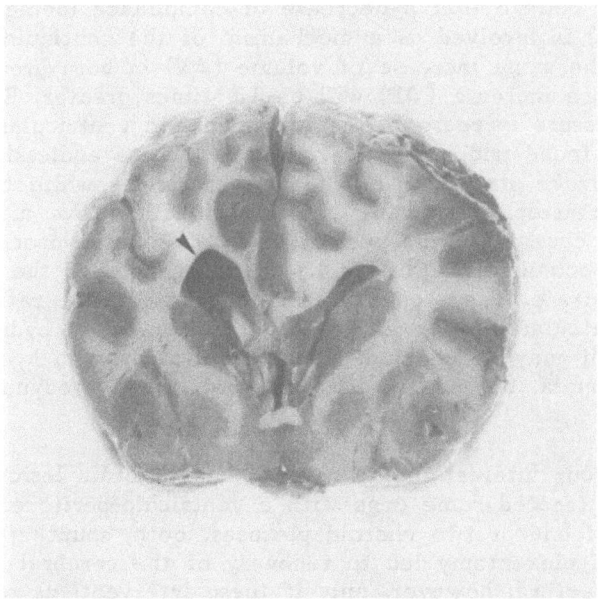

Fig.5: Coronal section through the frontal area of a dog treated by
plexectomy two months after the onset of hydrocephalus. The arrow
indicates the right lateral ventricle which remained dilated. The
left ventricle, however, returned to normal and the left periventricular white matter regained its normal diameter.

DISCUSSION

After intracisternal injection of kaolin in adult dogs hypertensive
obstructive hydrocephalus develops. After one month it changes into a
normotensive communicating hydrocephalus subsequent to recanalization of
the central canal of the spinal cord with outflow of CSF into the spinal
subarachnoid space. In the first phase, hydrocephalus develops as a
result of mechanical obstruction of the fourth ventricle leading to a
markedly increased outflow resistance. The resulting intraventricular
hypertension causes a pronounced dilation of the ventricles and atrophy
of the periventricular white matter. The latter is certainly not the result of a collapse of periventricular arterioles, as has been presumed,
e.g. by PENFIELD (1929), PENFIELD et al. (1932), and DeSN (1950). Instead of loss or collapse of vessels, we have always found a conservation
of blood vessels, or even a relative hypervascularization. Only some vessels, the rami medullares in particular, change their direction. The second phase of hydrocephalus is characterized by transformation of the obstructive hypotensive type into the normotensive communicating type due
to dilation and opening of the central canal, providing for CSF flow into
the lumbar subarachnoid space. A continued dilation of the ventricles
during the chronic stage in spite of a normal intraventricular pressure

is remarkable. We believe that a decrease of compliance (being 1.8 times lower than normal) is involved as a mechanism of the continuing ventricular dilation. For the same increase of volume (ΔV) in both groups of animals the pressure increase (ΔP) will be 1.8 times greater. Each systole causes a pressure increase in the hydrocephalic ventricular system of 1.8 times that found under normal conditions. These endlessly repeated pressure peaks provoke stretching of the periventricular white matter fibers, resulting in atrophy and ventricle dilation. The above mentioned angioarchitectonic changes in the normotensive phase of hydrocephalus can be explained by mechanical factors, e.g. by the changes of the underlying morphological substrate. These changes are the consequence rather than the cause of ventricular dilation and of the communicating hydrosyringo-bulbia. Therefore it may be concluded that brain atrophy in hydrocephalus is not due to ischemia, but due to the abnormal CSF hydrodynamic parameters.

Because of our interest in the reversibility of brain lesions in hydrocephalus, we treated some dogs with a ventriculo-peritoneal shunt, or with removal of one or two choroid plexuses. Both, shunting or unilateral or bilateral plexectomy led to recovery of the cerebral morphology and angioarchitecture, however, only if these interventions were carried out before the end of the eighth month of hydrocephalus. Thus, there is a time-limit, a point of no return, after which restoration of brain structures and of their blood supply to normal is no longer possible. Both kinds of operation are not entirely equivalent. Ventricular shunting is superior and brings about a quasi complete normalization of the morphological, hydrodynamic, and angioarchitectonic properties, if it is carried out within eight months after the onset of hydrocephalus. Uni- or bilateral plexectomy resulted only in partial recovery. Moreover, hydrocephalus continued to develop, although at a much slower rate. Plexectomy resulted in ischemia of the dorsolateral thalamus, which in turn led to atrophy and necrosis of this area. Ischemia developed because the choroidal arteries supply the dorsal part of the thalamus (Plets et al., 1970). Although the diameter of the periventricular white matter increased after early treatment of hydrocephalus, vascularization of these regions did not improve. Therefore postoperative angioarchitectonic transformation can not play a part in the restoration of brain structures, nor is the atrophy evolving after induction of hydrocephalus caused by the changes of the vascularization or by ischemia.

Our findings after uni- or bilateral plexectomy support the hypothesis that ventricular dilation in normotensive communicating hydrocephalus is caused by repetitive intraventricular pressure peaks during the systole, which are inadequately compensated by an impaired compliance. In the light of this hypothesis it is understandable, why after unilateral plexectomy asymmetrical ventricles develop with unilateral hydrocephalus of the unoperated hemisphere. In the latter instance the source of arterial pulsations remained in the lateral ventricle. After bilateral plexectomy, however, the ventricles are symmetrical, as the pulsations and systolic pressure peaks have been eliminated on both sides. After uni- or bilateral plexectomy the lateral ventricles did not permanently remain narrow but dilated again, since the choroid plexus was retained in the third and fourth ventricle. However, evolution of ventricular dilation was delayed, because the pulsations and pressure peaks were attenuated and reached the plexectomized ventricles from a distance only. Taken together, the data from untreated and treated hydrocephalic brains demontrate that the cerebral vascular changes in hydrocephalus are not responsible for the periventricular atrophy. The vascular changes are the consequence rather than the cause of atrophy and of ventricular dilation. The latter results from a decreased compliance of the CSF system.

REFERENCES

1. De SN, A study of the changes in the brain in experimental internal hydrocephalus, J Path Bacteriol 62:197 (1950).
2. Marmarou A, Shulman K, La Morgese J, Compartmental analysis of compliance and outflow resistance of the cerebrospinal fluid system, J Neurosurg 43:523 (1975).
3. Penfield W, Cerebral pressure atrophy, in: "Intracranial pressure in health and disease", Proc Ass Res Nerv Ment Dis, Williams and Wilkins Ltd, Baltimore (1929).
4. Plets C, De Reuck J, Van der Eecken H, Van den Bergh R, The vascularization of the human thalamus, Acta Neurol Belg 70:687 (1970).
5. Plets C, Van den Bergh R, Contribution des arteres chorioidiennes dans la vascularisation du thalamus humain, Bulletin de l`Association des anatomistes 58:163 (1974).
6. Plets C, De invloed van de experimentele hydrocephalus op de cerebrale bloedverzorging, 1-164, Acco, Louvain (1977).

AVOIDABLE MORTALITY, MORBIDITY AND SECONDARY BRAIN

DAMAGE AFTER RECENT HEAD INJURY

Bryan Jennett

Institute of Neurological Sciences
Glasgow G51 4TF, Scotland

It has long been recognised that various factors subsequent to the initial impact can cause death or disability after head injury. When a patient who has talked after head injury dies of unrelieved cerebral compression from an extradural haematoma, or he chokes to death from inhaled vomitus, or he bleeds to death from extracranial injuries, no one doubts that the patient might still be alive had these secondary events not occurred. That is not the same as declaring that they could have been avoided, because this depends on the when and where of the secondary events. When a solitary accident victim chokes or bleeds to death before anyone gets to him it is unfortunate but unavoidable. Even cerebral compression from clot can progress so rapidly that surgical relief is impractical.

In recent years neuropathologists have made us aware of less obvious secondary events - cerebral hypoxia and ischaemia, oedema, fat embolism and raised ICP. Their evidence comes from fatal cases but even with detailed dissection of the brain it can sometimes be difficult to ascribe with confidence the separate contributions to a patient's death of impact injury and secondary events - even sometimes to apportion the respective roles of intracranial and extracranial pathological events, whether primary or secondary. With survivors it is even more difficult - because it is seldom possible in life to distinguish between the contribution of various processes to the overall dysfunction of the brain, as reflected in the depth of coma and the extent of neurological signs. Raised ICP can certainly be measured, but cerebral hypoxia, ischaemia and oedema can be detected only intermittently and approximately. The relative contributions of primary and secondary brain damage to persisting disability is likewise usually difficult to discern, except when a patient who was little affected soon after injury suffers major disability from complications, or when the brain of a more severely injured patient is eventually retrieved for dissection after months or years of

disabled survival. CT scanning in long term survivors gives some insight
into the degree and distribution of atrophy; but this too is the net result of primary and secondary factors.

We do not know how to promote recovery of the damaged brain,
only how to prevent more damage occurring. Therefore reducing mortality
and morbidity depends mainly on minimising secondary brain damage by
appropriate management. I shall consider aspects of the organisation of
the care provided for head injuries that could reduce the risk of adverse
clinical developments that can have pathological consequences likely to
contribute to mortality and morbidity. My concern will be with clinical
events rather than with pathological processes. I leave it to others to
discuss therapeutic regimens designed to prevent, limit or reverse these
processes, e.g. by controlling ICP, manipulating CBF, reducing oedema and
so on. My concern is with the strategy of management rather than with
tactics of treatment.

It was in 1958 that papers written jointly by a neurosurgeon and an
anaesthetist from Newcastle-on-Tyne in England drew attention to the importance of dealing with respiratory problems after severe head injury
(1,2). They claimed that by doing tracheostomy on all unconscious head
injuries, and by administering a drug cocktail including chlorpromazine,
the mortality was dramatically reduced. There was no mention of controlled ventilation because this technology was not yet widely available.
The last 25 years have seen growing interest in this approach to the
treatment of severe head injury, and also of other kinds of acute brain
dysfunction such as those resulting from ischaemia and intoxication.
Whole books have been written on various aspects of therapy recommended
for the patient in coma (cf.:3,4). Anaesthesiologists and intensive care
physicians have figured prominently in these developments and in the
writings about them. In Europe, where neurosurgeons are nowhere near as
numerous as in the USA or Japan, there has been a tendency for these
specialists, together with accident or trauma surgeons, to play a major
role in the management of severe head injured patients, about a third of
whom have major extracranial injuries. In some European countries some
neurosurgeons have aided and abetted this conspiracy, fearful that traumatic cases might take up too much of their time or too many of their
beds they have been only too willing to shed the load of looking after
head injuries on to other willing shoulders. The coming of CT scanning
has further compounded this conspiracy. Its ease of performance and the
apparent simplicity of interpretation makes it seem less necessary to
consult a neurosurgeon where scanning is available outside regional
centres, even when an intracranial haematoma is suspected.

My thesis is that greater gains in reducing mortality and morbidity
from head injury are likely to come from earlier diagnosis and treatment
of remediable intracranial complications, and from minimising systemic
insults than will result from greater efforts to retrieve the brain that
is already badly damaged. Rather than concentrating attention on patients
in deep coma we should be more concerned that patients whose initial in-

jury was not very severe do not secondarily become comatose from factors that could have been prevented. In Europe, where both neurosurgeons
and CT scanners are in limited supply, this approach requires a review of
how patients are selected for scanning and also for transfer to regional
neurosurgical units. In countries where scanners are mostly located in
neurocentres the decision to scan usually implies transfer to a special
unit.

In 1975 the Glasgow team described a series of patients who had
"talked and died" (5). The implication was that if a patient had talked
soon after injury he had clearly not sustained overwhelming brain damage
and therefore he should not have died. This seemed an obvious group of
patients in whom to look for avoidable factors but we later studied these
also in patients who had remained continuously in coma until their death
(6,7). Avoidable factors that had contributed significantly to the patient's death were found in 54% of talk and die cases but in only 34%
of those who had remained continuously in coma. The overall rate in patients transferred to the Glasgow Neurosurgical Unit was 40%. A similar
incidence, both of talk and die cases and of avoidable factors, was reported some years later from the Merseyside Neurosurgical Unit (8).

About 70% of the avoidable factors discovered in this study were
related to intracranial lesions and 30% to extracranial, but some patients had both. The commonest single factor was delayed diagnosis or
treatment of an intracranial haematoma, accounting for two thirds of all
avoidable factors and about 90% of intracranial factors. Delay occurred
at various stages - before hospital was reached (23%), in the accident
and emergency department (9%) in the primary surgical ward (48%), and
after arrival at the neurosurgical unit (20%). Alcohol was a common factor when there was delay, usually because the patient's condition was
wrongly ascribed to alcohol rather than to head injury. In some patients
stroke rather than trauma was suspected and this likewise led to delay
(9). When a head injury was clearly identified as the main lesion the
commonest cause of delay was failure to anticipate the likelihood of
complications because the risks were underestimated. But sometimes the
failure was to observe clinical deterioration or to act promptly once it
had been recognised.

It might be expected that CT scanning would have improved the
situation - for it is widely acknowledged that one of its most useful
roles is the diagnosis of intracranial haematoma. However, the hospital
in London where the scanner was developed and that installed the first
prototype reported no reduction in head injury mortality - only that angiography and exploratory burr holes largely became procedures of the
past (10). Similarly disappointing results were reported from Liverpool
(11) and we found the same im Glasgow. It seemed to us that the reason
for this was the persistence of the now old-fashioned attitude that head
injuries should be transferred to neurosurgeons only when they are clinically deteriorating with obvious signs of an intracranial haematoma. By
then it is often too late for surgery to produce optimal results, or even

to save life. Moreover, CT scanning has taught that surgically significant intracranial haematomas can be detected long before they have produced the classical clinical syndrome of unilateral cerebral compression.
Indeed the textbook features of intracranial haematoma are uncommon
even in the later stages of development (12).

TABLE 1

Risk of intracranial haematoma in admitted patients

Skull fracture	Altered consciousness (not orientated)	Risk
0	0	1 : 900
0	+	1 : 70
+	0	1 : 29
+	+	1 : 4
+/0	coma	1 : 4

Data from (13)

The early detection of intracranial haematomas calls for identification of risk factors that indicate which patients are likely to develop
haematoma. In less severely injured patients we have found that the
interaction between the conscious state and the presence or absence of
skull fracture is a clear indication of the level of risk (Table 1). Because a skull fracture is such a significant risk factor it becomes important to undertake skull radiography in milder injuries in the casualty
department, contrary to the protests of many radiologists in recent years
(14,15). These studies on risk factors have also revealed how frequently
a haematoma is found in patients who have been continuously in coma
since the injury, often without any recorded deterioration. There is a
disturbing tendency to treat such patients in general intensive care
units, without specialised investigation and/or monitoring and without
the involvement of neurosurgeons (see below). A group of neurosurgeons
in Britain have therefore recently published guidelines for the initial
management of head injured adults (Table 2). These indicate which patients should have skull x-rays, which patients should be admitted for
observation to general surgical wards, and which patients should be
transferred to neurosurgeons (16).

TABLE 2a	**TABLE 2b**
Guidelines for skull x-ray	Guidelines for admission to hospital
Loss of consciousness or amnesia	Consciousness still impaired
Neurological symptoms or signs	Skull fracture
CSF leak	Neurological symptoms or signs
Suspected penetration	Difficult to assess
Scalp bruise	No responsible adult at home
	(Brief amnesia with recovery no sufficient indication)
Data from (16)	Data from (13)

TABLE 2c

Guidelines for neurological consultation

Fractured skull with:

- confusion or worse

- one or more fits

- any neurological symptoms or signs

Coma after resuscitation

Deterioration in conscious level

Depressed fracture

Suspected fracture of base

Data from (13)

Since using these guidelines in Glasgow we have significantly reduced mortality and morbidity from intracranial haematoma (17). Moreover
both talk and die patients and avoidable factors have occurred significantly less often. However, this improvement has been bought at the
price of doubling the number of head injured patients transferred each
year to the regional neurosurgical unit (Table 3). In order that neurosurgeons may calculate how their resources should be altered so that they
are able to offer a reasonable standard of care by modern methods to
head injured patients in their catchment population we have made estimates of the number of beds per million population required to adopt our
recommended policy (18). These calculations assume that patients will be
returned to general hospital beds as soon as they no longer have need for
the specialised facilities of a neurosurgical unit.

TABLE 3

Effect of new transfer policy for suspected intracranial haematoma

	1974-77	1978-80
Annual head injury transfers to neurosurgical unit	274	520
Annual operated haematomas*	76	126
Percent of cases in coma before operation	66	50
Percent of cases who talked but went into coma	48	24
Percent dead/vegetative	38	29

* No increase in incidence of head injury between two periods, yet 50
more haematomas per year operated in second period: presume overall
mortality from haematoma in first period was much higher, if undiagnosed
cases dying in primary hospital were included.   Data from (17)

There are several reasons why extracranial or systemic avoidable
factors occur. Hypotension may develop because extracranial injuries are
overlooked or it is wrongly believed that it is more important to rush
the patient to the neurosurgical unit than to undertake resuscitation. On
the other hand in some cases hypotension results from paying too much
attention to extracranial injuries and subjecting a patient to prolonged

TABLE 4

Treatment of severe head injuries in different centres

	Netherlands	Los Angeles	San Francisco	Glasgow
n	302	225	430	763
Steroids	34%	99%	49%	11%
Osmotics				
- Haematoma	31%	64%	85%	45%
- Non-haem	19%	28%	64%	25%
Bone flap	92%	93%	65%	12%
Tracheostomy	15%	66%	1%	6%
Controlled Ventilation	28%	19%	77%	35%

	Predicted deaths	Actual deaths
In San Francisco, based on Glasgow	236	237
In Glasgow, based on San Francisco	361	363

Data from (23)

surgical procedures under general anaesthesia during which hypotension can occur. Hypoxia is much more often the result of airway obstruction or inhalation than of ventilatory insufficiency,and this obstruction frequently occurs during transfer from the primary hospital to the neurosurgical unit. Failure to position the patient properly, to empty the stomach, and to ensure that a well-trained escort travelled with the patient were each found quite frequently in a study of 150 patients arriving at a neurosurgical unit in coma (19). Considering that these were inter-hospital transfers there is really no excuse for them. We also investigated the occurrence of avoidable factors in patients who died soon after injury without having reached hospital (7). Most of these patients had overwhelming multiple injuries or they had irrecoverable brain damage, e.g. the tearing of the brain stem or complete disruption of the brain. We found very few fatalities that might have been avoided had there been more skilled resuscitation at the roadside. However, it is possible that some patients who arrive alive at hospital may have suffered hypoxic or hypotensive incidents either at the roadside or en route to the first hospital which could have been avoided by more skilful intervention at that stage.

It is less easy to determine the frequency of avoidable mortality and morbidity in patients who are in coma from the outset and do not have an intracranial haematoma, or those who remain in coma after the removal of intracranial haematoma. Recent years have seen widespread adoption of various regimens including controlled ventilation, corticosteroids and osmotic agents and in some cases barbiturates. Others will deal with the relative roles of these treatment tactics but I wish to emphasise the definite risks of iatrogenic avoidable factors when some of these methods are used in patients whose condition is not sufficiently serious to justify them. Their use in patients too seriously damaged to benefit causes avoidable waste of valuable resources, but does no harm to patients destined to die.

My plea for triage for treatment of patients with acute brain damage could be applied in practice only if there were reliable means of predicting soon after injury which patients did need special care, which would recover anyway and which would die anyway. This can now be done by taking account of age and severity of injury, and there is increasing recognition that one of the most important new technologies now emerging in medicine is the power to predict - a power borne of advances in data processing by computers and in analysis by biostatisticians. Not only do such methods make it possible to reach practical decisions about individual patients more rationally. They also make it feasible to assess the relative value of different treatment regimens without the problems that are presented by randomised trials when these involve patients with life-threatening conditions. This method has been applied to comparing different general intensive care units within the US and also with those in France. In spite of differing mortality rates in the various US centres the numbers of deaths in each phase were correctly predicted from experience in one centre (20). Although there were differences in management between US and France the deaths in France were predictable from the American data (21).

Using statistical modelling and post-hoc stratification we have been able to show that it is possible to predict the deaths in one centre treating severe head injuries based on data collected from another where the details of management were markedly different (Table 4; 22,23). This pre-supposes excellent care directed at minimising avoidable mortality and morbidity along the lines discussed in this paper. It does not mean that intensive care is ineffective or unnecessary; what it does suggest is that variations in therapeutic details beyond the basic good care that everyone is agreed about do not seem to influence outcome.

REFERENCES

1. MacIver IN, Frew JCI, Matheson JG, The role of respiratory insufficiency, Lancet 2:390 (1958)
2. MacIver IN, Lassman LP, Thomson, CW, MacLeod I, Treatment of severe head injuries, Lancet 2:544 (1958)
3. Trubuhovich RV, Management of acute intracranial disasters, in: "International anaesthesiology clinics", Little, Brown & Co, Boston (1979).
4. Grenvik A, Safar P, "Brain failure and resuscitation", Churchill Livingstone, New York (1981).
5. Reilly PL, Adams HJ, Graham DI, Jennett B, Patients with head injury who talk and die, Lancet 2:375 (1975)
6. Rose J, Valtonen S, Jennett B, Avoidable factors contributing to death after head injury, Brit Med J 2:615 (1977).
7. Jennett B, Carlin J, Preventable mortality and morbidity after head injury, Injury 10:31 (1978).
8. Jeffreys RV, Jones JJ, Avoidable factors contributing to the death of head injury patients in general hospitals in Mersey region, Lancet 2:459 (1981).
9. Galbraith S, Misdiagnosis and delayed diagnosis in traumatic intracranial haematoma, Brit Med J 1:1438 (1976).
10. Ambrose J, Gooding MR, Uttley D, EMI scan in the management of head injuries, Lancet 1:847 (1973).
11. Jeffreys RV, Lozada L, The use of the CAT scanner in the management of patients with head injury transferred to the regional neurosurgical unit, Injury 13:370 (1981).
12. Jennett B, Teasdale G, "Management of head injuries", Davis FA, Philadelphia (1981).
13. Mendelow AD, Teasdale G, Jennett B, Bryden J, Hersett C, Murray G, Risks of intracranial haematoma in head injured adults, Brit Med J 287:1173 (1983).
14. Jennett B, Skull x-rays after recent head injury, Clin Radiol 31:463 (1980).

15. Mendelow AD, Campbell DA, Jeffrey RR, Miller JD, Hersett C, Bryden J, Jennett B, Admission after mild head injury: benefits and costs, Brit Med J 285:1530 (1983).

16. Guidelines for initial management after head injury in adults: suggestions from a group of neurosurgeons, Brit Med J 288:983 (1984).

17. Teasdale G, Galbraith S, Murray L, Ward P, Gentleman D, McKean M, Management of traumatic intracranial haematoma, Brit Med J, 285:1695 (1982).

18. Bryden JS, Jennett B, Neurosurgical resources and transfer policies for head injuries, Brit Med J 286:1791 (1983).

19. Gentleman D, Jennett B, Hazards of inter-hospital transfer of comatose head injured patients, Lancet 2:853 (1981).

20. Knaus WA, Draper EA, Wagner DP, Zimmermann JE, Birnbaum ML, Cullen DJ, Kohles MK, Shin B, Snyder JV, Evaluating outcome from intensive care: a preliminary multihospital comparison, Crit Care Med 10:491 (1982).

21. Knaus WA, Wagner DP, Loirat P, Cullen DJ, Glaser P, Mercier P, Nikki P, Snyder JV, Legall JR, Draper EA, Campos RA, Kohles MK, Granthil C, Nicolas F, Shin B, Wattel F, Zimmerman JE, A comparison of intensive care in the USA and France, Lancet 2:642 (1982).

22. Jennett B, Teasdale G, Fry J, Braakman R, Minderhoud J, Heiden J, Kune T, Treatment for severe head injury, J Neurol Neurosurg Psychiat 43:289 (1980).

23. Jennett B, Murray G, Pitts L, Using the international head injury data bank to compare outcome in two groups (centres or treatments), Abstract for AANS Meeting: 1984

RELEVANCE OF PRIMARY AND SECONDARY BRAIN DAMAGE FOR OUTCOME OF HEAD INJURY

J. Douglas Miller

Department of Surgical Neurology, University of Edinburgh
Edinburgh, Scotland

The outcome of head injury in the human depends upon the inter-action of pre-injury factors, the nature of the injury to the skull and brain and secondary factors that lead to further brain dysfunction and damage (Miller and Becker, 1982). The important pre-injury factors are age, the psychosocial status of the patient and state of health prior to injury, in particular previous brain injury, hydrocephalus or stroke.

THE NATURE OF PRIMARY BRAIN DAMAGE

The more closely it is examined, the more elusive the entity of primary brain damage becomes. From the pathological standpoint the ob-vious lesions are local contusions or lacerations of brain tissue under-lying the point of impact, polar contusions and lacerations affecting the frontal and temporal lobes of the brain, as a consequence of its motion within the skull upon acceleration or deceleration and diffuse axonal in-jury within white matter (Adams, 1975). This latter form of injury is identified by the presence of axonal retraction balls and microglial clusters in white matter. It has been held to represent tearing of axones with extrusion of axoplasm from the torn end. While this is an attractive explanation, it is probably an over-simplification because similar axonal retraction balls can be observed in conditions in which the axones have not been torn. The appearance can represent a localised block to axo-plasmic flow, the reasons for which have not yet been determined. The subtleties of the morphological basis of concussion remain to be ex-plained (Povlishock et al, 1979).

Clearly, many other events occur at the moment of impact. The physiological response to injury includes a tremendous surge of arterial hypertension, and it has been postulated by Kontos and his colleagues that this event triggers off the arachidonic acid cascade resulting in the formation of singlet oxygen and other free radicals, thromboxane,

323

prostacycline and the leukotrienes (Wei et al, 1980). These agents will in turn cause alterations in the calibre of cerebral vessels, their endothelial surface and areas of intravascular stasis resulting in multifocal cerebral ischaemia. The movements of the brain at impact may also produce vasospasm of central perforating branches of cerebral vessels by stretching and deformation. This local ischaemia has been suggested as a cause of loss of consciousness and impaired neurological function immediately after head trauma. The other major aspect of the physiological response to trauma is temporary cessation of spontaneous respiration. The extent of the hypoxic brain insult can seldom be determined because by the time medical help arrives on the scene spontaneous respiration has usually been resumed (Levine and Becker, 1979). It appears, therefore, as though a proportion of the damage to the brain normally included within the term "primary brain damage" may, in fact, be ischaemic or hypoxic rather than directly traumatic in origin.

SECONDARY BRAIN DAMAGE

While a wide variety of factors may lead to secondary brain damage, the two most important mechanisms by which neuronal death occurs are hypoxia and ischaemia with the addition of distortion of the brain due to mass and the toxic effects of CNS infection. The final common pathway for brain damage even for these mechanisms may be ischaemia. Graham and Adams (1971) have established that the incidence of ischaemic/hypoxic brain damage in fatal cases of head injury is more than 90%. Evidence from sequential CT scanning in survivors of head injury shows that large areas of cerebral infarction are not uncommon.

MECHANISMS OF SECONDARY BRAIN DAMAGE

These may be divided into systemic and intracranial factors (Table 1). Miller and his colleagues have drawn attention to the high incidence of arterial hypotension and hypoxia in severely head injured patients and the profound influence that this has upon mortality (Miller et al, 1978; 1981; 1982). They have also pointed out the high incidence of intracranial mass lesion formation and intracranial hypertension in these patients, both of which are associated with doubling of the mortality. In addition, the presence of moderate intracranial hypertension is associated with an increase in morbidity after head injury.

Arterial Hypotension

The principal cause of arterial hypotension observed soon after head injury is concomitant systemic injuries, particularly bleeding into the intra-abdominal cavity or pelvis. At a later stage, however, arterial hypotension may occur in the Intensive Care Unit as a result of inadequate fluid replacement after heavy diuresis due to osmotherapy, sepsis,

TABLE 1

Mechanisms of Secondary Brain Damage

SYSTEMIC	INTRACRANIAL
Arterial hypotension*	Extra/subdural haematoma*+
Hypoxaemia*+	Intracerebral haematoma/contusion*+
Hypercapnia+	Raised intracranial pressure*
Anaemia*	Cerebral oedema+
Hypoglycaemia*	Cerebral arterial vasospasm*
Hyperthermia+	Intracranial infection*+
Hyponatraemia+	Post-traumatic epilepsy*+

* associated with ischaemic/hypoxic brain damage; + associated with raised intracranial pressure

or injudicious use of drugs. In the severely injured patient, particularly those with any degree of intracranial hypertension, a fall in blood pressure below 90 mmHg is usually disastrous and startlingly rapid in the production of irreversible loss of brain function. It is difficult to investigate fully in the human patient why this should be so. In developing hypotheses to explain the severe effect of degrees of arterial hypotension in head injured patients that would be well tolerated during an elective neurosurgical operation, one must turn to evidence from animal models of brain injury and altered cerebral perfusion pressure.

Lewelt and his colleagues (1980) studied the effects of progressive reduction of arterial blood pressure by controlled haemorrhage in a series of cats that had been subjected to varying degrees of concussive head trauma. While non-injured animals tolerated a reduction of mean arterial pressure to 60 mmHg without a significant fall in cerebral blood flow, a pressure-passive reduction in flow was seen in all injured animals regardless of the severity of the injury. Related to the severity of the injury, however, was the level of blood pressure at which the cerebrovascular resistance began to increase during arterial hypotension, representing a relatively greater reduction in blood flow than in perfusion pressure. In severely injured animals, sustaining a level of injury that would produce prolonged coma, any reduction of arterial pressure below control levels resulted in an increase in vascular resistance. In less severely injured animals, the increase in vascular resistance occurred at a mean arterial pressure of 80 mmHg. After arterial pressure had been lowered to 30 mmHg, an attempt was made in most animals to restore arterial and cerebral perfusion pressure by replacement of the blood volume. In non-injured control animals the manoeuvre usually resulted in hyperaemia with obvious loss of autoregulation. In none of the injured animals, however, was it possible to restore cerebral blood flow once it had been reduced by haemorrhagic hypotension. These findings suggest that the injured brain may be particularly sensitive to any reduction in cerebral perfusion pressure, and when this is sufficiently severe to cause cerebral ischaemia it may not be possible to restore blood flow even when cerebral perfusion pressure is restored to normal levels.

Arterial Hypoxaemia

Similar studies of the changes in cerebral blood flow in response to alteration of arterial PCO_2 and PO_2 in cats subjected to concussional head injury have been carried out (Lewelt et al., 1982). Increasing levels of cerebral injury were associated with a progressive attenuation of the cerebrovascular response to changes in arterial PCO_2 . When the capacity of cerebral vessels to dilate to hypercarbia had been lost or severely depressed, a similar loss of the vasodilator properties to hypoxia was also present (Lewelt et al, 1982). Under such circumstances, therefore, the brain is subjected to an uncompensated reduction in oxygen delivery when hypoxia becomes so severe as to result in desaturation.

IMPLICATIONS FOR THE TREATMENT OF HEAD INJURY

After resuscitation, identification and treatment of any intracranial space-occupying haematoma, the intensive care of severely head injured patients is directed at maintenance of normal levels of relevant physiological variables (Becker et al, 1982). Artificial ventilation is widely practiced to preserve normal levels of arterial PO_2 while monitoring of arterial and intracranial pressure helps in the identification and preservation of normal levels of cerebral perfusion pressure. The value of information about the level of intracranial pressure can hardly be overstated. If there is not information about ICP, elevations of arterial pressure that are secondary to intracranial hypertension may mistakenly be treated by giving drugs to lower arterial pressure when the real problem is to reduce intracranial pressure. To reduce arterial pressure under these circumstances is disastrous.

It is best to think of the management of raised intracranial pressure in two stages, the first consisting of the application of safe proven treatments for raised ICP, the second of newer and unproven thera-

TABLE 2

Theoretical Benefits vs Practical Drawbacks of Barbiturate Therapy in Head Injury

Metabolic protection	Prolonged loss of observable neurological function
Stability of ICP	Loss of brain electrical activity
Mild hypothermia	Episodes of arterial hypotension
Reduced cerebral blood volume	Increased incidence of infection

pies. The first stage measures consist of hyperventilation, CSF drainage where appropriate, when ICP is being monitored via an intraventricular catheter, and intravenous infusion of hyperosmotic agents such as mannitol. The second group of therapies are not yet supported by solid evidence in favour of their effectiveness and their use has attendant risks. All of these drugs, barbiturates, althesin, gamma hydroxybutyrate and etomidate are anaesthetic agents. While claims have been made that each of these drugs can reduce the incidence of intracranial hypertension, improve cerebral perfusion pressure and lower mortality from severe head injury, these have not yet been supported by large-scale clinical trials. Ward and his colleagues have recently reported a randomised trial of pentobarbital treatment in severely head injured patients. They found no decrease in the incidence of intracranial hypertension in barbiturate treated patients, the requirements for intravenous mannitol therapy were the same, and there was no difference in overall outcome from injury. Within the barbiturate treated group, however, there was more systemic and intracranial infection and a significantly greater incidence of episodes of arterial hypotension despite the use of Swan-Ganz catheters to monitor pulmonary arterial pressure and dopamine infusions to support arterial pressure. It may be that a beneficial effect of barbiturate therapy in the management of severe head injury was offset by an increase in the incidence of brain ischaemia (Table 2). In the management of the patient with severe head injury the most important therapeutic measures are maintenance of stable and adequate levels of cerebral perfusion pressure and arterial PO_2 .

REFERENCES

Adams JH, The neuropathology of head injuries, in: "Handbook of clinical neurology", Vol 23, Vinken PJ, Bruyn GW, eds., Elsevier-North Holland Publ. Co., Amsterdam (1975).

Becker DP, Miller JD, Young HF, Selhorst JB, Kishore PRS, Greenberg RP, Rosner MJ, Ward DJ, Diagnosis and treatment of head injury, in: "Neurological surgery", Vol 4, Youmans J, ed., Saunders, Philadelphia (1982).

Graham DI, Adams JH, Ischaemic brain damage in fatal head injuries, Lancet 1:275 (1971).

Lewelt W, Jenkins LW, Miller JD, Autoregulation of cerebral blood flow after experimental fluid percussion injury of the brain, J Neurosurg 53:500 (1980).

Lewelt W, Jenkins LW, Miller JD, Effects of experimental fluid percussion injury of the brain on cerebrovascular reactivity to hypoxia and hypercapnia. J Neurosurg 56:332 (1982).

Levine JE, Becker DP, Reversal of incipient brain death from head injury apnoea at the scene of accidents, New Eng J Med 301:109 (1979).

Miller JD, Becker DP, General principles and pathophysiology of head injury, in: "Neurological surgery", Vol 4, Youmans J, ed., Saunders, Philadelphia (1982).

Miller JD, Becker DP, Secondary insults to the injured brain, J Roy Coll Surg Edin 27:292 (1982).

Miller JD, Butterworth JF, Gudeman SK, Faulkner JE, Choi SC, Selhorst JB, Harbison JW, Lutz H, Young HF, Becker DP, Further experience in the management of severe head injury, J Neurosurg 54:289 (1981).

Miller JD, Sweet RC, Narayan R, Becker DP, Early insults to the injured brain, J Am Med Ass 240:439 (1978).

Povlishock JT, Becker DP, Miller JD, Dietrich WD, The morphopathologic substrates of concussion? Acta Neuropath 47: 1 (1979).

Sullivan HG, Martinez J, Becker DP, Miller JD, Griffith R, Wist AO, Fluid percussion model of mechanical brain injury in the cat, J Neurosurg 45:520 (1976).

Ward JD, Becker DP, Miller JD, Choi SC, Marmarou A, Wood C, Newlon P, Kennan R, Failure of prophylactic barbiturate coma in the treatment of severe head injury, J Neurosurg 62:383 (1985).

Wei EP, Dietrich WD, Povlishock JT, Kontos HA, Functional, morphological and metabolic abnormalities of the cerebral microcirculation after concussive trauma in cats, Circ Res 46:37 (1980).

CAN SECONDARY BRAIN DAMAGE BE PREVENTED BY

PHARMACOLOGICAL OR OTHER MEANS?

Lawrence F. Marshall

Division of Neurosurgery, University of California
Medical Center, San Diego

Modern treatment of severe brain injury has centered on the hospital phase with neurosurgeons and others providing intensive care devoting much of their attention to the treatment of intracranial hypertension. This is appropriate since approximately 55% of severely head injured patients reaching the hospital alive will have an ICP in excess of 30 mmHg at some time during their intensive care. However, it is important to recognize that treatment of secondary insults to the already injured brain can not only reduce the number of patients who die prior to hospitalization, but may also reduce the incidence of elevated ICP, the most frequent cause of death in the severely head injured. Harr et al., have shown that severe hypoxia in the pre-hospital phase was associated with a substantially increased risk of elevated ICP during hospitalization (1). One might therefore conclude that early treatment of hypoxia should reduce the incidence of intracranial pressure rises and thus decrease mortality.

Recent data of San Diego County shown in Tables 1 and 2 indicate that the introduction of advanced life support services and helicopter evacuation of the severely injured results in a significant reduction of mortality (2). One must assume that this is due to the early treatment of shock, hypoxia and hypercarbia, the secondary insults so elegantly described by Miller and his colleagues (3). There was no substantial change of in-hospital mortality. Rather, the highly significant reduction appears to be almost entirely a product of treatment in the pre-hospital phase. This is in keeping with our own experience demonstrating a relatively stable in-hospital mortality during the past seven years. Thus, while the application of relatively simple principles of pre-hospital management has produced a dramatic decrease in mortality and morbidity in San Diego County, it appears appropriate at this time to focus on the treatment of additional, secondary insults during the pre-hospital

329

TABLE 1

San Diego County Head Injury

Death Rates by Year[*]

Year	Deaths	Population 1000's	Rate/ 100,000
1976	347	1623	21.3
1977	389	1667	23.3
1978	381	1710	22.3
1979	425	1786	23.8
1980	444	1862	23.8
1981	372	1920	19.4
1982	344	1965	17.5

[*] Included were all deaths for which the injury occurred within the county regardless of residence; gun shot wounds were excluded.

and hospital phase. During extensive epidemiologic surveys carried out over the past four years in San Diego County, we have demonstrated that severe arterial hypertension and bradycardia are associated with a significant excess risk of mortality. Thus, the immediately arising question is whether early treatment of these pathophysiologic responses to brain injury might reduces mortality. To date, no evidence can be produced to support or refute this hypothesis. Bradycardia and substantial elevations in SAP are known to be associated with an excess risk of mortality, not only in patients who have suffered severe devastating head injuries, but also if the initial Glasgow Coma Scale (GCS) is in excess of 8. This indicates bradycardia and arterial hypertension as a fruitful area of potential therapeutic intervention. As the treatment of bradycardia is relatively simple with atropine, small pilot studies in which physicians provide advanced life support at the scene appear appropriate. Systemic arterial hypertension would be extremely difficult to treat in the field, but can often be successfully dealt with in the hospital.

HOSPITAL PHASE

Others at this Conference have addressed the importance of treatment of brain tissue acidosis, intracranial hematomas, cerebral vasospasm as well as hypoxia. In these areas major attempts have and are being made to apply previously affirmed principles. There are other secondary injuries during the intensive care phase which appear to be targets for pharmacologic manipulation. These include cardio-pulmonary dysfunction, malnutrition and intracranial hypertension. It is proposed that the surge of catecholamines in serum and brain (Fig. 1) play a significant role in the production of all of these phenomena. The catecholamine storm can be seen as a potential unifying cause and might be labeled a "hyperdynamic state". Such a hypothesis first advanced by Wurtman and Zervas (4), although somewhat simplistic can serve as a starting point to rationally understand the mechanisms of the dynamic changes characterizing severe brain injury. Thus, several new areas have been identified where therapy could be directed to return the patient to a physiologically improved milieu. I have chosen to discuss two major areas: the hyperdynamic state and intracranial hypertension.

TABLE 2

Head Injury Death Rates

by Location and Year[*]

Death Location	Year	Deaths	Rate/ 100,000
Scene	1979	175	9.0
	1980	178	9.6
	1981	138	7.6
DOA	1979	56	3.1
	1980	53	2.8
	1981	20	1.0
ER	1979	50	2.8
	1980	46	2.5
	1981	65	3.3
Other	1979	144	8.1
	1980	167	8.9
	1981	149	7.7

[*] Includes non-county residents, excludes gun shot wounds; DOA: dead on arrival, ER: emergency room.

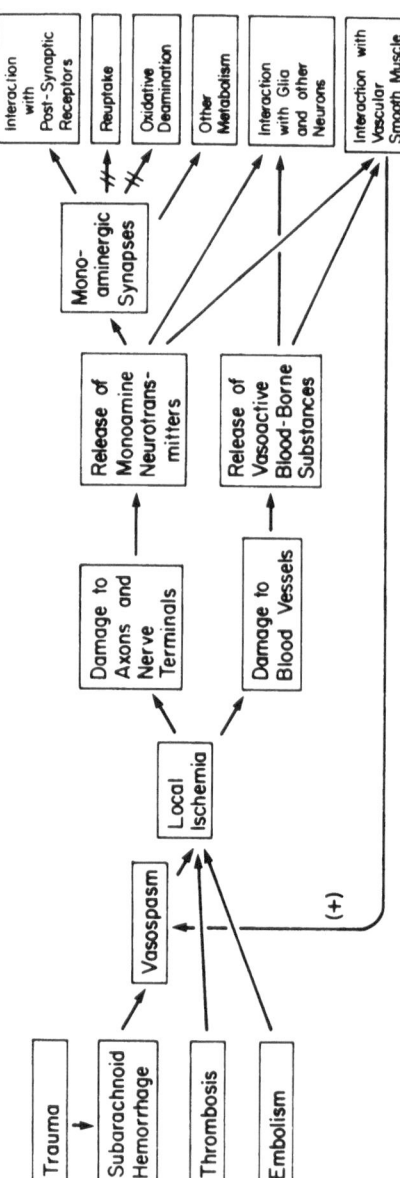

Fig.1: Role of brain injury in triggering intrinsic catecholamine storm. (From Wurtman and Zervas, 4).

THE HYPERDYNAMIC STATE

1. Cardio-pulmonary Performance

The increasingly sophisticated measures of cardio-pulmonary performance have become a subject of intense scrutiny in head injured patients. Pharmacological intervention to improve cardio-pulmonary dynamics of these patients, however, has been quite limited. One-half of the patients entered into the National Pilot Traumatic Coma Data Bank had one or more episodes of hypoxia, defined as a PaO_2 of < 65 mmHg some

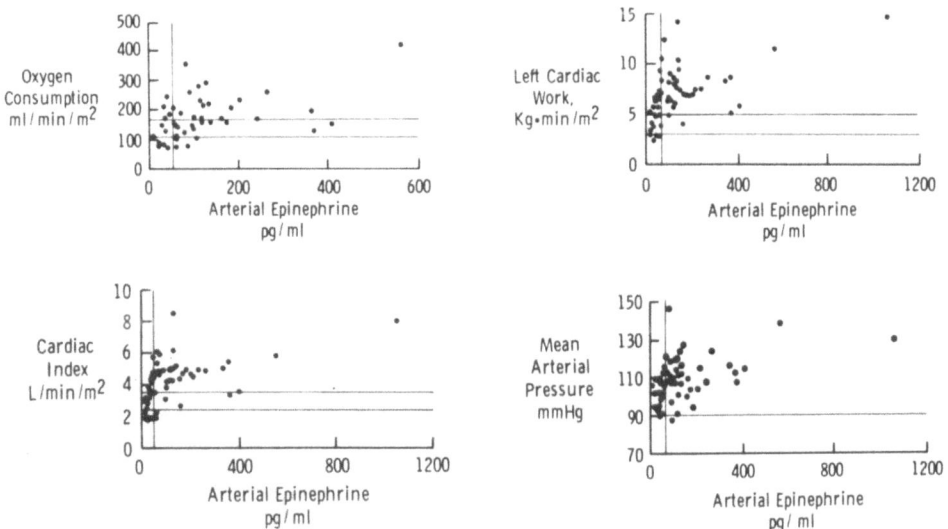

Fig.2: Correlations between arterial epinephrine levels and oxygen consumption, cardiac work, cardiac index and mean arterial pressure in patients with severe head injury (From Clifton, et al. 7; with permission).

time in their hospital course. Therefore, we began to manipulate cardio-pulmonary dynamics, as did Clifton and his colleagues at Baylor University. These studies were based on previous observations by Clifton and Ziegler, et al. that systemic arterial hypertension, right and left heart strain, and catecholamine storm all occurred in brain injured patients, and that the degree of cardiac dysfunction directly correlated with the serum catecholamine levels (5). A frequent association of pulmonary failure with systemic arterial hypertension was of considerable interest. The role of systemic arterial hypertension in the exacerbation of acute

brain injury is complex and unresolved. Obviously, under circumstances
with a defective autoregulation, increased cerebral blood volume must
follow systemic arterial hypertension and will produce elevations in ICP.
This can be easily demonstrated in many patients at the bedside simply by
allowing blood pressure to rise and watching ICP. While it was initially
thought by Cushing and others that the rise in blood pressure is a compensatory mechanism, Shalit postulated that this is perhaps a detrimental
response of the injured brain (6). We have adopted this philosophy as
well.

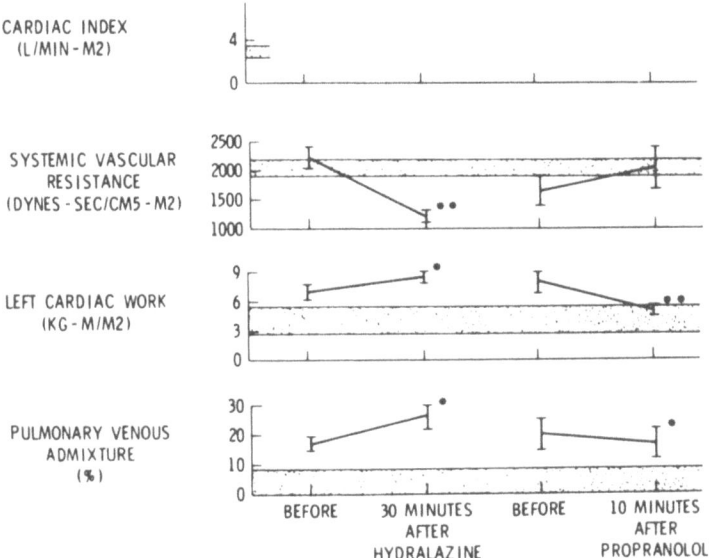

Fig.3: Deleterious effect of hydralazine and beneficial effect of propranolol on left cardiac work and pulmonary venous admixture.
(From Robertson, et al. 8; with permission).

Recently, Clifton and his colleagues expanded on their previous
work with Ziegler and described in detail the cardiovascular responses to
head injury and the treatment of systemic arterial hypertension associated with it (7,8). They have shown that elevations of serum norepi-
nephrine- and epinephrine levels are common, and that the levels of these
monoamines correlate with the increase in cardiac index, systemic
arterial pressure, and oxygen consumption (Fig. 2). While abnormalities
of cardiovascular function have long been recognized in patients suffering from head injury and subarachnoid hemorrhage, much less atten-

tion has been paid to the potential mechanisms and to the possibility of therapy. We do not know yet, whether the hyperdynamic state is useful, maladaptive, or both. Evidence that it may be maladaptive includes (1) ischemic cardiac damage detected pathologically and by changes of the electrocardiogram, (2) pulmonary failure characterized by hypoxia, and (3) marked increases in oxygen consumption.

Robertson and Clifton utilized two regimens to treat this state. The effective one employing propranolol is discussed here. They were able to show an improvement of pulmonary venous admixture, cardiac work (Fig. 3), and a decrease of plasma catecholamine levels following the administration of propranolol (Fig. 4). Moreover, systemic oxygen consumption was substantially reduced. Since severe brain injury may be characterized by a local, i.e. intracerebral release of catecholamines, it is tempting to speculate that part of these effects might be blocked directly by propranolol taken up into the brain. Propranolol is lipophilic and therefore penetrates the blood-brain barrier. Substantially higher concentrations were indeed demonstrated in the brain following intravenous administration.

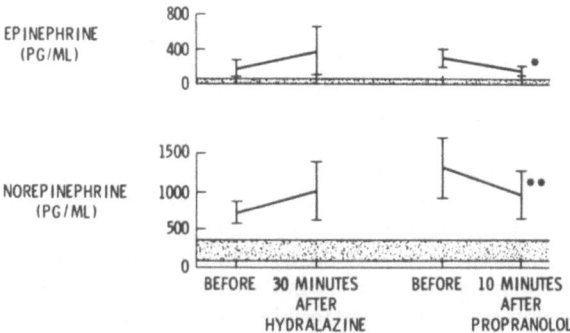

Fig.4: Effects of hydralazine and propranolol on plasma epinephrine- and norepinephrine levels; p < .01, and p < .001, respectively. (From Robertson, et al. 8; with permission).

2. Catabolism in the Hyperdynamic State

The second major component of the hyperdynamic state is enhancement of catabolism which is often associated with severe head injury. Decerebrate patients may require a caloric intake in excess of 4,000 kilocalories per 24 hours. It has been generally accepted that attempts should be made to restore nitrogen balance and avoid catabolism. There is a severe depression of the immune competence and a substantial increase in the incidence of infection in patients in whom severe catabolism is allowed to continue for more than a few days. As elevated serum catecholamine levels may play a role in the development of hypermetabolism, it is tempting to consider pharmacologic depression rather than total parenteral nutrition. This concept deserves consideration because ischemic brain injury probably is exacerbated by hyperglycemia (9, 10). Early enteral feeding which usually does not produce hyperglycemia is impossible if narcotic sedation and muscle relaxants are administered. Total parenteral nutrition may produce hyperglycemia and requires large fluid volumes which may exacerbate intracranial hypertension. Rapp and his colleagues have suggested, however, that early parenteral feeding has a favorable effect on the outcome from severe head injury (11). However, the sample size of this study was small and interpretations of the results must be cautious. Nevertheless, the mortality of the patients with total parenteral nutrition was not increased.

Although the study might allay some concerns regarding early parenteral nutrition, the question remains whether it would not be better to suppress the caloric needs by propranolol while not entirely interfering with the breakdown of certain proteins and fats required for recovery of the brain. Robertson et al. studied oxygen consumption during administration of propranolol and demonstrated a salutary effect on the hypermetabolic state. If elevated levels of catecholamines in both brain and serum are deleterious and produce secondary brain injury, several objectives could be achieved by propranolol or similar agents in the early treatment of severe head injury. First, the hyperdynamic state would be substantially reduced and the pulmonary function improved. Second, the metabolic needs of the patient would be lessened and, at least for the first days, total parenteral nutrition could be avoided with its attendant hazards of hyperglycemia, major swings in serum osmolality, and the potential for increasing ICP. Finally, interference with synaptic transmission due to the release of free monoamines in the brain might be ameliorated. Brain tissue acidosis which characterizes the natural history of acute brain injury interferes with the reuptake of monoamines. Therefore, it appears appropriate to reverse the metabolic derangement and to suppress the level of free catecholamines in the brain. Such an approach may be naive given the complexity of acute brain injury. Nevertheless, it may provide a rationale to manipulate the cerebral milieu towards recovery.

INTRACRANIAL HYPERTENSION

It is apparent that pharmacologic intervention can substantially reduce ICP. Mannitol and ventricular drainage, if instituted promptly and in concert with mechanical hyperventilation, yields an overall incidence of uncontrolled ICP which is quite low compared to our earlier experience. As we have become more sophisticated in the implementation of these conventional methods and have instituted earlier surgery for patients with intracranial hematomas and intracerebral contusions producing shift, the need to utilize other methods to reduce ICP has diminished. It is not the purpose of this discussion to debate the value of barbiturates or althesin, for example, in intracranial hypertension. It is our own experience and that of the national multicenter trial that ICP control can be improved if barbiturates are superimposed on the previously described more conventional therapies.

1. Should Modest Elevations of ICP be Treated?

It is not apparent whether early intervention should be instituted at ICPs considerably lower than those treated just five years ago. Our observation is that pupillary abnormalities indicative of early transtentorial herniation can occur at ICPs below 30 mmHg. At least in some patients, pressures between 20 and 30 mmHg are poorly tolerated and require therapy (12). Furthermore, although reduction of the incidence of uncontrollable ICP was hypothesized earlier as a function of prompt treatment of shock and hypoxia, it is also possible that more aggressive therapy of mild or modest elevations, in the 15-20 mmHg range, reduce uncontrolled ICP. Ducker has attributed the reduction of the incidence of uncontrolled ICP at the Maryland Institute of Emergency Medicine to earlier intervention (13). In their initial series, 43 out of 127 patients, or 34% had an ICP equal to or in excess of 25 mmHg despite conventional therapy. In the latest series, the incidence of such elevations was reduced to 12 out of 97, or 12%. As the initial ICP can be used to predict the likelihood of a high ICP later (14), a therapeutic trial which randomizes high-risk patients into two groups, one receiving treatment when the ICP exceeds 15 mmHg, and the other when it exceeds 30 mmHg, would not be difficult to design. It is important, however, that such trials should be carried out only in patients with head injury or other diseases, where ischemia is generally secondary to ICP elevation rather than primary, as e.g. occurring by drowning or cardiac arrest. In the latter group of patients, intracranial hypertension has in our experience been associated with universal fatality, indicating irreversible and untreatable primary injury of the brain parenchyma, although ICP control can be achieved in some instances. The question whether earlier intervention for elevated ICP at lower pressures is useful should focus not only on mortality but also on morbidity. The National Traumatic Coma Data Bank in the United States could serve as a mechanism for undertaking such a trial.

The influence of aggressive treatment of elevated ICP on outcome is clouded by the fact that mortality statistics are reported from university centers and institutions with a substantial interest in clinical head injury. In reviewing outcome in community hospitals, we have found a wide variance. In some hospitals with a dedicated neurosurgeon and good support, similar outcomes to those reported from major centers have been accomplished. On the other hand, we have identified circumstances where mortality rates in excess of 70% have occurred in large series of patients. In one sample, two-thirds of the patients had diffuse injuries, while one-third was initially talking or obeyed commands at admission. Yet, many died within several days with a clinical picture strongly suggestive of diffuse brain swelling. It is difficult to argue under these circumstances that such patients had irreversible injuries from inception, and that rational therapy for intracranial hypertension would not be meaningful.

2. Iatrogenic Intracranial Hypertension

The debate regarding aggressive therapy for elevated ICP has tended to obscure the substantial advances in our understanding of the pathophysiology of brain injury and steps that have been taken to avoid iatrogenic errors. As one looks back at the 1970s we recognize the importance of an improved understanding of drug action in avoiding iatrogenic ICP rises. One can only crudely estimate how many patients have avoided significant secondary ischemic brain injuries because halothane was not chosen as anaesthetic agents, or sodium nitroprusside was not administered when the blood pressure rose in response to an elevation of ICP. The avoidance of secondary insults must have the avoidance of iatrogenic insults as one of its underlying assumptions. The multiplicity of insults which might lead to secondary brain damage makes the management of the severely brain injured patients terribly complex. Cerebral vasospasm is frequent in patients suffering from severe brain injury, or spontaneous subarachnoid hemorrhage. Neuropathologic and neuroradiologic studies have demonstrated the relationship between vasospasm and irreversible ischemic damage (15). In the treatment of patients with signs of ischemia from vasospasm following subarachnoid hemorrhage, elevated ICP is a relatively infrequent problem. Treatment of cerebral vasospasm includes a standardized regimen of fluid expansion, blood pressure elevation and, more recently, calcium channel blockers in some centers. What, however, happens if a calcium channel blocker is superimposed on a patient with acute brain injury who has either an intracranial mass lesion or diffuse brain swelling?

Figure 5 shows the ICP following administration of verapamil to a patient in the operating theater prior to removal of a supratentorial tumor.In this patient with low compliance verapamil caused a dramatic rise in ICP (16). The illustration serves to demonstrate the problem of how to avoid secondary insults, and the potential for exacerbating them with pharmacologic intervention. Under circumstances with elevated ICP or low compliance, it is inappropriate to uncritically expand the intravascular volume, raise blood pressure or administer calcium channel blockers. These substances might be appropriate, if ICP is normal and cerebral vasospasm develops late.

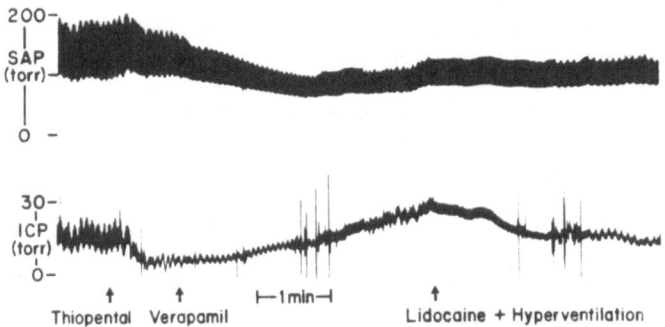

Fig.5: Effect of the calcium channel blocker verapamil on ICP in a patient with a supratentorial tumor. The ICP rose from 10 to 30 mmHg within 30 minutes of verapamil administration.

We cannot become discouraged by the complexity of secondary brain injury, rather we need to consistently develop more rational forms of treatment based on a better understanding of the basic pathophysiology. Avoidance of iatrogenic insults to the nervous system is just as important as pharmacologic or other manipulation of the patient's milieu to protect the already injured brain. Those of us with an aggressive philosophy of treatment of acute brain injury must recognize the limitation of our efforts under circumstances where salvage appears possible. At the same time, it is a substantial advance that we begin to rationally apply our understanding of cerebral flow, metabolism, neural activity in using pharmacologic regimens to manipulate those variables. One recognizes, of course, that barbiturates, althesin or other "wonder drugs" are no panaceas but represent the application of basic principles of neurophysiology and neurochemistry to the treatment of a complex brain injury. Their role in the improvement of survival is perhaps not as important as that we can use them to manipulate these variables. At least in the operating theater, their use helps to avoid intracranial catastrophes that formerly occurred with substantial frequency.

REFERENCES

1. Harr FL, Phillips S, Huchton JI, The incidence and significance of
   early hypoxemia in head injury patients, Trans Amer Assoc
   Neurosurg, Boston (1981.
2. Klauber MR, Marshall LF, Toole BM, Knowlton SL, and Bowers S.A.,
   Cause of decline in head-injury mortality rate in San Diego
   County, California, J Neurosurg 62:528 (1985).
3. Miller JD, Sweet RC, Narayan R, Becker DP, Early insults to the
   injured brain. JAMA 240:439 (1978).
4. Wurtman RJ, Zervas NT, Monoamine neurotransmitters and the pathophysiology of stroke and central nervous system trauma.
   J Neurosurg 40:34 (1974).
5. Clifton GL, Ziegler MG, Grossman RG, Circulating catecholamines and
   sympathetic activity after head injury, Neurosurgery 8:309
   (1981).
6. Shalit MN, Cotev S, The cushing response - a compensatory mechanism
   or a dangerous phenomenon? in: "ICP II", Lundberg N, Ponten U,
   Brock M, eds., Springer Verlag, Berlin (1975).
7. Clifton GL, Robertson CS, Kyper K, Taylor AA, Dhekne RD, Grossman
   RG, Cardiovascular response to severe head injury, J Neurosurg
   59:447 (1983).
8. Robertson CS, Clifton GL, Taylor AA, Grossman RG, Treatment of
   hypertension associated with head injury, J Neurosurg 59:455
   (1983).
9. Myers RE, Lactic acid accumulation as a cause of brain edema and cerebral necrosis resulting from oxygen deprivation. in: "Advances
   in perinatal neurology", Korobskin R, Guilleminault G, eds.,
   Spectrum Publishers, New York (1977).
10. Rehncrona S, Rosen I, Siesjö BK, Excessive cellular acidosis: An im-
    portant mechanism of neuronal damage in the brain? Acta
    Physiol Scand 110:435 (1980).
11. Rapp RP, Young B, Tayman D, Birins BA, Haack D, Tibbs PA, Bean
    JR, The favorable effect of early parenteral feeding on survi-
    val in head injured patients, J Neurosurg 58:906 (1983).
12. Marshall LF, Barba D, Toole BM, Bowers SA, The oval pupil: Clinical
    significance and relationship to intracranial hypertension,
    J Neurosurg 58:566 (1983).
13. Bellegarrigue R, Ducker RB, Control of intracranial pressure in
    severe head injury, in: "ICP V", Ishii S, Nagai H, Brock M,
    eds., Springer Verlag, Berlin (1983).
14. Klauber MR, Toutant SM, Marshall LF, A model for predicting delayed
    intracranial hypertension following severe head injury (submit-
    ted)
15. MacPherson P, Graham DI, Correlation between angiographic findings
    and the ischemia of head injury, J Neurol Neurosurg Psychiat
    41:122 (1978).
16. Bedford RF, Dacey R, Winn HR, Lynch C III, Adverse impact of a
    calcium entry blocker (verapamil) on intracranial pressure in
    patients with brain tumors, J Neurosurg 59:800 (1983).

CONTUSIONAL HEMORRHAGE - PROGNOSTIC SIGNIFICANCE

OF PRIMARY AND SECONDARY BRAIN DAMAGE

Karl E. Richard, K. Radebold, and R.A. Frowein

Dept. Neurosurgery, University of Cologne, Cologne, FRG

Traumatic contusional hemorrhage is a severe form of brain injury with a high mortality of 40 to 70% (1). The following prognostic factors must be taken into consideration:

- Primary damage of brain tissue as, e.g. reflected by the magnitude of contusional hemorrhage,
- General and focal secondary reactions of brain tissue, such as brain swelling and focal edema of the surrounding brain,
- Brain shift and disturbances of the intracranial volume-pressure relationship.

CLINICAL MATERIAL

Neurologic conditions, computer-tomographical findings and ICP-courses were analyzed in 36 patients with traumatic contusional hemorrhage during the first three weeks after head injury. All patients were unconscious at the time of admission. Four levels of coma were differentiated:

- **Coma I:** Without neurologic disturbances

- **Coma II:** With unilateral pupillary disturbance and/or hemiparesis

- **Coma III:** With uni- or bilateral extension spasms

- **Coma IV:** With bilaterally dilated and fixed pupils and general muscular flaccidity.

The volumes of contusional hemorrhages (CH), environmental reactions (ER), and the degree of brain swelling were estimated by CT taken on the day of admission and on subsequent days. Three degrees of brain swelling were differentiated:

- **I:** Small ventricles and small subarachnoid spaces,
- **II:** Collapse of ventricles, or of subarachnoid spaces, or of the cisterns,
- **III:** Collapse of the entire cerebral fluid spaces

ICP was measured epidurally during a period of six days on average. Every day, the peaks of maximal pressure (ICPmx) were evaluated. Four outcome groups were considered:

- Full recovery
- Slight disability
- Severe disability
- Lethal outcome

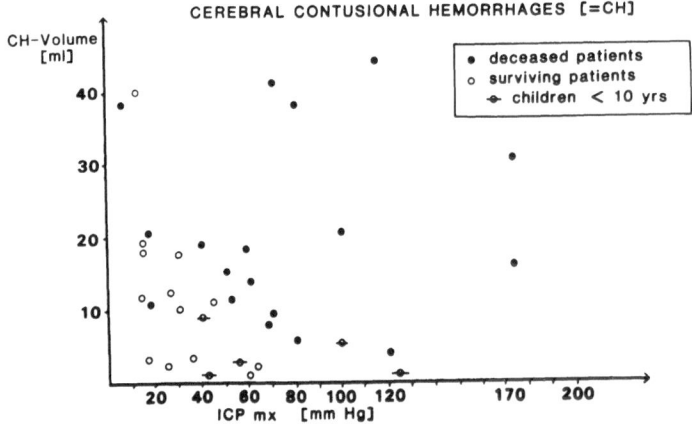

Fig.1: Relationship of contusional hemorrhage volumes and maximum of intracranial pressure peaks (ICPmx) in patients who survived or died from severe head injury. Only with small CHs patients could survive high pressure peaks and, conversely, large CH volumes, if pressure peaks remained below 60 mmHg. Children with small CHs tolerated epidural pressure peaks of up to 120 mmHg.

RESULTS

1. Contusional Hemorrhage (CH)

The size of the CHs varied considerably reaching volumes of up to 50 ml. An increase in size of CHs was seen in 36% of the patients on the first day post trauma. Patients with CH volumes of more than 5 ml survived only, if the ICP did not exceed 45 mmHg (Fig. 1). In 17 cases with lethal outcome (47%) CH volumes were 10 ml higher on average than in patients who survived (Table 1). On the other hand, the mean CH volume was highest in those patients (11%) who had a complete recovery. Two patients in this group were 74 or 76 years old, respectively.

TABLE 1

Outcome	N	%	Age [x̄]	CH-Volume [ml]	ER-Volume [ml]	Brain Swelling [Grade]	Coma [Grade]	ICP mx [mm Hg]
Full recovery	4	11	43	20	59	2	1.8	26
Moderate disability	5	14	39	8	18	2.4	2.3	35
Severe disability	10	28	15	3	10	2.9	2.8	52
Lethal outcome	17	47	33	16	30	2.9	2.7	77
Σ	36	100	29					

Hemorrhage (CH-volume), perifocal edema (ER volume), brain swelling, level of consciousness and intracranial pressure peaks (ICPmx) of all patients in relation to the clinical outcome.

2. Environmental Reaction (ER)

The magnitude of the secondary, peri-contusional reaction of the brain tissue varied greatly. In the majority of cases the maximum of ER became apparent only several days or even 2-3 weeks later. After the third week, space-occupying reactions could no longer be recognized on CT. Cases with ERs of more than 30 ml survived only, if the ICP did not exceed 50 mmHg (Fig. 2). In patients who died the average ER volume was higher than in patients who survived with either slight or severe disability (Table 1). An exception were four patients who fully recovered.

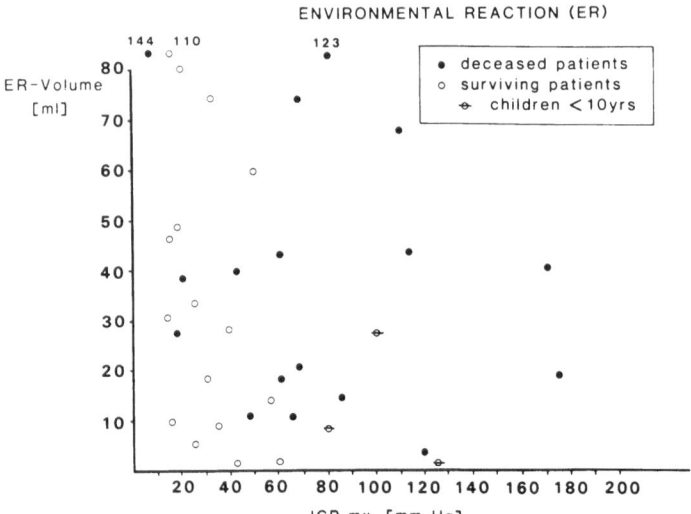

Fig.2: Relationship of ER volume and ICPmx (cf: Fig.1) in patients with severe head injury. The effect of ER volume or ICPmx, respectively, on outcome is obvious.

3. Brain Swelling

Severe brain swelling (IIIrd degree) was a very frequent (62%) and characteristic finding on the day of trauma, and nearly always seen on CT taken at the first day post trauma. Brain swelling disappeared by the end of the third week. The extent of brain swelling was of critical significance for the outcome (Table 1).

4. Neurological Dysfunction

Both, the severity and the duration of disturbances of consciousness, especially coma were significantly related to recovery (Table 1).

5. Intracranial Pressure

The effect of ICP on recovery was particularly evident. Patients who survived with full recovery or with only slight deficits had significantly lower pressure peaks than patients who died or survived with severe disabilities (Table 1). Only children were found to survive an increase of the intracranial pressure of up to 100 mmHg, juveniles of

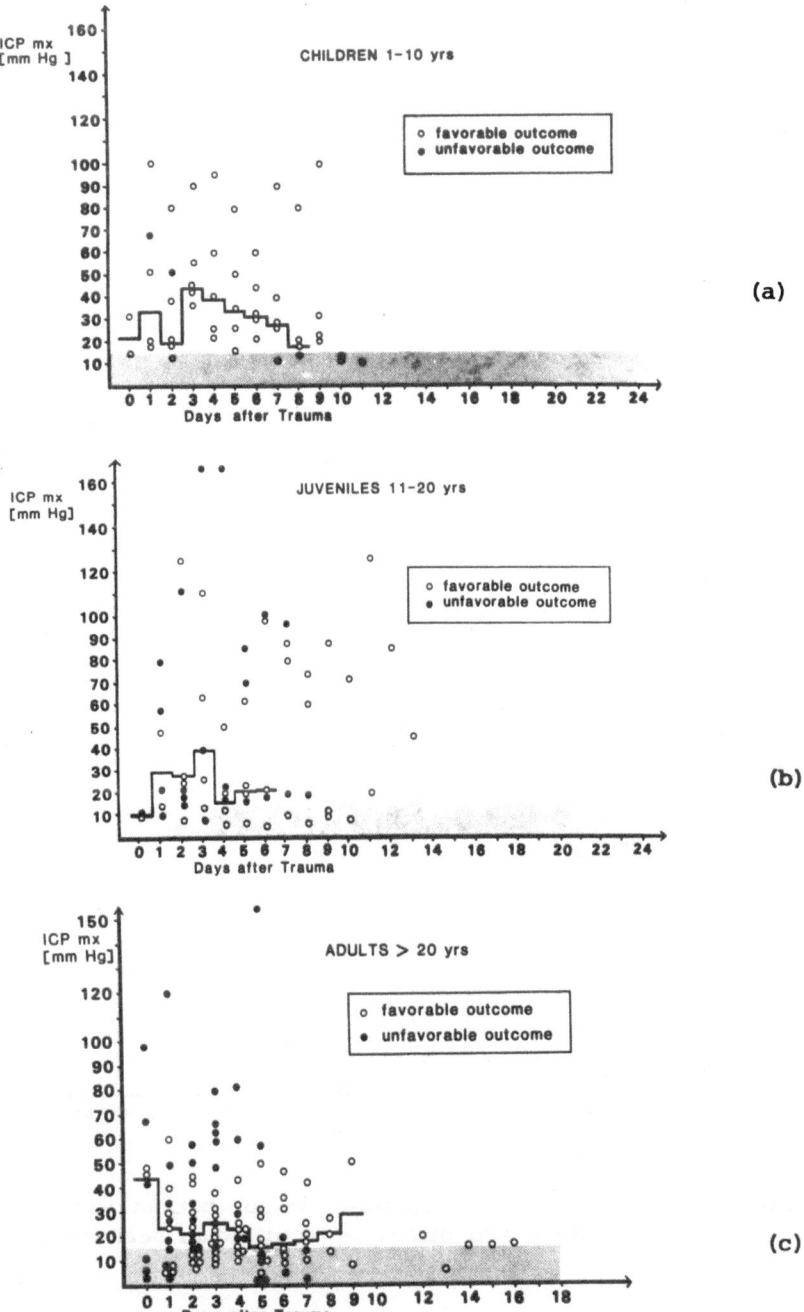

Fig.3: Intracranial peak pressures (ICPmx) of patients with contusional hemorrhage. As compared to adults children and juvenile patients tolerated higher ICP peaks.

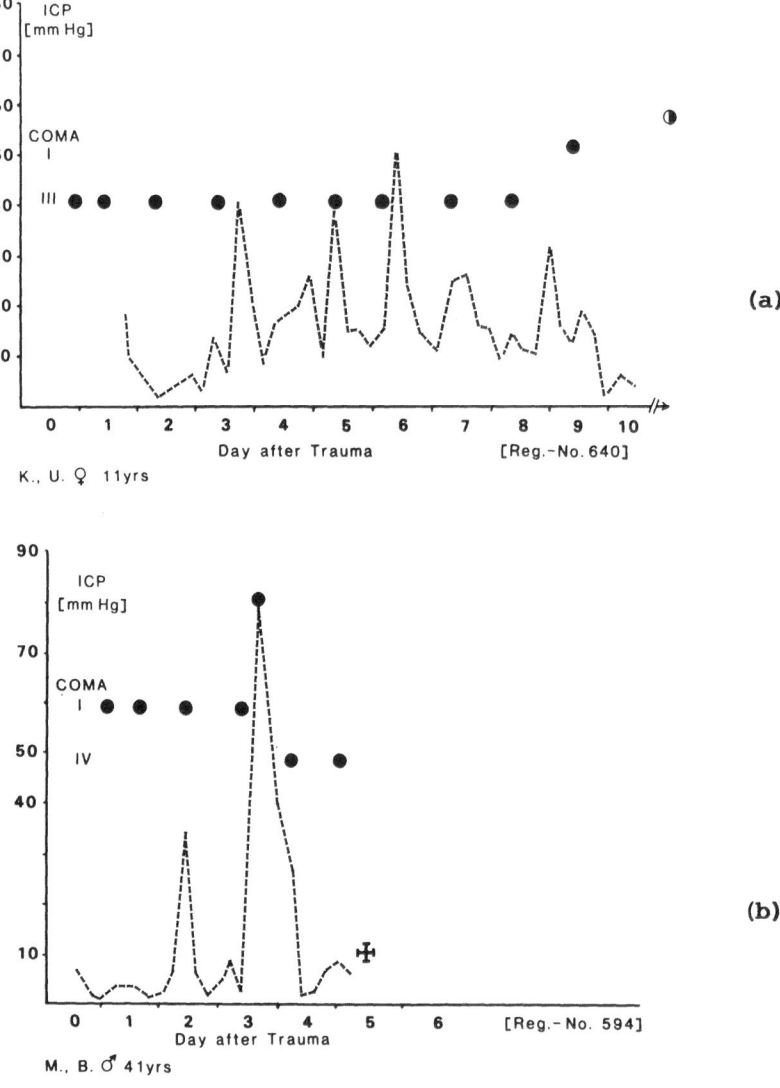

Fig.4: ICP and level of consciousness in one patient who survived (top), and in another who died from contusional hemorrhage (bottom).

even up to 130 mmHg, while in adults elevations of ICP above 60 mmHg resulted in death (cf: Fig. 3, 4). On the other hand, patients with an extreme increase of ICP could survive, if the contusional hemorrhage volume was relatively small. Children and juveniles survived with severe disability, if the intracranial pressure exceeded 60 mmHg.

CONCLUSION

Traumatic contusional hemorrhage can be considered as a clinical model of brain damage with primary and secondary components. Primary consequences are laceration of the tissue and hemorrhage. A secondary increase of the hemorrhagic area on CT most likely is the result of a continuing gradual bleeding. It is also possbile that release of agressive serum factors, such as kinins play a role (3; cf: Unterberg et al., this volume). An early manifestation of a secondary reaction is brain swelling caused by traumatic vasoparalysis combined with an increase of intracranial blood volume. A few hours later the environmental reaction, i.e. vasogenic brain edema appears in the perifocal brain. Contusional hemorrhage and edema of the surrounding brain attain critical proportions for recovery, if they exceed a certain extent and size, in particular, if they lead to an intracranial pressure rise.

Therefore treatment must be directed at limiting these negative factors, particularly brain swelling, brain edema and the intracranial pressure rise. Contusional hemorrhages should be removed, if the epidural pressure exceeds 40-50 mmHg.

REFERENCES

1. Lanksch W, Grumme T, and Kazner E, "Computed tomography in head injuries", Springer, Berlin-Heidelberg-New York (1979).
2. Richard KE, and Frowein RA, The value of ICP-monitoring in the treatment of traumatic bilaterally or medially situated intracerebral contusional hemorrhages, in: "Intracranial pressure V", Ishii S, Nagai H, and Brock M, eds., Springer, Berlin-Heidelberg (1983).
3. Sicuteri F, Fanciullacci M, Bavazzano A, Franchi G, and Del Bianco PL, Kinins and intracranial hemorrhages, Angiology 21:193 (1970).

SYSTEMATIC SEARCH FOR BRAIN RESUSCITATION POTENTIALS

AFTER TOTAL CIRCULATORY ARREST

Peter Safar and Per Vaagenes

Resuscitation Research Center and Department of Anesthesiology/Critical Care Medicine, and the Presbyterian-University Hospital, University of Pittsburgh, Pittsburgh, PA, 15260 USA

INTRODUCTION

In the 1970s, the concepts and techniques of cardiopulmonary resuscitation (CPR) were extended to cardiopulmonary-cerebral resuscitation (CPCR; 81). A systematic search for improved CPCR methods after cardiac arrest is important for socio-economic, medical and scientific reasons. About 25% of all deaths are the results of acute potentially reversible dying processes, in the absence of end-stage incurable disease and before natural dying from old age. The majority of such dying processes represent indications for CPCR attempts. Moreover, systematic CPCR research should not only lead to more effective therapies, but also to reliable information early post-arrest to predict persistent vegetative state and thereby help in deciding on whom to let die. Brain resuscitation research on complete isolated global brain ischemia (GBI) and total body circulatory arrest (TCA) might also benefit patients with severe shock, stroke, brain injury, brain hemorrhage, encephalitis, and organ preservation for transplantation. This paper is not a report of the authors' new research data, but rather summarizes some methods, results and ideas primarily, but not exclusively of the authors' groups.

Some visionary laboratory resuscitation researchers envisioned already at the turn of this century an extension of CPR to CPCR (95). Until 1970, most brain resuscitation related research had focussed on mechanisms, on focal ischemia, and on brain trauma. In the 1970s in Pittsburgh, we developed animal models, studied mechanisms, and evaluated treatment potentials for GBI and TCA; the treatments were selected semi-empirically (12, 31-35, 41,42,58, 67-69, 80-84, 92,102,103). These studies involved rats, and short-term and long-term (3-7d) experiments in dogs and monkeys. In 1979, with the initiation of the Resuscitation Re-

search Center (RRC), GBI and TCA animal studies were designed with a
more systematic approach (17-19, 31-34, 102). The RRC's first five years
(1979-1984) gathered data from acute (< 12 h), short-term (< 24 h) and
long term (48-168 h) experiments in rats, monkeys and dogs. A systematic
study mechanism was developed consisting of therapy screening studies in
rats (41, 42, 50) and dogs (80,84), and definitive longterm outcome studies in dogs (80,102) and monkeys (31-35). Moreover, a multi-institutional randomized clinical study mechanism was established for evaluating
cardiac arrest patients with, versus without, novel CPR clinical treatment protocols (1, 2, 15). The clinical study completed its first phase
1979-1983. Extensive data were collected from 280 patients, one-half with
standard therapy only, and one-half with additional thiopental loading.
The second phase of this study (1983-1987) is evaluating the effects of a
calcium entry blocker.

We have urged clear terminologies to differentiate between the
following:
(1) Types of insults, e.g., focal versus global ischemia, ischemia versus
trauma, complete versus incomplete ischemia, temporary versus permanent
ischemia
(2) Transient functional changes versus permanent brain damage (nonvia-
bility factors)
(3) Survival with cerebral dysfunction versus mortality
(4) Changes produced by the primary insult versus secondary post-insult
derangements of the brain
(5) Cerebral versus extracerebral derangements
(6) Pre-insult treatment (protection), versus during-insult treatment
(preservation), versus post-insult treatment (resuscitation)
(7) Short-term mechanism-oriented studies versus long-term outcome studies, and
(8) Experiments with post-insult intensive care life support versus experiments with observation of spontaneous recovery.

Our working hypotheses on the post-cardiac arrest situation, which
have evolved since the 1960s are as follows:
(1) Cerebral damage is the result of the initial insult plus post-reper-
fusion changes in the brain and in extracerebral organs all of which can
add further injury to the postischemic brain (65, 82-84).
(2) The resulting changes in the brain consist of three types of processes, which can worsen each other:
    (a) Reperfusion failure leading to impairment of reoxygenation (4, 44,
       68, 92);
    (b) Reperfusion injury, i.e., intracellular cascades of chemical reac-
       tions which lead to membrane damage and necrosis of neurons
       triggered by the interaction of iron, free chemical radicals
       (7,24,71,103a) and calcium (89,106); and
    (c) Self-intoxication from post-anoxic extracerebral organs.
These processes, which lead to membrane lipid peroxidation (86) and feedback loops leading to cell necrosis (89), include multifocal hypoperfusion, hypermetabolism, acidosis, abnormal electrolyte shifts, tissue ede-

ma, neurotransmitter derangements, and toxic factors from the brain and malfunctioning extracerebral organs.

(3) The multifactorial pathogenesis of postischemic-anoxic encephalopathy requires etiology-specific multifaceted therapies (31, 82-84).

(4) The key to maximally effective brain resuscitation after prolonged cardiac arrest or severe shock state lies in an understanding and amelioration of the multi-organic mechanisms of the post-resuscitation syndrome (PRS; 17-19, 48, 65, 82, 91). The concept of the PRS was pioneered by Negovsky (65).

After a decade of semi-empirical animal trials (1973-1983), post-arrest general "brain-oriented extracerebral life support" protocols proved to ameliorate brain damage as compared with "conventional care" (12, 31-34, 69, 80-83). Several special treatments proved to have adjunctive value: reflow promotion (84), barbiturates (2, 12, 33, 100, 110), calcium entry blockers (35,94,102,109), and seizure prophylaxis (3,23). Their beneficial effects are inconsistent, however. A breakthrough has not yet occurred. We believe that it is more likely to come from systematic multi-center research than from empirical trials. Although the problem is very complex, the potentials and limits of reversibility of global ischemic brain damage after reversal of clinical death must and can be clarified.

BACKGROUND

Brain ischemia and brain resuscitation have been reviewed (44,65, 72,74,82,83,89,106). Circulatory arrest of more than five min duration leads to cerebral ATP loss (60) and death or permanent brain damage (20). Retinal neurons of rabbits in vitro can tolerate up to 20 min of normothermic anoxia (5). Some cerebral neurons can recover electric and biochemical activity after up to 60 min of normothermic global ischemia (44, 45). After total circulatory arrest (TCA), brain damage can be ameliorated by post-insult therapy (84, 102). Novel therapies were investigated which started immediately upon reperfusion and continued for days. We have occasionally been able to produce full neurologic recovery in monkeys after 16-18 min of global head ischemia by neck tourniquet (12, 31-35, 69), and in exceptional clinical cases after up to 15-20 min of normothermic cardiac arrest (1, 2, 15), or up to 40 min of hypothermic arrest (88). In dogs after up to 10-12 min of TCA by ventricular fibrillation (VF) post-arrest reflow promotion (84), or a calcium entry blocker (102) were used. In these experiments we have achieved neurologic but not histologic recovery. All these beneficial results were inconsistent because of the numerous interacting variables mentioned above (80). Pre-insult and insult variables can be controlled experimentally. Known post-insult variables are partially controllable, but some, particularly extracerebral factors, are still unexplored (19, 65).

It should be possible to consistently achieve complete recovery after up to 20 min TCA. This would require control of the post-arrest no-reflow phenomenon (4), of the delayed postischemic hypoperfusion (92), and of the protracted multi-organ systems post-resuscitation syndrome (PRS; 65). Special therapies were tested by us and others. Transient moderate hypertension (4, 28, 84), hemodilution (31, 84), heparinization (84, 96), streptokinase (57), barbiturates (12, 33, 59, 100, 110), phenytoin (3, 23), or calcium entry blockers (35, 94, 102, 107, 109) have produced some amelioration of brain damage, but not consistently. Only some of these therapies enjoy documentation by improved long-term neurologic outcome (12, 84, 102). We believe that a wise selection of treatment potentials (Table 1) for systematic studies in animal models will result in consistent and maximally effective therapy results. Before such an approach is possible, more knowledge is required on interactions of the multi-organ systems failure in PRS (65).

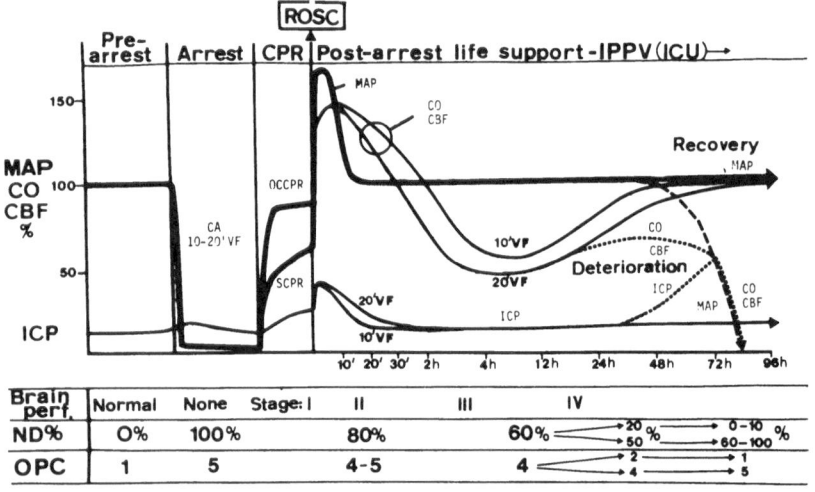

Fig.1: Cardiac arrest of 10-20 min ventricular fibrillation and CPCR in dogs. Diagramatic presentation of cerebral and cardiovascular changes observed by authors' and Negovsky's groups. MAP = mean arterial pressure. CO = cardiac output. CBF = cerebral blood flow. OCCPR = open-chest CPR. SCPR = standard (external) CPR. ROSC = restoration of spontaneous circulation. ICP = intracranial pressure. ND = neurologic deficit (ND score 100% = brain death, 0% = normal). OPC = overall performance categories (#1 = normal; #5 = death).

We recognize the symptoms and signs of the PRS, but do not know its exact mechanisms. The known symptomatology is as follows (Figure 1): After prolonged cardiac arrest, in spite of normotension, a delayed protracted severe reduction in cardiac output, concomitant with the reduction in CBF is common in dogs and humans (19, 48, 65, 91). Evidence suggests a therapy-resistant combination of central cardiogenic and peripheral microcirculatory failure. Pulmonary insufficiency post-arrest seems to be primarily related to a lack of advanced respiratory intensive care, rather than arrest-induced adult respiratory distress syndrome (ARDS; 17, 80). After long-lasting arrests, however, the metabolic functions of lung tissue might be impaired. Conversely, isolated brain ischemia without TCA can trigger dysrhythmias (12,31-35,69), renal problems (32,33) and pulmonary edema and consolidation (12,32,33,69). Acute renal and hepatic failure, while common and investigated after shock (10,21) are not obvious and are relatively unexplored after cardiac arrest (65). Negovsky's and our group (17-19, 65, 80, 102) have acquired only suggestive evidence so far of post-cardiac arrest subtle renal and hepatic derangements, hyper-hypocoagulability (44,65), and self-intoxication by substances released from anoxic tissues. These substances are probably not detoxified by the sick liver, kidneys, and lungs (65). These and other still unknown factors might enhance the cascades of autotoxic reactions, which occur post-TCA in all organ systems, but particularly in the brain. Transient extracerebral organ systems failure post-TCA may have deleterious effects on the brain. This is suggested by the fact that global head ischemia alone (12,31-35,44,45,69) is better tolerated than the same period of TCA (58,80,84,102). Toxic substances may enter the blood from tissues subjected to anoxia. These and endotoxin penetrating through a damaged gastrointestinal wall might reach the brain, if the blood-brain barrier is damaged. The same substances found in shock could be involved, namely kinins, myocardial depressant factor, angiotensin and other peptides, prostaglandins, amino acids, ammonia, and bacterial toxins. Bacteria and white blood cells might also be considered (21). Some of these substances damage endothelial, glial and neuronal cell membranes. Others known to enhance microcirculatory obstruction, may precipitate energy failure of the brain. Possible mechanisms of microcirculatory obstruction include vasospasm, decreased red cell deformability, aggregation of red and white blood cells and of thrombocytes, increased viscosity of blood and plasma, capillary compression by glial, endothelial and interstitial edema, and clotting.

Confusion and controversies have resulted from unclear terminologies and definitions,use of different, often clinically non-relevant animal models, and lack of appreciation of the many pre-arrest-, arrest-, and post-arrest factors, which influence ultimate brain damage. These are species, sex, age, general health and vigor of the animal, pre- and post-ischemic state of nutrition, blood sugar, temperature, hematocrit, perfusion pressure - the immediate post-arrest perfusion pressure pattern is a most important aspect - PaO_2 , $PaCO_2$, pHa, BE, serum osmolality and drugs, anesthesia before, during and after the insult. Equally important are drugs used for life support (vasopressors), complications such as polyuria, electrolyte disturbances, uremia or infections, type, dose and timing of the special post-arrest therapy to be evaluated, and methods and observers used for assessing the outcome. Researchers must try to control as many of these variables as possible. This is easier in animals than in patients. In patients only randomized assignment to a treatment versus control group and large numbers may balance out the many unknown variables.

We and others have used a variety of animal models to identify potentially treatable cerebral and extracerebral derangements post-TCA. We have grouped these derangements into:
(A) Cerebral and overall perfusion-oxygenation problems
(B) Intracellular necrotizing cascades
(C) Malfunctioning neurons, and
(D) Extracerebral organ failure (Table 1, left).

All four occur concurrently. The most important derangements, cerebral and overall perfusion-oxygenation problems, and their effects on brain damage are illustrated schematically in Figure 1. Briefly, in dog experiments (18,19,80,102) TCA by ventricular fibrillation (VF) of 10-20 min was followed by a few minutes of hypoperfusion produced by CPR until successful defibrillation. External CPR could not be counted on (11,55,81) to produce a CBF above the critical threshold of 20 % of normal required for viability of neurons (8). Open-chest CPR, however, was more reliably to produce CBF values above the threshold (93) and to restore spontaneous circulation after prolonged arrest and CPR (11). External CPR raises intrathoracic, venous, and intracranial pressures and thus produces low cerebral and myocardial perfusion pressures. Open-chest CPR produces high perfusion pressures due to low venous pressures (81). After myocardial reoxygenation by CPR, epinephrine and defibrillation, a spontaneous pulse returned usually with 5-10 min hypertension. During this period of cerebral and overall hyperemia, however, some brain areas may not be reperfused at all (4, 68). Then, with mean arterial pressure (MAP) normal, first with fluid load and vasopressors later spontaneously, and in spite of a normal ICP, global CBF and cardiac output decreased gradually reaching 50 % of pre-arrest values at 3-4 h post-arrest (18,19,44,89,92). The cerebral and peripheral delayed hypoperfusion continued for 12-24 h and was reflected in increased arterio-venous O_2 gradients (18). The signs of increased demand/supply relationships for the whole organism lasted over 12 h (19), and for the brain over 24 h (18). After 10 min TCA by ventricular fibrillation, CBF and cardiac output gradually recovered at 12-24 h, while after 15 or 20 min TCA by ventricular fibrillation, CBF and cardiac output deteriorated further toward progressive vasopressor and fluid resistant hypotension in spite of advanced life support (18,19).

Neurologic recovery was quantitated as neurologic deficit scores (ND score 100% = brain death, 40-80% = vegetative state, 0% = normal) (32,69,80). Overall recovery (behavior, disability) was stated as overall performance categories (OPC #1 = normal, #2 = moderate disability, #3 =

severe disability, #4 = coma, vegetative state, #5 = brain death, death) (32,47,80-82). Neurologic deficit and OPC improved at 12-24 h depending on the duration of sedation, paralysis and controlled ventilation. After 10 min ventricular fibrillation and during special therapy, significant improvement occurred at 48-72 h, and maximal improvement by 4-7 d. After longer periods of ventricular fibrillation and during standard life support there was, starting at 24-48 h, secondary deterioration in about 1/2 of animals (18,19,31-35,69,80,102). The mechanisms are still obscure.

Histopathologic damage (HD; 16,30) examined semi-quantitatively in our animals by Moossy et al. (12,31-34,69,102) correlated well with neurologic deficit scores and OPC, both in dogs at 96 h after 10 min ventricular fibrillation and CPR, and in monkeys at 96 or 168 h after 16-20 min head ischemia. Using light microscopy, 14-20 brain regions were scored for ischemic neuronal changes, microinfarcts and edema. Edema was not seen at 3-7 d post-arrest. Ischemic neuronal changes were always seen in the hippocampus, even when OPC was normal. Similar changes occurred in the cerebellum and brain stem but were less common (12,31-34,69,102). Ischemic neuronal changes and microinfarcts in the neocortex, however, were seen only in animals which remained stuporous or comatose. These had neurologic deficit scores over 30% and OPC of #3-5. Similar histologic results were observed in patients after cardiac arrest (103). Monkeys with head ischemia of longer duration (16-18 min) had less neocortical damage than dogs with TCA of only 10 min (12,31-33,69).

Brain-specific cytosolic enzyme activity in the CSF correlated well with OPC, neurologic deficit- and histologic deficit-scores in patients (103) and dogs (102). Studies in rats and dogs indicate loss of cytosolic enzymes during and after ischemia from brain into CSF (50,80,102). CSF brain-specific creatine phosphokinase (CPK-BB) usually peaked at 48 h post ischemia, but earlier after cerebral cold injury.

MODELS AND RESULTS

A systematic search for brain resuscitation potentials after total circulatory arrest requires controlled study mechanisms in animals and patients. We have developed and are using:
(1) An acute rat model for therapy screening
(2) Short-term dog experiments for the study of early pathophysiologic mechanisms
(3) Long-term dog and monkey models for the evaluation of the post-resuscitation syndrome (PRS) and outcome, and
(4) A clinical, multi-institutional international, randomized mechanism for the study of new CPCR methods in cardiac arrest patients - the "ultimate measure" of new therapies.

Experimental designs of animal and patient studies should include randomized controls, and life support by the same team. Blinded administration of special therapy versus placebo should be considered as well as outcome evaluation by the same observer who is not involved in life support and does not know the treatment given. We are trying to control the above mentioned pre-arrest, arrest and post-arrest variables, and have achieved reproducible outcomes with the same protocol. When studying treatment potentials for patients, the animal model including insult and life support should be clinically relevant. The post-resuscitation-syndrome is rather complex, and acute experiments fail to expose the irreversibility of brain damage. Therefore, therapy-focussed and mechanism-focussed studies of PRS should be conducted in long-term animal models with prolonged intensive care. Short-term (< 24 h) recovery studies can give erroneous results, since they miss secondary deterioration. They can, however, be useful for therapy screening. Knowledge of mechanisms will help select therapeutic potentials.

Studies in small animals like mice, rats and gerbils do not permit monitoring and support of vital function. We have favoured dogs and monkeys because of the feasibility to carry out CPR and intensive care as in patients. Dogs have a larger blood volume and larger vessels than monkeys, facilitating blood sampling and monitoring. Contrary to monkeys, dogs usually do not defibrillate spontaneously. The fact that dogs more readily develop gastrointestinal necrosis is a major drawback in shock studies. However, we have not found this to be a serious problem after cardiac arrest. Our post-arrest studies of the brain include chemical changes in acute (<12 h) experiments, cerebral blood flow and metabolism in short-term (< 24 h) experiments, and outcome evaluation as overall performance categories (OPC), neurologic deficit (ND) scores, CSF enzymes, and histopathologic damage (HD) in long-term (3-7 d) experiments. All resuscitation animal studies require continuous control of extracerebral variables. For studies longer than 12 h this is provided in an animal research intensive care unit. To correctly interpret changes in cerebral variables, one must also monitor and control cardiovascular and other vital organ variables.

Early markers of irreversible brain damage are needed. Early EEG recovery time and neurologic test results (1,82,99) correlated fairly well with the severity of the insult (37), but not always with outcome (32,102). Early post-arrest elevation of brain lactate and decrease in energy charge do not reflect the degree of irreversible brain damage (44,65,89). Free fatty acids continue to increase during ischemia while lactate concentrations in brain tissue eventually reach a plateau (66). However, free fatty acids are difficult to measure and have not yet been established as a reliable marker of irreversible brain damage after reperfusion. We found that activities in CSF of cytosolic brain-specific enzymes peak at 48 h post-arrest (26,64,102,103), and that this is correlated better with neurologic deficit and histopathologic damage at 96 h than CSF lactate.

The rat model of asphyxial cardiac arrest (41,42,50,80) includes control of MAP, ventilation and oxygenation. Apnea plus airway obstruction results in asystole within 4 min. Asphyxia times of 5-15 min were studied. Restoration of spontaneous circulation was attempted by i.v. epinephrine and external CPR/O_2 ; it was successful within 60 sec in 60% of the rats. Blood pressure support with norepinephrine and controlled ventilation followed as long as necessary. Long-term survival was not achieved with asphyxia longer than 6 min mainly because of pulmonary failure after extubation. A long-term rat model would be desirable, although it is not suitable for screening of drug effects because long-term life support is too difficult. Asphyxia of 6 min causing arrest of 2 min produced only mild ischemic neuronal changes scattered in the hippocampus. We are using now asphyxia of 10 min, followed by $IPPV/O_2$ and MAP control for therapy screening in two short-term models: (a) Cardiac resuscitability, observation of return of reflexes and EEG activity. After 20-120 min IPPV, decapitation follows for brain freezing and determination of brain lactic acid levels and activity of cytosolic enzymes. Brain cytosolic enzymes, particularly CPK-BB was found decreased at 20 min post-anoxia (50). Brain lysosomal enzymes might increase post-anoxia, but not constantly. (b) Cardiac resuscitability, observation of return of reflexes and EEG activity, and after continuance of $IPPV/O_2$ for 2 h, crude neurological recovery during an attempt to wean from IPPV.

In model (a) we found a significantly beneficial effect of methylprednisolone (30 mg/kg i.v.) given immediately post-reflow, but not when given pre-arrest (50). Because of the clinical relevance, we have selected this asphyxial cardiac arrest model over the Levine model (56), the Pulsinelli fore-brain ischemia model (73), and global ischemia induced by raising ICP (14,54,89). The new model of Siesjö of incomplete global fore-brain ischemia consisting of temporary hemorrhagic hypotension plus bilateral carotid artery clamping looks promising even for long-term recovery.

For the first time a <u>monkey model</u> was developed for long-term intensive care (3-7d) and outcome to study global brain ischemia without total circulatory arrest, and to evaluate treatment potentials without problems of cardiac resuscitability (12,31-35,69). Ventilation is maintained during ischemia via a noncompressible tracheal tube. The high pressure neck tourniquet occludes all blood flow into the cranium, even through the vertebral arteries (32,69). Brain ischemia triggered in some animals dysrhythmias, arterial pressure changes, pulmonary dysfunction and polyuria. Continuous long-term life support in animals proved difficult and expensive, but nevertheless feasible. Standard therapy resulted in survival with brain damage. Details of life support influenced outcome (80,82). Global head ischemia of 16-18 min in most animals resulted in a vegetative state at 3-7 d post-ischemia. However, in most standard treatment groups of 10 monkeys, one or two developed secondary brain death, while one or two recovered almost completely. The reperfusion pressure pattern proved crucial. We have tested about ten treatment potentials with this model (34,82). <u>Barbiturate</u> studies in rhesus monkeys looked promising (12) and sparked interest in brain resuscitation research by others (3,23,35,44,57,94,100,106). Better controlled experiments in pigtail monkeys on barbiturates led to the same outcome as standard thera-

py (33). The model is very expensive, variable in outcome, and does not
simulate a clinical insult. Therefore, we will reserve it in the future
for special indications. We have not chosen clamping of intrathoracic
head-arteries (44,58,62) or laminectomy plus neck tourniquet in dogs -
the species has many intra-osseous arterial connections to the brain (58)
- because the surgical trauma may complicate outcome evaluation.

Dog cardiac arrest (TCA) models. In the 1950s and 1960s Safar and
associates developed clinically relevant short-term models of drowning
(75), cardiac arrest by exsanguination (51), normovolemic hemodilution
(98), asphyxia (75,76) and ventricular fibrillation in dogs (77). These
were recently extended to long-term life support models of ventricular
fibrillation (80,84,102) and asphyxial asystole (80). Electrically induced ventricular fibrillation in the dog has been used for animal resuscitation research since the turn of the century (95). Ventricular fibrillation in dogs with a 10 min arrest time and standard post-arrest intensive
care (80,102) resulted in a more reproducible outcome than the neck tourniquet in monkeys (32,69). OPCs at 96 h post-arrest were #2-4. None developed brain death and none recovered completely (102). Restoration of
spontaneous circulation with standard external CPR succeeded within 5 min
in 80% of the dogs, and with life support resulted in 96 h survival with
brain damage. Ventricular fibrillation arrest of 15 and 20 min had higher
failure rates with external CPR. Experimental ventricular fibrillation of
over 10-12 min should best be reversed by open-chest CPR or cardiopulmonary bypass (13,17). Asphyxial arrest in dogs with an asphyxiation
time of 7 min followed by an arrest time of 7 min using the same life
support, resulted in easier cardiac resuscitability, but worse cerebral
outcome than 10 min ventricular fibrillation (80).

During 1982-84 several treatment potentials were studied using
these cardiac arrest dog models. The clinically most relevant results
were obtained in a 10 min ventricular fibrillation study with 96 h standard life support (n=11) versus the same life support plus i.v. administration of the calcium entry blocker lidoflazine (1 mg/kg immediately
post-arrest and at 8 and 16 h; n=11). 5 of 11 lidoflazine treated dogs
recovered to OPC#1, none developed OPC#4-5, while none of the 11
control dogs reached OPC#1, and 3 of 11 developed OPC#4 ($p < 0.05$).
Neurological deficit scores were also significantly better after lidoflazine. Histological damage scores and CSF enzymes correlated with ND
and OPC. Secondary deterioration after 24h occurred in 6 of 11 controls
while in none of 11 lidoflazine treated dogs. Treatment with lidoflazine
after 7 min asphyxial cardiac arrest in dogs did not result in neurological improvement (102). Asphyxia-induced TCA is a different insult with
low-flow and greater hypoxia and acidosis before no-flow. Significant but
also inconsistent neurologic improvement by post-ischemic calcium blocker
therapy was found in three other studies (35,94,109).

A ventricular fibrillation model would be most relevant in which
the insult is not caused by electric shock but by transient coronary
ischemia (70). This, however, requires major surgery. We have not chosen

cats for ventricular fibrillation, since they tend more easily to develop post-anoxic seizures than man (101). Monkeys and cats also defibrillate spontaneously and require continued stimulation through a cardiac lead. We have not chosen ascending aorta occlusion via thoracotomy (58, 92,94) because of the surgical trauma. Neither did we study venae cavae and ascending aorta balloon occlusion (46) because of an uncertain quality of ischemia and a lack in clinical relevance. Alveolar hypoxia (110), or potassium-induced arrest (55, 107, 109) were also found inappropriate as it is either clinically seldom the cause of cardiac arrest, is not clinically relevant, and reverses spontaneously.

Randomized controlled patient studies are the final tests for expensive or risky therapies. They require large numbers of patients and are tedious and expensive. We have developed the first cardiac arrest brain resuscitation clinical trial (BRCT) study mechanism in 12 hospitals of 7 countries. The first phase (1979-83) proved that thiopental loading after cardiac arrest is feasible and safe in skilled hands, but cannot be counted on to improve cerebral outcome (1,2). The data suggest that only in a subgroup with long arrest times thiopental might have been of borderline benefit. Standard therapy by protocol without thiopental gave better results than usual post-arrest therapy reported by others. This patient study and many laboratory studies indicate that treatment with barbiturates is not a breakthrough, as was initially hoped. We have begun the second phase of BRCT, which is designed to test the brain damage ameliorating effect of post-arrest calcium-entry antagonist (lidoflazine).

FUTURE EVALUATIONS OF TREATMENT POTENTIALS

Brain resuscitation potentials should satisfy the following criteria before they are clinically tested:
(a) The chemical structure of drugs and their pharmacological effects, also on extracerebral organ systems, should be known.
(b) The distribution, half-life and elimination should be known, taking into account that this may be different in different species and vary between health and disease.
(c) The method(s) should not exert uncontrollable negative side effects on vital parameters.
(d) They should be tested in animal brain ischemia models and given post-insult, starting with the reperfusion period. Direct influence on the brain may be different when given post- versus pre-arrest, since the blood-brain barrier may be more permeable post-arrest.
(e) The insult should be of moderate degree, e.g. 10 min ventricular fibrillation or 7 min asphyxial cardiac arrest. With standard therapy, the animals should survive at least 96 h with brain damage, and if the special therapy is effective, recover completely.
(f) The treatment should have a beneficial effect in a short-term screening model of a small or large species such as dog or monkey with short-term (24-48 h) intensive care.

(g) Finally, promising treatment should be evaluated by long-term (< 96h) experiments with prolonged life support. This is to observe whether the therapy prevents secondary deterioration, which may occur after 24h, and to detect possible long-term side effects on other vital organ systems.
(h) Pharmacologic or physical agents, which individually appear promising should be tested as "tailored" combination therapies.

Secondary derangements in the brain and other organs which have been demonstrated or are expected to occur after ischemic-anoxic insults, are listed in Table 1. Treatment potentials for evaluation of their efficacy after cardiac arrest should be tailored to the most important pathophysiologic processes (s. Table 1 (A) to (D)).

(A) Perfusion Oxygenation Problems

All four stages of hypoperfusion may be therapeutically influenced.

Stage I the multifocal no-reflow phenomenon, has been acutely ameliorated by hypertension (4,28,68,84). It might be overcome with carotid flush plus hypertension (63,84). Standard external CPR does not provide a sufficient perfusion pressure to overcome the no-reflow phenomenon after arrest times longer than 3-5 min (11,55). Trials with open-chest CPR (11, 93) and cardiopulmonary bypass (CPB) are indicated (13,17). CPB permits control of perfusion pressure, flow, blood composition and temperature, irrespective of concomitantly used cardiac depressant drugs. Acute animal studies on reperfusion, e.g. by in-vivo microscopy, should map out the best reperfusion pattern to overcome the primary microvascular obstruction. Against sludging and increased viscosity we suggest prolonged carotid flushing with oxygen carriers, such as stroma-free hemoglobin (10,25) or fluorocarbons (90). Blood cell disaggregation may be promoted by hemodilution (84,98), aspirin and indomethacin (38). In prolonged arrest when clotting may already have occurred, streptokinase should be tried (57). Heparin would prevent secondary clotting (84,96). The use of osmotherapy early post-arrest should be evaluated (104). The brain edema sequence occurring post-global ischemia and its relevance, needs study and documentation (9,44,52). The tissue hypoxia due to microvascular obstruction may benefit from enhancing PaO_2 (by FIO_2 of 1.0 and PEEP) or even from hyperbaric oxygenation. The latter is clinically usually not feasible. One may also try ATP containing solutions (21,53). Tissue acidosis (78) may be reversed by hyperventilation and THAM. Lipid peroxidation (86,89) might be prevented by a calcium-entry antagonist (102,106), free radical scavengers, or antioxidants (7,24), such as vitamin C and E, superoxide dismutase (SOD), catalase, mannitol and the xanthine oxidase inhibitor allopurinol (22).

Stage II the transient hyperemia (vasodilation), is not understood as to its possible deleterious effects on outcome. Hyperventilation and vasoconstricting drugs may be helpful (81) No-reflow therapy should continue.

Stage III the delayed and protracted cerebral and peripheral hypoperfusion represents the greatest challenge and potential for brain resuscitation trials. Neither global nor regional CBF measurements (97) may reflect fully the pathogenetic mechanisms involved. Initial moderate hypertension may be prolonged by a vasopressor and its effects augmented by hyperpulsatility (JW Severinghaus, personal communication). The same methods listed for Stage I may be used against sludging, cell aggregation, clotting, edema and vasospasm. In Stage III, there may be secondary disseminated intravascular coagulation. One might consider also defibrination by ancrod (87). DIC may be enhanced by disturbances of proteolytic enzyme systems, which may be influenced by a kallikrein inhibitor such as trasylol (6). Endothelial damage calls for a trial of steroids. In spite of the fact that steroids have profound protective effects on cell membranes and the intracellular milieu, clinical trials of steroids after various brain insults have given mixed results. Nevertheless, our positive results with post-arrest methylprednisolone in rats should be confirmed (50).

Stage IV represents an evolution from Stage III. One is progressive recovery. Another is partial recovery with secondary deterioration or progressive deterioration due to irreversible microvascular complications, leading to multi-organ systems failure in spite of advanced therapeutic efforts. In trying to reverse the irreversible stage, one should continue all the above treatments, treatments (B), (C) and (D) below, plus ICP control. Secondary deterioration may be accompanied by intracranial hypertension (18,32,69,102). Systematic steps to reduce intracranial hypertension have been worked out for other brain insults (39, 61, 54, 59, 82, 83). It is not clear whether edema, vasodilation and ICP rise are cause or result of the intracellular necrotizing processes listed below.

(B) Cell Necrotizing Cascades

The intracellular events post-reflow, which occur simultaneously with and are enhanced by the previously discussed perfusion problems, consist of metabolic derangements and cell swelling. These lead to cell death and to interstitial edema of the whole brain in addition to cell swelling. Treatment potentials would start with trying to put the injured brain at rest, i.e. combating hypermetabolism and seizures. Barbiturates have been shown to exert adjunctive beneficial effects but are no breakthrough (2,12,15,33,59,66,89,100,110). Their evaluation has been carried through the entire chain of systematic studies to controlled clinical trials (1,2,15). Other CNS depressants listed still need controlled animal studies. They include diazepam, phenytoin (3,23) and GABA (67). In spite of the clear-cut protective effects of hypothermia when induced pre-ischemia, useful studies with optimized post-arrest hypothermia have not been conducted yet. There are suggestions to re-evaluate immediate post-arrest hypothermia (31,79,108). Short-term hypothermia should be combined with hemodilution to counteract its increased blood viscosity. Inhibition of glucose metabolism such as deoxyglucose (85), or administration of glucose-ATP (53) have been suggested.

Mitochondrial energy failure leads to ion pump failure with loss of potassium into the extracellular fluid. Moderate increases of potassium cause vasodilation, while high concentrations vasospasm. This and the influx of calcium and sodium into cells may be ameliorated by phenytoin and local anesthetics (8), in addition to reoxygenation and re-establishment of the cell energy charge. Enhanced post-ischemic loading of mitochondria with calcium may be reduced by calcium-entry blockers and magnesium (106). This effect, however, has not yet been documented (44). Trials of calcium-entry antagonists post-TCA in animals (35,94,102,109) appear promising, although these compounds do not prevent histologic damage (35,102). As in the case of barbiturates, Ca^{++}-entry antagonists will be shown to exert benefits without representing a breakthrough.

Free iron is suspected to trigger free radical generation, which may enhance tissue destruction. A trial of an iron chelator such as desferrioxamine seems justified (40). A free radical scavenger solution significantly improved cerebral outcome after asphyxial cardiac arrest (103a). Membrane lysis is reflected by increased free fatty acid concentrations, particularly of arachidonic acid (66). Studies of free fatty acids after reperfusion with and without membrane stabilizers would be of interest. Leakage of lysosomal enzymes may be beneficially influenced by steroids (50). The cell-necrotizing cascades result in a release of prostaglandins and thromboxanes which may have deleterious vascular effects, such as cell aggregation and vasospasm. This might be ameliorated with prostacyclin (38). Secondary neurologic deterioration may also be attributed to aggregation and endothelial adhesion of white blood cells. However, this has not been documented by histologic studies. Finally, leukotriene antidotes might be considered (89,106).

(C) Malfunctioning Neurons

Neurotransmitter derangements during and following ischemic anoxia have not been studied systematically (67). Whether or not neurotransmitter derangements play a role in the process toward cell death has not been clarified. Modification of the functional balance between dopamine and GABA, or intermediates might restore normal behavior. Abnormal behavior post-ischemia may occur even in the absence of histologic or neurologic deficits. The role of brain endorphins is equally speculative (43).Yet, patients with brain- or spinal cord-injury or shock showed - at least transient - functional recovery upon administration of naloxone. Again, no data are available that opiate antagonists beneficially influence irreversible neuronal damage. Could these agents merely act as analeptics?

(D) Extracerebral Organ Failure

Deleterious effects of cardiovascular-pulmonary, hematologic, renal and hepato-gastrointestinal derangements on the postischemic brain have been partly established and partly suggested (19, 48, 65, 91, 80-84). 1-2 hours after prolonged total circulatory arrest cardiac output becomes markedly reduced, which parallels the reduction in CBF (Figure 1; 19, 102). Cardiac failure should be corrected by appropriate drugs, and perhaps by assisted circulation. Peripheral vascular failure can be influenced by drugs, fluids, and a transient use of anti-shock trousers (49). For both combined, cardiopulmonary bypass (CPB) should be considered (13,17). CPB is a promising new technology for emergency resuscitation and also for assisted circulation post-TCA. Hyperpulsatile moderate hypertension may help all organs. Excessive and prolonged use of vasoconstrictors worsens the process (19,102). Pulmonary failure can usually be prevented by advanced respiratory care. Established ARDS is rare post-arrest if advanced respiratory therapy is used. It may call for steroid treatment. Most exciting is the idea of controlled trials of blood detoxifying measures, such as cross circulation (65), blood and plasma exchange (36,65) to combat self-intoxication, which is suspected to originate from all tissues, and is enhanced by post-anoxic renal and hepatic failure. There is suggestive evidence that gastrointestinal wall leakage seen in shock (10,21) is also present albeit mildly after cardiac arrest (19,41,65). Endotoxin or bacteria entering the bloodstream might further hurt an already injured brain (27). The hepato-gastrointestinal failure may lead to an increase of aromatic (i.e. toxic) over branched-chain (i.e. beneficial) amino acids. This may enhance cerebral damage if the blood-brain barrier permeability is increased. Therapy with branched-chain amino acids has been suggested (29). The endotoxemia and bacteremia, secondary to an enhanced intestinal permeability and liver failure, would call for antibiotics (27). A hepatic RES failure may be influenced by cryoprecipitate to increase fibronectin necessary to restore RES function (21).

CONCLUSIONS AND RECOMMENDATIONS

Since 1970, several research centers have evolved worldwide, which concern themselves primarily with global brain ischemia or total circulatory arrest (TCA). Much has been learned about the mechanisms following these insults. Although this has helped to improve somewhat the efficacy of therapies, a breakthrough has not been reached so far. Post-TCA brain oriented intensive care life support of extracerebral organs "by protocol" increases cerebral and general recovery in animals and patients, as compared to "usual care". Post-TCA special treatments tested semi-empirically in addition to standard intensive care have so far been of only inconsistent benefit. These are intracarotid hemodilution plus moderate transient hypertension, barbiturates, calcium-entry antagonists, seizure prophylaxis, free radical scavangers, and multifaceted therapy protocols. No beneficial cerebral effect post-TCA was produced by i.v. heparinization or hemodilution alone, or repetitive and prolonged severe hypertension.

In order to optimize clinical brain resuscitation, one must learn more about the multifactorial nature of the secondary cerebral changes. This should include effects of secondary extracerebral organ systems failure on the brain. Long-term animal models are required with post-TCA intensive care and outcome evaluation in terms of irreversible damage (non-viability). We recommend to study the mechanisms of the multiorganic PRS for determining the limits and potentials of CPCR for cardiac arrest. Then, etiology-specific treatment potentials should be evaluated. These should first be screened, then pilot-tested, and then confirmed by long-term animal outcome studies. Tests of single treatments should be followed by "tailored" combination therapies. Expensive and hazardous treatment protocols should be finally confirmed in a randomized controlled clinical study. Simultaneously with systematic mechanisms-oriented studies of treatment potentials, empirical trials seem justified, to enhance the likelihood of achieving a breakthrough. History has shown that knowledge of the mechanisms of the disease process and of treatment is not essential for therapeutic breakthroughs. Anesthesia, antibiotics, cancer treatments, steroids, insulin may serve as examples.

There should be joint planning and communication for individual planning between research groups working on TCA, shock, focal brain ischemia, brain trauma and neurosurgery-neuroanesthesia. We have initiated such communication and joint planning of multi-disciplinary, multi-center laboratory and patient studies. The former include a laboratory program to study the PRS in a dog TCA outcome model over the next five years. The results shall lead to the development and evaluation of titrated combination therapies. The most promising and clinically feasible protocol will be channeled to our multi-center institutional Brain Resuscitation Clinical Trial (BRCT) of cardiac arrest patients. Other TCA-related studies of the Resuscitation Research Center in Pittsburgh concern themselves with CPR techniques and with hemorrhagic-traumatic shock. The systematic study mechanism developed for TCA permits not only quantification of resuscitation effects, but also identification of early measurements with which one can reliably predict outcome with severe brain damage. We hope that systematic plus empirical studies will lead to a therapeutic breakthrough for secondary brain damage after cardiac arrest. If not, they will establish the potentials and limits of CPCR.

ACKNOWLEDGEMENTS

Research support of the authors has been from the National Institutes of Health (USA), the Pennsylvania State Department of Health, the Asmund Laerdal Foundation, the Pittsburgh Foundation and the Pharmacia Laboratories. Mrs. Gale Foster, Ms. Tancy Crawford and Ms. Fran Mistrick helped with the preparation of the manuscript.

TABLE 1

Post-Total Circulatory (Cardiac) Arrest

Multi-Organ Derangements of Post-Resuscitation Syndrome (PRS) and Etiology-Specific Treatment Potentials (suggested for systematic studies)

Derangements (partially proven)	Treatment Potentials (s. Text)

A. PERFUSION-OXYGENATION PROBLEMS
Brain > extracerebral organs

Stage I: Reperfusion-no reflow
(immediate)

CPR (minimal flow)	Optimal CPR, OCCPR, CPB
Post-CPR hypotension-hypoxemia	Brief hypertension, IPPV/O$_2$
Primary (arrest-induced) microvascular failure	
sludging, incr. viscosity	Carotid flush, hemodilution (electrol., colloid, SFH, FC)
cell aggr.(RBC, WBC, Plat.)	Hemodilution, aspirin, indomethacin, SOD
clotting	Streptokinase (treatment), Heparin (prevention)
edema	Hyperosmolality
tissue hypoxia	Hyperoxia, ATP
tissue acidosis	Hyperventilation, THAM
Beginning lipid peroxidation → membrane breakdown	Ca^{++} blockers Free rad. scavengers, antioxidants (Vit. C, E; SOD; allopurinol, mannitol)
Stage II: Hyperemia vasoparalysis (5-30) min	Hyperventilation (cerebral) Vasoconstrictor drugs (extracerebral)

(cont'd)

TABLE 1
(cont'd)

Post-Total Circulatory (Cardiac) Arrest

Multi-Organ Derangements of Post-Resuscitation Syndrome (PRS) and Etiology-Specific Treatment Potentials (suggested for systematic studies)

Derangements (partially proven)	Treatment Potentials (s. Text)

A. PERFUSION-OXYGENATION PROBLEMS (cont'd)
  Brain > extracerebral organs

Stage III: Delayed prolonged hypoperfusion
            (30 min - 12 h or longer)

Secondary (PRS-induced) micro-vascular failure	
sludging, cell aggregation clotting, edema	Same as in Stage I
vasospasm	Ca^{++} blockers, indomethacin a.o. vasodilators
secondary clotting (DIC)	Heparin (prevention) → Ancrod (defibrination)
multifactorial occlusion	Hyperpulsatile hypertension
increased vasoactive kinins	Kallikrein inhibitors
endothelial damage	Steroids

Stage IV: (12-48 h →)

(a) progressive recovery	Continue above
(b) partial recovery → deterioration	Continue above, ICP control, try (B, D) cf. below
(c) progressive deterioration irreversible microvasc. damage → brain death	Continue above, ICP control, try (B, D) cf. below

TABLE 1
(cont'd)

Post-Total Circulatory (Cardiac) Arrest

Multi-Organ Derangements of Post-Resuscitation Syndrome (PRS) and Etiology-Specific Treatment Potentials (suggested for systematic studies)

Derangements (partially proven)	Treatment Potentials (s. Text)

B. CELL NECROTIZING CASCADES
(simult. with A) (inhomogenous, multifocal)

(1) Metabolic dysfunction, acidosis

(a) Hypermetabolism, seizures	Barb.-diazepam-phenytoin-GABA hypothermia (+ hemodilution)
(b) Increased tissue glucose	deoxyglucose
(c) Decreased O_2 ,glucose, ATP	Glucose-ATP-Insulin-K^+ ; hyperbaric O_2
(d) Mitochondrial energy failure	
→ ion pump failure, K^+ efflux	Phenytoin, local anesthetics
→ Intracellular calcium load of mitochondria	Ca^{++} blocker, $MgSO_4$,$MgCl_2$
→ free radical generation	Free rad. scavengers (Vit. E,C;SOD; catalase, allopurinol, mannitol)
triggered by Fe^{++}	Fe^{++} chelators (desferrioxamine)
→ free fatty acid generation	(Result or cause of membrane breakdown?)
→ lysosomal enzyme leakage	Steroids
(e) Prostaglandins (thromboxane) release	Prostacyclin
(f) Leucotrienes release (from WBCs)	Anti-leucotrienes

(2) Intracellular edema (ion pump failure, incr. osmolality)	Osmotherapy
	Phenytoin, local anesthetics
(3) Cell pyknosis (cell death)	
(4) Interstitial edema	ICP control, steroids, hypothermia, hyperosmolality

(cont'd)

TABLE 1
(cont'd)

Post-Total Circulatory (Cardiac) Arrest

Multi-Organ Derangements of Post-Resuscitation Syndrome (PRS) and Etiology-Specific Treatment Potentials (suggested for systematic studies)

Derangements (partially proven)	Treatment Potentials (s. Text)

C. MALFUNCTIONING NEURONS
(simult. with A)

Neurotransmitter failure	Neurotransmitters (e.g.dopamine,GABA)
Endogenous opiates increased	Naloxon

D. EXTRACEREBRAL ORGAN FAILURE
(simult. with A)
(may augment brain failure)

(1) Cardiac failure	Drugs, assisted circulation, CPB
Peripheral vasc. fail.	Drugs, fluids, hyperpulsatility, MAST
(2) Pulmonary failure	IPPV, PEEP, FIO_2 control, steroids
(3) Hematologic derangements	Cross circulation
(4) Hepato-renal failure	Blood exchange, plasma exchange
Self-intoxication	Hemodialysis, hemofiltration
(5) Gastro-intestinal failure	
Endotoxemia, bacteremia	Antibiotics
Aromatic amino acids incr.	Branched-chain amino acids
(6) RES-failure	RES promoting fibronectin (cryoprecipitate)

OCCPR:	open-chest cardiopulmonary resuscitation	SOD:	superoxide dismutase
CPB:	cardiopulmonary bypass	ATP:	adenosine triphosphate
SFH:	stroma free hemoglobin	MAST:	med. antishock trousers
FC:	fluorocarbons	ICP:	intracranial pressure
RBC:	red blood cells	GABA:	gamma amino butyric acid
WBC:	white blood cells	Fe++:	free iron
DIC:	disseminated intravascular coagulation	RES:	reticuloendothelial system

REFERENCES

1. Abramson NS, Safar P, Detre K, et al, An international collaborative clinical study mechanism for resuscitation research, Resuscitation 10:141 (1982).
2. Abramson NS, Safar P, Detre K, et al, Results of a randomized clinical trial of brain resuscitation with thiopental, Anesthesiology 53:A101 (1983).
3. Aldrete JA, Romo-Salas F, Jankowsky L, et al, Effect of pretreatment with thiopental and phenytoin on postischemic brain damage in rabbits, Crit Care Med 7:466 (1979).
4. Ames A, Wright RL, Kowada M, et al, Cerebral ischemia. II. The no-reflow phenomenon, Am J Pathol 52:437 (1968).
5. Ames A, Earliest irreversible changes during ischemia, Am J Emerg Med 1:139 (1983).
6. Amundsen E, Aspects of plasma kinin physiology, Scand J Clin Lab Invest 107:65 (1969).
7. Arfors KE, Hillered L, Oxygen radicals and biological injury, in: "Protection of tissue against hypoxia", Wauquier A, et al., eds., Elsevier, Amsterdam (1982).
8. Astrup J, Symon L, Branston NM, Lassen NA, Cortical evoked potential and extracellular K^+ and H^+ at critical levels of brain ischemia, Stroke 8:51 (1977).
9. Baethmann A, Oettinger W, Rothenfusser W, Geiger R, Biochemical aspects of cerebral edema, in: "Pathophysiology of cerebral energy metabolism", Mrsulja BB, Rakic LM, Klatzo I, Spatz M, eds., Plenum Press, New York (1979).
10a. Bar-Joseph G, Safar P, Stezoski WS, et al, A realistic hemorrhagic shock model in the monkey, Circ Shock 8:206 (1981; Abstract).
10b. Bar-Joseph G, Safar P, Stezoski WS, et al, Survival after severe hemorrhagic shock in monkeys: CNS, pulmonary, hepatic and renal outcome, Anesthesiology 57:A97 (1982; Abstract).
10c. Bar-Joseph G, Safar P, Stezoski WS, et al, Stroma-free hemoglobin, hydroxyethyl starch, lactated Ringer's, or blood, for severe prolonged hemorrhagic shock in monkeys, Crit Care Med 11:219 (1983; Abstract).
11. Bircher NG, Safar P, Cerebral preservation during cardiopulmonary resuscitation in dogs, Crit Care Med 13:185 (1985).
12. Bleyaert AL, Nemoto EM, Safar P, et al, Thiopental amelioration of brain damage after global ischemia in monkeys, Anesthesiology 49:390 (1978).
13. Bozhiev AA, Tolova SV, Trubina IE, Peculiar features of resuscitation with the use of extracorporeal circulation, Kardiologiia 14:101 (1974).
14. Bradford K, Experimental increase in intracranial pressure, Diseases Nervous System 25:46 (1964).
15. Breivik H, Safar P, Sands P, et al, Clinical feasibility trials of barbiturate therapy after cardiac arrest, Crit Care Med 6:228 (1978).

16. Brierley JB, Meldrum BS, Brown AW, The threshold and neuropathology of cerebral anoxic-ischemic cell change, Arch Neurol 29:367 (1973).

17. Cantadore R, Vaagenes P, Safar P, Stezoski W, Cardiopulmonary bypass for resuscitation after prolonged cardiac arrest in dogs, Ann Emerg Med 13:398 (1984; Abstract).

18. Cantadore R, Vaagenes P, Safar P, Oxygen utilization (cerebral arterio-venous oxygen gradients) after prolonged cardiac arrest in dogs, Circ Shock 13:69 (1984; Abstract).

19. Cantadore R, Vaagenes P, Safar P, Alexander H, Prolonged cardio-vascular failure after cardiac arrest and cardiopulmonary re-suscitation in dogs, Circ Shock 13:70 (1984; Abstract).

20. Cole S, Corday E, 4-minute limit for cardiac resuscitation, JAMA 161:1454 (1956).

21. Cowley RA, Trump BF, eds., "Pathophysiology of shock, anoxia, and ischemia", Williams & Wilkins, Baltimore (1982).

22. Crowell JW, Jones CE, Smith EE, Effects of allopurinol on hemorrhagic shock, Am J Physiol 16:744 (1969).

23. Cullen JF, Aldrete JA, Jankowsky L, et al, Protective action of phe-nytoin in cerebral ischemia hypoxia, Anesth Analg 58:165 (1979).

24. Del Maestro RF, An approach to free radicals in medicine and biology, Acta Physiol Scand 492:153 (1980).

25. DeVenuto F Acellular oxygen-delivering resuscitation fluids, Crit Care Med 10:237 (1982).

26. Edgren E, Terent H, Hedstand U, et al, Cerebral spinal fluid markers in relation to outcome in patients with global cerebral ische-mia, Crit Care Med 11:4 (1983).

27. Ekstrom-Jodal P, Haggendal E, Larsson LE, Cerebral blood flow and oxygen uptake in endotoxin shock. An experimental study in dogs, Acta Anaesth Scand 26:163 (1982).

28. Fisher E, Ames A, Studies on mechanisms of impairment of cerebral circulation following ischemia: effect of hemodilution and per-fusion pressure, Stroke 3:538 (1973).

29. Freund H, Hoover HC, Atami S, Fischer JD, Infusion of branched chain amino acids in post-operative patients, Ann Surg 190:18 (1979).

30a. Garcia JH, Kalimo H, Kamijyo Y, et al, Cellular events during par-tial cerebral ischemia. Part I: Electron microscopy of feline cerebral cortex after middle-cerebral artery occlusion, Virchows Archiv B Cell Pathology 25:191 (1977).

30b. Garcia JH, Kalimo H, Kamijyo Y, et al, The ultrastructure of "brain death", Virchows Archiv B Cell Pathology 25:207 (1977).

31. Gisvold SE, Safar P, Rao G, et al, Multifaceted therapy after global brain ischemia in pigtail monkeys, Stroke 15:803 (1984).

32. Gisvold SE, Safar P, Rao G, et al, Prolonged immobilization and controlled ventilation after global brain ischemia in monkeys, Crit Care Med 12:171 (1984).

33. Gisvold SE, Safar P, Hendrickx H, et al, Thiopental treatment after global brain ischemia in pigtail monkeys, Anesthesiology 60:88 (1984).

34. Gisvold SE, Safar P, Systematic studies of cerebral resuscitation po-
    tentials after global brain ischemia, Crit Care Med 10:466
    (1982).
35. Gisvold SE, Steen PA, et al, Nimodipine improves neurologic recovery
    after global head ischemia in monkeys, Personal communication
    (see Steen PA et al, Anesthesiology 62:406 (1985)).
36. Gurland JH, Heinze V, Lee HA, eds., "Therapeutic plasma exchange",
    Springer, New York (1981).
37. Gurvitch AM, Determination of the depth and reversibility of post-
    anoxic coma in animals, Resuscitation 3:1 (1974).
38. Hallenbeck JM, Leitch DR, Dutka AJ, et al, Prostaglandin I_2 , indo-
    methacin, and heparin promote postischemic neuronal recovery
    in dogs, Ann Neurol 12:145 (1982).
39. Hayes RL, Pechura CM, Becker DP, A head injury animal model:
    physiological studies of mechanical brain injury in the cat,
    J World Assoc Emerg Disaster Med 1:Suppl. (1985).
40. Hedenberg L, Studies on iron metabolism with desferrioxamine in man,
    Scand J Haematol 6, (1969).
41. Hendrickx H, Rao GR, Safar P, et al, Asphyxia, cardiac arrest and re-
    suscitation in rats. I. Short-term recovery, Resuscitation 12:97
    (1984).
42. Hendrickx H, Asphyxia, cardiac arrest and resuscitation in rats. II.
    Long-term behavioural changes. Resuscitation 12:117 (1984).
43. Hosobuchi Y, Baskin DS, Woo SK, Reversal of postischemic neurologic
    deficit in gerbils by the opiate antagonist naloxone, Science
    215:69 (1982).
44. Hossmann KA, Treatment of experimental cerebral ischemia,
    J Cereb Blood Flow Metabol 2:275 (1982).
45. Hossmann KA, Kleihues P, Reversibility of ischemic brain damage,
    Arch Neurol 29:375 (1973).
46. Jackson DL, Dole WP, McGloin J, et al, Total cerebral ischemia:
    Application of a new model system to studies of cerebral micro-
    circulation, Stroke 12:66 (1981).
47. Jennett B, Bond M, Assessment of outcome after severe brain damage:
    a practical scale, Lancet 1:480 (1975).
48. Kampschulte S, Smith J, Safar P, Oxygen transport after cardiopulmo-
    nary resuscitation, Anaesthesiology and Resuscitation 30:95
    (1969).
49. Kaplan BC, Civetta JM, Nagel EL, et al, The military anti-shock trou-
    ser in civilian pre-hospital emergency care, J Trauma 13:843
    (1973).
50. Katz L, Vaagenes P, Safar P, et al, Protective and resuscitative
    properties of methylprednisolone in asphyxial cardiac arrest in
    rats, Anesthesiology (1985, in press).
51. Kirimli B, Kampschulte S, Safar P, Cardiac arrest from exsanguination
    in dogs. Evaluation of resuscitation methods, Acta Anaesth Scand
    (suppl.) 39:183 (1969).
52. Klatzo I, Neuropathological aspects of brain edema, J Neuropathol
    Exper Neurol 26:1 (1967).

53. Kraven T, Rush BF, Gosh A, et al, Correlation of survival and metabolic response produced by ATP-MgCl$_2$ in hemorrhagic shock, Circ Shock 6:186 (1979).
54. Langfitt TW, Tannenbaum HM, Cassell NF, et al, Acute intracranial hypertension, cerebral blood flow and EEG, Electroenceph Clin Neurophysiol 20: 139 (1960).
55. Lee SK, Vaagenes P, Safar P, et al, Effect of cardiac arrest time on the cortical cerebral blood flow generated by subsequent standard external cardiopulmonary resuscitation in rabbits Anesthesiology (1985, in press).
56. Levine S, Anoxic-ischemic encephalopathy in rats, Am J Pathol 36:1 (1960).
57. Lin SR, O'Connor MJ, Fischer HW, et al, The effect of combined dextran and streptokinase on cerebral function and blood flow after cardiac arrest: an experimental study in the dog, Invest Radiol 13:490 (1978).
58. Lind B, Snyder J, Kampschulte S, Safar P, A review of total brain ischemia models in dogs and original experiments on clamping the aorta, Resuscitation 4:19 (1975).
59. Marshall LF, Shapiro HM, Rauscher A, et al, Pentobarbital therapy for intracranial hypertension in metabolic coma: Reye's syndrome, Crit Care Med 1:293 (1973).
60. Michenfelder JK, Theye RA, The effects of anesthesia and hypothermia on canine cerebral ATP and lactate during anoxia produced by decapitation, Anesthesiology 33:430 (1970).
61. Miller JD, Barbiturate and raised intracranial pressure, Ann Neurol 6:189 (1979).
62. Miller JR, Myers RE, Neuropathology of systemic circulatory arrest in adult monkeys, Neurology 22:888 (1972).
63. Miyake T, Kinoshita K, Ishii N, et al, First report of an experimental study in dogs of cerebral cardiopulmonary resuscitation (CCPR), Resuscitation 10:105 (1982).
64. Mullie A, Lust P, Penninckx J, et al, Monitoring of cerebrospinal fluid enzyme levels in post-ischemic encephalopathy after cardiac arrest, Crit Care Med 5:399 (1981).
65. Negovsky VA, Gurvitch AM, Zolotokrylina ES, "Postresuscitation Disease", Elsevier, Amsterdam (1983).
66. Nemoto EM, Shiu G, Alexander H, Brain free fatty acids during decapitation ischemia in awake and pentobarbital anesthetized rats, Fed Proc 39:407 (1980).
67. Nemoto EM, Pathogenesis of cerebral ischemia-anoxia, Crit Care Med 6:203 (1978).
68. Nemoto EM, Frinak S, Rat brain tissue PO$_2$ after 16 min global ischemia and thiopental therapy, Crit Care Med 6:113 (1978).
69. Nemoto EM, Bleyaert AL, Stezoski SW, et al, Global brain ischemia: A reproducible monkey model, Stroke 8:558 (1977).
70. Otto CW, Yakaitis RW, Ewy GA, Spontaneous ischemic ventricular fibrillation in dogs: a new model for the study of cardiopulmonary resuscitation, Crit Care Med 11:883 (1983).

71. Parks DA, Bulkley GB, Granger DN, Role of oxygen derived free radicals in digestive tract diseases, Surgery 94:415 (1983).
72. Plum F, Symposium on the threshold and mechanisms of anoxic-ischemic brain injury, Arch Neurol 29:359 (1973).
73. Pulsinelli W, Brierley J, Plum F, Temporal profile of neuronal damage in a model of transient forebrain ischemia, Ann Neurol 11:491 (1982).
74. Raichle ME, The pathophysiology of brain ischemia, Ann Neurol 13:2 (1983).
75. Redding J, Cozine RA, Voigt GC, et al, Resuscitation from drowning, JAMA 178:1136 (1961).
76. Redding JS, Pearson JW, Resuscitation from asphyxia, JAMA 182: 283 (1962).
77. Redding JS, Pearson JW, Resuscitation from ventricular fibrillation, JAMA 203:93 (1968).
78. Rehncrona S, Rosen I, Siesjö BK, Excessive cellular acidosis: an important mechanism of neuronal damage in the brain?, Acta Physiol Scand 110:435 (1980).
79. Rosomoff HL, Shulman K, Raynor R, et al, Experimental brain injury and delayed hypothermia, Surg Gynecol Obstet 110:27 (1960).
80. Safar P, Gisvold SE, Vaagenes P, et al, Long-term animal models for the study of global brain ischemia, in: "Protection of tissues against hypoxia", Wauquier, et al, eds., Elsevier, Amsterdam (1982), also: Crit Care Med (1985, in press).
81. Safar P, Cardiopulmonary cerebral resuscitation. A manual for physicians and paramedical instructors. World Federation of Societies of Anaesthesiologists. Laerdal, Stavanger; W.B. Saunders, Philadelphia (1981).
82. Safar P, Resuscitation after brain ischemia, in: "Brain failure and resuscitation", Grenvik A, Safar P, eds., Clinics in Critical Care Medicine. Churchill Livingstone, New York (1981).
83. Safar P, eds., "Special symposium issue. Brain resuscitation", Crit Care Med 6:199 (1978).
84. Safar P, Stezoski W, Nemoto EM, Amelioration of brain damage after 12 minutes cardiac arrest in dogs, Arch Neurol 33:91 (1976).
85. Schuier F, Orzi F, Sokoloff L, Brain edema and mortality after cerebral ischemia in the gerbil, J Cereb Blood Flow Metabol 3:S339 (1983).
86. Schultz HW, Day EA, Sinnbauer RO, eds., "Lipids and their oxidation", Avi, Westport CN (1962).
87. Sharp AA, Warren BA, Paxton AM, et al, Anticoagulant therapy with a purified fraction of Malayan pit viper venom, Lancet 1:493 (1968).
88. Siebke H, Rod T, Breivik H, et al, Survival after 40 minutes submersion without cerebral sequelae, Lancet 1:1275 (1975).
89. Siesjö BK, Cell damage in the brain: A speculative synthesis, J Cereb Blood Flow Metabol 1:155 (1981).
90. Sloviter HA, Petokovic M, Ogoshi S, et al, Dispersed fluorochemicals as substitutes for erythrocytes in intact animals, J Appl Physiol 27:666 (1969).

91. Smith J, Penninckx JJ, Kampschulte S, et al, Need for oxygen enrichment in myocardial infarction, shock and following cardiac arrest, Acta Anaesth Scand 29:127 (1968).

92. Snyder JV, Nemoto EM, Carroll RG, et al, Global ischemia in dogs: Intracranial pressure, brain blood flow and metabolism, Stroke 6:21 (1975).

93. Stajduhar K, Steinberg R, Sotosky M, Safar P, Stezoski SW, Cerebral blood flow and common carotid artery blood flow during open chest cardiopulmonary resuscitation in dogs, Anesthesiology 59:A117 (1983).

94. Steen PA, Newberg LS, Milde JH, et al, Nimodipine improves cerebral blood flow and neurologic recovery after complete cerebral ischemia in the dog, J Cereb Blood Flow Metabol 3:38 (1983).

95. Stewart GN, Guthrie C, Burns RL, Pike H, The resuscitation of the central nervous system of mammals, J Experimental Med 8:289 (1906).

96. Stullken EH, Sokol MD, The effects of heparin on recovery from ischemic brain injuries in cats, Anesth Analg 55:683 (1976).

97. Symon L, et al, Physiological responses of local areas of the cerebral circulation in experimental primates determined by the method of hydrogen clearance, Stroke 4:632 (1973).

98. Takaori M, Safar P, Treatment of massive hemorrhage with colloid and crystalloid solutions, JAMA 199:297 (1967).

99. Teasdale G, Jennett B, Assessment of coma and impaired consciousness. A practical scale, Lancet 2:81 (1974).

100. Todd MM, Chadwick HS, Shapiro HM, et al, The neurologic effects of thiopental therapy following experimental cardiac arrest in cats, Anesthesiology 57:76 (1982).

101. Todd MM, Dunlop BJ, Shapiro HM, et al, Ventricular fibrillation in the cat: a model for global cerebral ischemia, Stroke 12:808 (1981).

102. Vaagenes P, Cantadore R, Safar P, et al, Amelioration of brain damage by lidoflazine after prolonged ventricular fibrillation cardiac arrest in dogs, Crit Care Med 12:846 (1984).

103. Vaagenes P, Kjekshus JK, Torvik A, The relationship between cerebrospinal fluid creatine-kinase and morphological changes in the brain after transient cardiac arrest, Circulation 61:1194 (1980).

103a. Vaagenes P, Safar P, Cantadore R, et al, Amelioration of brain damage with a free radical scavenger solution after asphyxial cardiac arrest in dogs, (submitted).

104. Wise BL, Chater M, Use of hypertonic mannitol solution to lower CSF pressure and decrease brain bulk in man, Surg Forum 12:398 (1961).

105. Welsh SA, O'Connor MJ, Marcy VR, et al, Factors limiting regeneration of ATP following temporary ischemia in cat brain, Stroke 13:234 (1982).

106. White B, Winegar CD, Wilson RF, et al, Possible role of calcium blockers in cerebral resuscitation: a review of the literature and synthesis for future studies, Crit Care Med 11:202 (1983).

107. White BC, Gadzinski DS, Hoehner PJ, et al, Effect of flunarizine on canine cerebral cortical blood flow and vascular resistance post cardiac arrest, Ann Emerg Med 11:119 (1982).
108. White RJ, Hypothermic preservation and transplantation of brain, Resuscitation 1:197 (1975).
109. Winegar CP, Henderson O, White BC, et al, Early amelioration of neurologic deficit by lidoflazine after fifteen minutes of cardiopulmonary arrest in dogs, Ann Emerg Med 12:471 (1983).
110. Yatsu FM, Diamond I, Graziana C, et al, Experimental brain ischemia: protection from irreversible damage with a rapid acting barbiturate (methohexital), Stroke 3:726 (1972).

NON STEROIDAL AND NON BARBITURATE TREATMENT OF

SECONDARY BRAIN DAMAGE

Petter A. Steen, S.E. Gisvold, L.A. Newberg, W.L. Lanier, B.W. Scheithauer, J.D. Michenfelder

Department of Anesthesiology, University of Oslo, Ulleval Hospital, Oslo 1, Norway

Drug treatment in cerebral hypoxia is after decades of research still controversial, and no treatment appears to be established at the moment. Not even the mechanisms causing the primary or possible secondary damage are established (Siesjö, 1981), only that they are obviously linked to a disturbance in the oxygen demand/oxygen supply ratio. Many hypotheses have been recently advanced involving increased intracellular concentrations of free calcium as responsible for large parts also of the secondary brain damage (Raichle, 1982; Siesjö, 1981; White et al., 1983a). In Michenfelders's laboratory at the Mayo Clinic we have especially studied complete cerebral ischemia. Secondary brain damage in the post-ischemic period due to a disturbed calcium homeostasis could include decreased CBF due to vasospasm (Somlyo and Somlyo, 1968; Van Neuten and Vanhoutte, 1980; Vanhoutte, 1981), reduced blood cell deformability (DeCree, et al. 1979), and platelet aggregation (Vanhoutte and Van Neuten, 1980). These factors could either be due to direct Ca^{++} effects or indirectly to a calcium stimulated breakdown of arachidonic acid to products such as prostaglandins, leukotrienes or thromboxanes (Raichle, 1982; Siesjö, 1981).

It is well established that after complete cerebral ischemia a brief 10-15 min hyperemia period is followed by delayed hypoperfusion lasting probably more than 18 and less than 48 hours (Steen et al., 1979; White, 1982). The hypoperfusion is severe with CBF down to 20-30% of control (Steen et al., 1978), but - at least until recently - clear evidence has not been provided yet that this hypoperfusion is indeed detrimental. If no-reflow occurs or the microcirculation is impaired by massive tissue edema this would obviously explain further damage (Siesjö, 1981), but complete cerebral ischemia with a dismal outcome has been described with normalization of the tissue energy state, and thus apparently without a no-reflow phenomenon (Steen et al., 1979).

To investigate if delayed hypoperfusion is indeed deleterious, we infused nimodipine into dogs in an attempt to improve post-ischemic blood flow. The neurologic outcome was correlated with the post-ischemic CBF in a separate group of dogs. Cerebral ischemia was induced by temporary occlusion of the aorta and both venae cavae for 10 minutes. The treated group received nimodipine in a dosis of 10 µg/kg i.v. 2 min pre-ischemia, followed by 1 µg/kg min for 2 hours. Nimodipine nearly doubled cerebral blood flow in the delayed hypoperfusion period compared to untreated dogs approaching 45% of the pre-ischemic control value vs. 25 % in untreated animals. Parallel to the enhancement of CBF nimodipine significantly improved the neurologic outcome (Steen et al., 1983). It is impossible to start treatment of, for instance cardiac arrest before arrest occurs, and even though nimodipine improved post-ischemic CBF and outcome, it is possible that this was attributable to protection during the period of complete ischemia rather than to alleviation of hypoperfusion.

A similar series of experiments was therefore performed with nimodipine given only post-ischemia. Again, post-ischemic CBF was significantly improved whereas $CMRO_2$ or cerebral metabolites 2 hours post-ischemia were not affected. The neurologic results 48 hrs post-ischemia were, however, intermediate and not significantly different from neither the controls nor from dogs given nimodipine pre-ischemia (Steen et al., 1984). What conclusion could be drawn? We knew that nimodipine can increase CBF in the post-ischemic hypoperfusion period, both when treatment was started before and after the ischemic episode. This was in agreement with most of the literature. The manufacturer's laboratory reported that hypoperfusion was completely abolished by nimodipine after complete cerebral ischemia in cats (Hoffmeister et al., 1982). White et al. (1982, 1983b) reported similar results after 20 min of circulatory arrest in dogs for flunarizine, lidoflazine, verapamil, and $MgSO_4$.

Some uncertainty applies to the latter results as the control CBF was 2-3 times what is normally reported for dogs. None of these studies analyzed the regional distribution of CBF, however. Smith et al. (1983) from Siesjö's laboratory found in rat brain that although global CBF was improved post-ischemia by nimodipine, there were great regional differences. Thus it was possible that the calcium blockers only increased blood flow to already well perfused areas of the brain without improvement of really ischemic areas. Our neurologic results gave no clear answer. The outcome was improved if treatment was started pre-ischemia, but this could be due to effects occurring during the ischemic event, while post-ischemic treatment only gave an intermediate result. Before considering human application it was furthermore necessary to do studies on primates.

In Michenfelder's laboratory we have therefore given nimodipine in a controlled, blind study to pig-tailed monkeys after 17 min of complete cerebral ischemia. Ischemia was obtained with a neck-cuff inflated to 1500 mmHg combined with arterial hypotension to 50 mmHg by infusion of trimetaphan. Completeness of ischemia was confirmed by blood flow measurements using ^{133} Xenon. Nimodipine was given in a dosis of 10 ug/kg 5 min post-ischemia followed by 1 μg/kg min for 10 hrs. The monkeys remained under intensive care for 96 hrs followed by neurologic and pathologic evaluation. 27 monkeys were studied, 6 were discarded for various reasons by a blinded observer. One had incomplete ischemia, one received accidentally a large glucose load pre-ischemia, one was very old, and three had severe cardiopulmonary problems. Of the remaining 21 monkeys, 11 received nimodipine which significantly improved outcome when compared to the placebo group. Lidoflazine has also been reported to improve the outcome after complete cerebral ischemia. In a controlled, blind study by Winegar et al. (1983), lidoflazine 1 mg/kg was given as part of the resuscitation after 15 min of cardiac arrest induced by potassium injection into dogs. The study indicated that lidoflazine could improve neurologic recovery in that all 5 treated dogs had spontaneous ventilation, reactive pupils, voluntary movements and responded to tactile stimuli 12 hours post-resuscitation, while 4 of 5 control dogs had maximum neurologic deficit scores. Lidoflazine 4 mg/kg given over a 16 hrs period also appeared to improve outcome when given after 10 min of ventricular fibrillation in dogs, while 9 mg/kg given over 72 hours had no effect in dogs after 7 or 10 min of asphyxial cardiac arrest (Vaagenes et al., 1983).

Two conclusions must be drawn from these studies:

1. Secondary brain damage occurs after a period of complete cerebral ischemia.
2. The calcium entry blockers nimodipine and lidoflazine protect the brain, at least against part of this damage. Both drugs appear to increase cerebral blood flow in the post-ischemic hypoperfusion period. Thus it seems possible that post-ischemic hypoperfusion is involved in secondary brain damage, although it can not be excluded that improvement of CBF by these drugs is only incidental to the protection seen.

REFERENCES

1. Hoffmeister F, Benz U, Heise A, Krause HP, Neuser V, Behavioral effects of nimodipine in animals, Arzneim Forsch 4:347 (1982).
2. Raichle ME, The pathophysiology of brain ischemia, Ann Neurol 13:2 (1982).
3. Siesjö BK, Cell damage in the brain: A speculative synthesis, J Cereb Blood Flow Metabol 1:155 (1981).
4. Smith ML, Kagström E, Rosen I, Siesjö BK, Effect of the calcium antagonist nimodipine on the delayed hypoperfusion following incomplete ischemia in the rat, J Cereb Blood Flow Metabol 3:543 (1983).
5. Somlyo AP, Somlyo AV, Vascular smooth muscle. 1. Normal structure, pathology, biochemistry and biophysics, Pharmacol Rev 20:197 (1968).

6. Steen PA, Milde JH, Michenfelder JD, Cerebral metabolic and vascular effects of barbiturate therapy following complete global ischemia, J Neurochem 31:1317 (1978).

7. Steen PA, Milde JH, Michenfelder JD, No barbiturate protection in a dog model of complete cerebral ischemia, Ann Neurol 5:343 (1979).

8. Steen PA, Newberg LA, Milde JH, Michenfelder JD, Nimodipine improves cerebral blood flow and neurologic recovery after complete cerebral ischemia in the dog, J Cereb Blood Flow Metabol 3:38 (1983).

9. Steen PA, Newberg LA, Milde JH, Michenfelder JD, Cerebral blood flow and neurologic outcome when nimodipine is given after complete cerebral ischemia in the dog, J Cereb Blood Flow Metabol 4:82 (1984).

10. Vaagenes P, Cantadore R, Safar P, Alexander H, Effect of lidoflazine on neurologic outcome after cardiac arrest in dogs, Anesthesiology 59:A100 (1983).

11. Van Neuten JM, Vanhoutte PM, Improvement of tissue perfusion with inhibitors of calcium ion influx, Biochem Pharmacol 29:479 (1980).

12. Vanhoutte PM, Calcium entry blockers and cardiovascular failure, Federation Proc 40:2882 (1981).

13. Vanhoutte PM, Van Neuten JM, The pharmacology of lidoflazine, Proc R Soc Med Int Congr Symp Ser 29:61 (1980).

14. White BC, Gadzinski DS, Hoehner PS, Krome C, Hoehner T, White JD, Trombley JH,Jr, Effect of flunarizine on canine cerebral cortical blood flow and vascular resistance post cardiac arrest, Ann Emerg Med 11:119 (1982).

15. White CB, Winegar CD, Wilson RF, Hoehner PJ, Trombley JH, Jr, Possible role of calcium blockers in cerebral resuscitation: a review of the literature and synthesis for future studies, Crit Care Med 11:202 (1983a).

16. White CB, Winegar CD, Wilson RF, Krause GS, Calcium blockers in cerebral resuscitation, J Trauma 23:788 (1983b).

17. Winegar CP, Henderson O, White BC, Jackson RE, O'Hara T, Krause GS, Viga DN, Kontry R, Wilson W, Shelby-Lane C, Early amelioration of neurologic deficit by lidoflazine after fifteen minutes of cardiopulmonary arrest in dogs, Ann Emerg Med 12:471 (1983).

FEASIBILITY OF PREVENTIVE BRAIN PROTECTION IN PATIENTS AT EXCESSIVE RISK OF STROKE

Cesare Fieschi*, N. Battistini**, S. Passero**, M. Rasura*

1st Department of Neurology*, University of Rome,
Department of Neurology and Psychiatry**, University of Siena, Italy

Medical therapy of acute focal cerebral ischemia in man has made no substantial progress thus far (1,6,13).

The remarkable decline in stroke mortality observed in several countries during the past decades appears to be related to a declining incidence of new cases rather than to a decrease of case fatality. In the Rochester study, both the decline in incidence and mortality have been greater than 50% since 1950 (12), and the long term survival after stroke has gradually improved as well, more than 70% from 1945-1949 to 1970-1974. However, the 30-days death rate was only slightly lower in 1979 than in 1950, and there was virtually no improvement in survival up to three days for cerebral infarction between these periods (4,5). In other words, in the past 30 years prevention of stroke and care for associated diseases seem increasingly effective, while management of the severe cases, who die in the very acute phase after cerebral infarction is no more effective nowadays than it was 30 years ago.

Prevention remains, therefore, the main goal of stroke research in the clinic. Yet, it will be difficult to achieve a complete success, and the problem will remain of patients who develop a stroke in spite of the best health care and individual preventive measures. Research will have to continue to find new guidelines for treatment. Therapeutical methods must be identified to protect the brain from some of the devastating effects of ischemia, and to limit or counteract the factors leading to secondary postischemic brain damage. A list of factors potentially involved in this process, and forms of treatment suggested or tested thus far, has been discussed both in experimental procedures and in clinical work in this symposium and elsewhere (8,10) (Table 1). However, the time delay between occurrence of stroke and arrival of patients in the appropriate clinical setting (6 hours on average in our institutions)

TABLE 1

Treatment of Cerebral Ischemia

1. Restore / Increase Perfusion

    - Surgery
    - Fibrinolysis
    - Hypertension
    - Prostacyclin

2. Prevention of Brain Edema

    - Steroids
    - Osmotic Agents

3. Protect Microcirculation and Delayed Hypoperfusion

    - Reduce Viscosity
    - Reduce Intravascular Coagulation
    - Ca^{2+} Blocking Agents

4. Metabolic Protection

    - Anesthetics
    - Steroids
    - Ca^{2+} Blocking Agents
    - Oxygen Radical Scavengers
    - Hyperbaric Oxygen (?)

TABLE 2

Scopes of Preventive Treatment of Cerebral Ischemia

- Reduce occurrence of ischemia: Prevention proper
- Reduce consequences of ischemia: Preventive brain
  protection

TABLE 3

Preventive Correction of Risk Factors for Secondary Brain Damage

- Hypertension
- Hyperviscosity
- Hyperglycemia
- Coagulation factors

TABLE 4

Italian Multicenter Study of Reversible Ischemic Attacks

(462 pts followed for 4 years)

Arterial hypertension	38.3%
Diabetes	14.9%
Hematocrit > 46	16.3%
Platelet aggregate ratio (10)	81.0%

TABLE 5

Patients at "Excessive Risk" of Stroke: Follow up at 6 Months

Groups	No.	RIAs	Stroke	MI* or other death
RIA within 2 weeks, with carotid steno- sis or cardiac em- bolism	45	8	5	2
Recurrent carotid RIAs within 2 weeks	9	0	0	1
Surgical risk for bilateral or mul- tiple angiographic lesions	38	8	5	1
Open heart surge- ry in pts with RIAs and/or carotid stenosis	8	0	1	0

*myocardial infarction

limits the chances of a successful postischemic treatment. Therefore, neurologists have developed a new strategy in stroke prevention comprising: "...protection of the brain from the effects of imminent ischemia in patients at excessive risk in spite of preventive measures" (Table 2).

It is in this light that the correction of factors should be considered known to determine the final size of the lesion, if the patient at risk is not adequately treated (Table 3). In a large consecutive case-record of RIAs collected between 1977 and 1978, (reversible focal ischemic attacks, including TIAs as well as TIAs with incomplete recovery, TIA-IR) (2) we have noticed the following frequencies of factors (Table 4). All patients were carefully treated and monitored during 4 years. No case was missing, and the occurrence of stroke was indeed very limited (37 strokes vs 189 patients with new RIAs, 2% and 10% per year, respectively). There was a strong bias in the selection of patients towards subjects of younger age than the average TIA cases, and towards patients with less cardiac involvement. Therefore, no extrapolation can be made regarding the efficacy of the treatment employed, yet the provoking possibility exists that some potential strokes have been converted into RIAs. In fact, compared to the natural history of TIAs it is the incidence of strokes, rather than that of new TIAs, which appears to be at variance (7).

Apart from these measures, pharmacological "brain protection" can be attempted of a type that is effective if administered prior, or within minutes of occurrence of ischemia as suggested by Steen (this Symposium). Such a trial must consider a population in which the expected frequency of events is high enough to allow the measurement of any effects. Disappointingly, however, RIAs - let alone asymptomatic carotid stenoses - do not fulfill these criteria. In fact, stroke occurrence in treated unselected RIAs, is now of the order of 4% per year (11), or 3.7% according to an other population study of 300 consecutive cases that we are following since 1980 (3). The first or last reversible attack was observed within two months before entry implying recent RIAs.

This and similar studies have provided sufficient data to stratify subgroups with excessive risk of stroke in spite of current medical or surgical prevention. Accordingly, in a new prospective study we have observed up to now with a 6-months follow-up 11 strokes and 16 RIAs in 100 patients (Table 5) belonging to the category at excessive risk. This population lends itself to analysis of the effects of pharmacological "preventive brain protection", both in terms of stroke occurrence and severity.

REFERENCES

1. Buonanno F, Toole JF, Management of patients with established ("completed") cerebral infarction, Stroke 12:7 (1981).

2. Fieschi C, Mariani F, Brambilla GL, Prencipe M, Tomasello F, Argentino C, Bono G, Candelise L, De Zanche L, Inzitari D, Nardini M, Italian multicenter study on reversible cerebral ischemic attacks: population characteristics and methodology, Stroke 14:424 (1983).

3. Fieschi C, Carolei A, Bastianello S, D'Aloja E, Possibilita di prevenzione farmocologica e brain protection nell'ischemia cerebrale reversibile, Rassegna Clinico Scientifica 59:33 (1983).

4. Garraway WM, Whisnant JP, Furlan AJ, The declining incidence of stroke, New Engl.J.Med. 300:449 (1979).

5. Garraway WM, Whisnant JP, Drury I, The changing pattern of survival following stroke, Stroke 14:699 (1983).

6. Hatchinson EC, Management of cerebral infarction, in: "Vascular disease of the central nervous system", Ross Russel RW, 2nd ed., Churchill-Livingstone, London (1983).

7. Mc Dowell FH, Prevention of subsequent infarction and transient ischemic attacks, in: "Cerebrovascular disorders and stroke", Goldstein M, Bolis L, Fieschi C, eds., Raven Press, New York (1979).

8. Plum F, What causes infarction in ischemic brain? Neurol 33:222 (1983).

9. Prencipe M, Buttinelli C, Paolucci S, Fieschi C, Lucignani G, Lenzi GL, Circulating platelet aggregates in 191 RIA patients and 117 control subjects, in: "Cerebral vascular disease 4", Meyer JS, Lechner H, Reivich M, Ott EO, eds., Excerpta Medica, Amsterdam, Oxford, Princeton (1983).

10. Raichle ME, The pathophysiology of brain ischemia, Ann Neurol 13:2 (1983).

11. Warlow C, Transient ischemic attacks, in: "Research advances in clinical neurology", Matthews WB, Glaser GH, eds., Churchill-Livingstone, New York (1982).

12. Whisnant JP, The role of the neurologist in the decline of stroke, Ann Neurol 14:1 (1983).

13. Yatsu F, Acute medical therapy of stroke, Stroke 13:524 (1982).

CONTRIBUTORS

J. Adams,
Univ. Dept. Neuropathol.,
Inst. Neurol. Sci.,
Southern General Hospital,
Glasgow, G51 4TF
SCOTLAND

J. Astrup,
Dept. Neurosurg., Rigshospitalet
University of Copenhagen,
2100 Copenhagen
DENMARK

R.N. Auer,
Health Sci. Center
Univ. Calgary
Calgary, Alberta, T2N 1N4
CANADA

A. Baethmann,
Inst. Surg. Res.,
University of Munich,
Klinikum Großhadern
8000 München 70
FRG

D.P. Becker,
Dept. Neurol. Surgery, UCLA
School of Medicine,
Los Angeles, Ca., 90024
USA

J. Cervos-Navarro,
Institute of Neuropathology,
Klinikum Steglitz, Free University
Hindenburgdamm 30
D-1000 Berlin 45
FRG

R.A. Clasen,
Dept. Pathol. Diagn. Radiol.,
Rush-Presbyt.-St.Luke's Med. Ctr.
Chicago, Il., 60612
USA

F. Cohadon,
Lab. Neurosurg. Exp. Neurobiol.,
University of Bordeaux II,
Service de Neurochirurgie A,
Hopital Pellegrin-Tripode,
33076 Bordeaux
FRANCE

C. Fieschi,
1st Dept. Neurol.,
University of Rome,
00185 Rome
ITALY

J.H. Garcia,
Dept. Pathol. Neuropath.
University of Alabama,
Birmingham, Alabama, 35233
USA

Th.A. Gennarelli,
Dept. Neurosurgurgery,
University of Pennsylvania,
Philadelphia, Pa., 19104
USA

K.G. Go,
Dept. Neurosurg.,
University of Groningen,
9700 RB Groningen
THE NETHERLANDS

K.-A. Hossmann,
Max-Planck-Institute
for Neurological Research,
5000 Köln 91
FRG

H.E. James,
Div. Neurosurg., UCSD,
University Hospital,
San Diego, Ca., 92103
USA

L.W. Jenkins,
Div. Neurol. Surg.,
Medical College of Virginia,
Richmond, Va., 23298
USA

B. Jennett,
Inst. Neurol. Sci.,
University of Glasgow,
Glasgow G51 4TF
SCOTLAND

O. Kempski,
Inst. Surg. Res.,
University of Munich,
Klinikum Großhadern
8000 München 70
FRG

L.F. Marshall,
Div. Neurosurg., UCSD,
University Hospital,
San Diego, Ca., 92103
USA

G. Mchedlishvili,
Beritashvili Inst. Physiol.,
Georgian Academy of Sciences,
Gotua Street 14
Tbilisi, 380060
USSR

J.D. Miller,
Dept. Surg. Neurol.,
Western General Hospital,
University of Edinburgh,
Edinburgh EH2 2XU
SCOTLAND

E.M. Nemoto
Dept. Anesthesiol. Crit. Care Med.,
University of Pittsburgh,
Pittsburgh, Pa., 15261
USA

H.M. Pappius,
Donner Lab. Exp. Neurochem.,
Montreal Neurol. Inst.,
McGill University,
Montreal, Quebec, H3A 2B4
CANADA

Ch. Plets,
Univ. Ziekenhuis, St. Rafael,
Kapucijnenvoer 33
3000 Louvain
BELGIUM

K.E. Richard,
Dept. Neurosurgery,
University of Cologne,
5000 Köln 41
FRG

P. Safar,
Resuscitation Res. Ctr.,
Univ. Pittsburgh,
Pittsburg, Pa., 15260
USA

M. Spatz,
Lab. Neuropathol. and
Neuroanat. Sci, NINCDS,
Natl. Inst. Health,
Bethesda, Md., 20892
USA

L. Symon,
Gough Cooper Dept. Neurol. Surg.,
Inst. Neurology,
Queen Square Natl. Hospital,
London WC1E 3BG
UNITED KINGDOM

P. Steen,
Dept. Anesthesiol.,
University of Oslo,
Ulleval Hospital,
0407 Oslo 4
NORWAY

A. Unterberg,
Inst. Surg. Res.,
University of Munich,
Klinikum Großhadern
8000 München 70
FRG

R. Vara-Thorbeck,
Catedra de Patol. y Quirurgicas II
Universidad de Granada
Granada,
SPAIN

T. Yamaguchi,
Lab. Neuropath. Neuroanat. Sci,
NINCDS, Natl. Inst. Health,
Bethesda, Md., 20892
USA

K.J. Zülch,
Max-Planck-Institute
for Neurological Research
5000 Köln 91
FRG

SUBJECT INDEX